Frontier Studies in Soil Science

Series Editor

Avelino Núñez-Delgado, Lugo, Spain

This series aims to illuminate the latest research, insights, and advancements in the ever-evolving world of soil science. This series is dedicated to fostering a comprehensive understanding of soils, from their micro to macro characteristics, their role in ecosystems, to their significance in global challenges. Our commitment is to present groundbreaking studies, carried out by esteemed experts and emerging scholars, ensuring that readers are consistently provided with the most up-to-date and relevant information in the realm of soil science.

Éder de Souza Martins · Suzi Huff Theodoro
Editors

Soil Remineralizers and Silicate Fertilizers

Regional Solutions to a Healthy Agriculture

Springer

Editors
Éder de Souza Martins
Agrogeological Laboratory
Embrapa Cerrados
Brasília, Brazil

Suzi Huff Theodoro
Postgraduate Program in Environment
and Rural Development (PPG-MADER)
Universidade de Brasília
Brasília, Brazil

ISSN 3004-9865 ISSN 3004-9873 (electronic)
Frontier Studies in Soil Science
ISBN 978-3-032-14655-7 ISBN 978-3-032-14656-4 (eBook)
https://doi.org/10.1007/978-3-032-14656-4

This Springer imprint is published by the registered company Springer Nature Switzerland AG
The registered company address is: Gewerbestrasse 11, 6330 Cham, Switzerland

Foreword

The planet cries out for solutions to ensure the continuity of life as we know it today. Recurrent climatic disasters, hunger, mass migrations, and the overconsumption of natural resources are among the causes and consequences of climate collapse. We live with the contradiction of producing food, commodities, and disposables for a population that consumes at a faster rate than the Earth's carrying capacity. The frantic expansion in search of new arable land advances and devastates biodiversity reserves and forests. Furthermore, the excessive use of soluble fertilizers, employed to produce commodities and bioenergy, neglects the necessary replenishment of all essential inorganic elements present in naturally fertile soils, derived from the cycling of organic nutrients and geological sources that ensure a continuous supply of nutrients.

In lateritic soils, common in tropical regions, it becomes necessary to replenish this constant nutrient loss through additions of multi-nutrient sources, such as those derived from ground rocks. It is important to remember that the use of comminuted rocks has been proposed for over 200 years and has always proven suitable for restoring lost fertility due to weathering processes. One of the most iconic examples of fertilization through geological and biological renewal is the Nile River Valley and Delta, often called the Fertile Crescent. Sediments from its diverse drainage basin, along with organic matter from vegetation, were carried downstream and deposited along the riverbanks and delta. This combination of minerals and organic material replenished the floodplains, sustaining the region's fertility for millennia.

Similarly, the Incas harnessed natural processes, blending traditional knowledge with simple technology to enhance soil fertility in the Andes. Their irrigation channels distributed nutrient-rich water, supporting agriculture in the highlands. Centuries later, in eighteenth-century Scotland, James Hutton advocated using rocks to enrich the nutrient-poor soils of the Highlands. This idea was further developed in the nineteenth century by Julius Hensel, who promoted crushed rock as a soil amendment in his pamphlet Stone Meal. By the early 20th century, Russian geochemist Wladimir Vernadsky revived and expanded these concepts, inspiring further scientific exploration worldwide.

However, since the mid-twentieth century, with the implementation and spread of the green revolution, the greed and the urgency to serve a market specialized in a few types of agricultural products have relegated the option provided by nutrient replenishment derived from rocks to secondary plans. Although such materials exhibit less expressive solubility than soluble chemical fertilizers, hundreds of modern studies and research prove the nutritional potential of various types of rocks and minerals, with surprisingly rapid solubility rates.

Several case studies, conducted in different countries and agroecosystems with various rock types, show the effects of using ground rocks as an effective and smarter alternative compared to NPK fertilizers. These studies report improved harvests in soils that received ground rocks, increased soil pH and cation exchange capacity (CEC) indicators, reduced aluminum toxicity, better soil aggregation, among other parameters.

Considering these urgent needs and important considerations, I am pleased to present this book titled *Soil Remineralizers and Silicate Fertilizers: Regional Solutions to a Healthy Agriculture*, organized by Éder de Souza Martins and Suzi Huff Theodoro. This work represents a collaborative effort aimed at exploring and highlighting the significance of soil remineralizers and silicate-based fertilizers in agriculture, particularly in a country like Brazil, which faces significant challenges related to soil fertility and reliance on imported inputs.

The chapters in this book cover a wide range of topics, beginning with the agrogeological principles of soil remineralization, which establish the foundation for understanding the crucial role that remineralizers play in enhancing soil quality and agricultural productivity. Through a comprehensive analysis, the authors discuss the chemical and mineralogical classification of silicate agrominerals, detailing their potential for soil fertility management and plant nutrition.

Moreover, the studies presented demonstrate how these products can contribute to correcting soil acidity, promoting fertility, and increasing the productivity of various crops, including soybeans, maize, sugarcane, coffee, and others. Research on the impact of silicate rocks (remineralizers) in the Cerrado and their implications for charge generation and carbon accumulation highlights the potential of these materials in mitigating climate change.

The chapters also address the agronomic characterization of remineralizers and silicate fertilizers under tropical conditions, emphasizing the need for innovative and sustainable strategies to ensure food security and preserve natural resources. The integration of management practices and the use of regional technologies are discussed as ways to maximize the benefits of remineralizers.

What makes this work particularly noteworthy is the strong academic foundation of its organizers, Éder de Souza Martins and Suzi Huff Theodoro, both of whom I have had the pleasure of guiding throughout their academic journeys. Under my mentorship, they have acquired the knowledge and expertise necessary to address these pressing agricultural challenges. This book not only contributes to the advancement of scientific knowledge but also serves as a practical guide for researchers, academics, and agricultural professionals seeking sustainable solutions to the issues faced in the field.

I am confident that the information contained in this volume will be immensely valuable to all those dedicated to promoting sustainable agriculture and improving soil quality in Brazil and beyond.

October 2025 Prof. Othon Henri Leonardos

Preface

Book Presentation

In the face of pressing global and local challenges, such as climate change, food insecurity, and geopolitical upheavals, science is constantly seeking innovative solutions. Among these, the use and development of fertilizers have attracted renewed focus. While fertilizers have historically played a pivotal role in boosting agricultural productivity, their environmental, economic, and health impacts have also raised concerns. Recently, the revival and modernization of rock dust application, using ground rock powders, have emerged as promising strategies within sustainable agriculture.

Since the 1970s, environmental concerns rooted in the green revolution's production model motivated the search for more comprehensive and safer fertilizers, capable of providing a wider spectrum of nutrients while being more accessible. It was only in the early 21st century that mineral sources derived from ground silicate rocks, known as soil remineralizers (REMs), began to gain traction globally. Particularly in Brazil, REMs have gained prominence as a technological alternative to increase productivity and reduce dependence on highly soluble, chemical fertilizers controlled by a few countries. With significant investments driven by research and supported by a clear regulatory framework (Brazil's Law 12,890/2013, MAPA Normative Instruction No. 05/2016, and inclusion in the National Fertilizer Plan Decree No. 10,991/2022), REMs are becoming safer, more accessible, and increasingly adopted by farmers. Beyond enhancing productivity, REMs, sourced from regional minerals, contribute to a more sustainable and resilient agricultural paradigm aligned with the United Nations Sustainable Development Goals (SDGs).

In Brazil, the growth in REM production and commercialization has been exponential, doubling annually. Over the past five years, more than 7 million metric tons have been sold, now covering approximately 10% of the country's agricultural land. Product registrations with MAPA increased by 400% between 2019 and 2023.

Given Brazil's vast mineral potential, projections suggest that the target of 18 million tons annually by 2050 could be reached much sooner, unlocking further opportunities for sustainable development.

This publication, *Soil Remineralizers and Silicate Fertilizers: Regional Solutions to a Healthy Agriculture*, centers on this transformative potential. Its chapters explore various aspects of soil remineralizers, including new proposals for classifying silicate rocks and their direct and indirect benefits, as well as practices that mitigate climate impacts associated with agriculture.

The introductory chapter ("Agrogeological Principles on Soil Remineralization and Brazilian Protagonism") emphasizes silicate rock powders as strategic tools for countries relying on imported soluble fertilizers, particularly in tropical regions with weathered soils like Brazil. As a country endowed with rich geodiversity and a robust agricultural sector, Brazil is ideally positioned to adopt these technologies, supported by a strong regulatory framework that ensures their agronomic effectiveness and environmental safety, based on scientific principles. Nonetheless, realizing the full potential of REMs demands further public policies focused on financing and technical training, which could accelerate large-scale adoption, boosting productivity, food security, and sustainability.

Chapter "Changing Time Scales: From Silicate Rocks to Agrominerals Through Bioweathering" examines the process of bioweathering, which involves the biological transformation of Earth's lithosphere through interactions between plants, microbes, and minerals. It highlights how bioweathering accelerates mineral dissolution, especially of silicate rocks, beyond the slow pace of abiotic weathering, occurring on timescales relevant to agriculture and climate mitigation. The rhizosphere, as a biogeochemical hotspot, plays a crucial role by releasing organic acids, enzymes, and phytosiderophores that enhance nutrient release and mineral transformation. Experimental evidence from greenhouse studies demonstrates rapid mineral alterations, such as vermiculitization of biotite within months. The research underscores the potential to use silicate agrominerals not only to supply vital nutrients like potassium but also to strengthen soil properties, increase cation exchange capacity, and sequester atmospheric CO_2. Future directions focus on optimizing mineral application, fostering microbe-mineral interactions, and advancing technologies to accelerate weathering, ultimately supporting sustainable agriculture and climate change mitigation strategies.

Chapter "Chemical and Mineralogical Classification of Silicate Agrominerals" introduces a chemical and mineralogical classification of silicate agrominerals (ASi). These rocks and industrial materials contain over 50% silicates and are rich in bases, calcium, magnesium, and potassium. Designed for regional use, ASi directly influence soil fertility through nutrient supply, pH elevation, and new mineral formation, as well as indirectly aiding in inorganic carbon sequestration and organic matter stabilization. The classification delineates eight classes, such as ASiMg, ASiCa, ASiK, and their combinations, based on mineral reactivity and potential effect duration. Despite their abundance, systematic studies of ASi are recent but show promising impacts, especially in tropical soils.

Chapter "Impact of Silicate Agrominerals on Charge Generation in Cerrado Soil" examines the effects of silicate agrominerals on the chemical and mineralogical properties of Brazilian Cerrado soils, focusing on carbon accumulation and charge generation. Data from long-term field trials using various agrominerals indicate accelerated mineral weathering, formation of organo-mineral complexes, enhanced nutrient release, organic matter stabilization, and increased microbial activity. These processes boost soil cation exchange capacity and reduce nutrient leaching, creating a promising pathway for greenhouse gas mitigation through long-term carbon retention.

Chapter "Potential of Soil Remineralization to Carbon Dioxide Removal in Agricultural Soils" explores the potential of Enhanced Rock Weathering (ERW) as a mechanism for Carbon Dioxide Removal (CDR) in agriculture. It details three primary pathways: (A) bicarbonate leaching into oceans, (B) formation of mineral-associated organic matter (MAOM), and (C) carbonate mineral precipitation. The chapter discusses the feasibility of each process, emphasizing that in tropical soils, bicarbonate leaching and MAOM formation show the greatest promise, while carbonate precipitation is less viable due to unfavorable chemical conditions. The authors highlight ongoing efforts to develop reliable measurement and verification protocols to quantify CDR and advocate for further interdisciplinary research to optimize these processes, especially in tropical regions.

Chapter "Agronomic Characterization of Calcium Magnesium Remineralizers in Tropical Conditions" discusses REMs derived from Ca- and Mg-rich aluminosilicates, such as basalt. Containing 9–20% alkaline oxides, these REMs increase soil macronutrients and micronutrients, raise pH, decrease exchangeable Al^{3+}, and enhance crop yields, often with the added benefit of reducing CO_2 emissions. When combined with microbial inoculants, REMs can produce even better agronomic responses, supporting more resilient and sustainable soils.

In Chapter "A Protocol for Assessing the Potential of Soil Acidity Correction by Employing Silicate Agrominerals Through Short-Term Incubation", a protocol for assessing soil acidity correction using silicate agrominerals in short-term incubation experiments is detailed. It explores dissolution mechanisms, the role of oxidation (especially in iron-containing rocks), and incubation time effects. This methodology aims to provide practical guidance for evaluating the reactivity and neutralization potential of these materials, bridging scientific understanding and field application.

Chapter "Assessment of Soil Acidity Correction Potential Using Metabasalt" evaluates the use of metabasalt as a soil remineralizer in the Brazilian Cerrado, an area challenged by acid soils. The study assesses its capacity to improve soil chemical properties and increase yields of maize and soybean. Results demonstrate that metabasalt effectively corrects acidity, enhances nutrient availability, and boosts crop productivity, presenting a sustainable alternative to traditional liming materials and offering promising potential for soil management in tropical environments.

Chapter "Agronomic Characterization of Magnesium Soil Remineralizers and Silicate Fertilizers in Tropical Conditions" focuses on the factors influencing the solubilization of magnesium silicates, such as dunite and thermomagnesium (TM). It reviews greenhouse and field studies demonstrating that these rocks can increase Mg and Si availability, improve photosynthesis, and promote crop yields in soybeans and maize. Dunite consistently elevates soil pH and releases Mg gradually, while finer TM particles show superior nutrient release and agronomic performance. These findings support the role of Mg-silicate agrominerals as environmentally friendly fertilizers that could reduce reliance on synthetic inputs in tropical soils.

Chapter "Agronomic Characterization of Potassium Soil Remineralizers and Silicate Fertilizers in Tropical Conditions" discusses the agronomic characterization of potassium (K) soil remineralizers and silicate fertilizers under tropical conditions. It emphasizes K's vital role in processes such as enzyme activation, photosynthesis, osmotic regulation, and stress response. The chapter details how K exists in multiple forms within soils and how factors like mineralogy, particle size, pH, and microbial activity influence its availability. Different technologies, physical, chemical, and biological, aim to enhance K release from silicate minerals, with ongoing research highlighting the need for long-term validation under field conditions to improve efficiency in nutrient-depleted soils. Implementing sustainable cultivation systems can maximize the benefits of these products.

Chapter "The Potential of Remineralizers and Bioinputs as a Strategy for Sovereignty in Tropical Agriculture" addresses Brazil's heavy dependence on imported soluble fertilizers, emphasizing the importance of regional solutions like REMs and bio-inputs. Utilizing locally available resources could reduce reliance on imports, increase agricultural resilience, and support the country's role as a leading global food and commodity producer. The chapter discusses how integrating these alternative inputs improves food quality, productivity, and conservation of natural resources, offering a comprehensive approach to soil management suitable for tropical, input-dependent economies.

In Chapter "An Overview of Research and Innovation in Stonemeal Technology in Brazil", the authors review the evolution of research involving remineralizers in Brazil, highlighting key contributions from the four editions of the Brazilian Rock Congresses. These events catalyzed scientific discussion and shaped the regulatory framework supporting the use of crushed silicate rocks for soil enrichment, with the potential to revolutionize tropical agriculture and promote sustainable development.

Finally, Chapter "Optimizing Research Protocols for Evaluating the Agronomic Efficiency of Soil Remineralizers" emphasizes that effective strategies for researching soil remineralizers (REMs) must account for their unique properties, being insoluble, slow-release, multi-nutrient sources. Conventional experimentation with short-term or soluble fertilizers often fails to accurately assess REM efficacy. The chapter advocates integrating biological processes, such as bio-weathering and mineralogical interactions, into modern research designs. It underscores that a comprehensive evaluation, considering REM's intrinsic properties, realistic field conditions, and simpler systems, will produce practical recommendations for application, including optimal granulometry, dosing, crop selection, and combination with other inputs. This approach ultimately aims to optimize REM use across diverse agricultural environments, supporting more sustainable soil management.

This publication offers an updated synthesis on soil remineralizers and silicate fertilizers, emphasizing sustainable approaches to global soil fertility challenges. It provides valuable insights and practical guidance to researchers, policymakers, and practitioners dedicated to advancing environmentally friendly and resilient agricultural systems.

Enjoy all its content!

Brasília, Brazil
October 2025

Éder de Souza Martins
Suzi Huff Theodoro

Acknowledgements

We express our deep gratitude to the Brazilian Agricultural Research Corporation (Embrapa) and the University of Brasília (UnB) for their steadfast institutional support, which has sustained research on soil remineralizers for more than two decades. We also extend our sincere appreciation to all colleagues from various public institutions who comprise the Interinstitutional Group of Remineralizers. Their dedicated efforts have been instrumental in bridging the diverse, and often complex, interests of Brazil's agricultural and mineral sectors.

We sincerely acknowledge the valuable contribution of the external reviewer, Dr. Pedro Maciel de Paula Garcia, from the Universidade Federal de Mato Grosso, whose thoughtful review and constructive feedback greatly improved the quality of this manuscript.

The authors gratefully acknowledge the financial support provided by Embrapa and by FINEP—Fundo Setorial Mineral through the project *"Rede de PD&I sobre remineralizadores e agrominerais (REMAg)"* (Project No. 1541/2022), and the project *"Intemperismo acelerado de agrominerais silicáticos em solo sob pastagem no Distrito Federal"* (Edital FAP-DF 09-2022), which made possible the development of this book Chapters "Changing Time Scales: From Silicate Rocks to Agrominerals Through Bioweathering", "Chemical and Mineralogical Classification of Silicate Agrominerals", "Impact of Silicate Agrominerals on Charge Generation in Cerrado Soil", "A Protocol for Assessing the Potential of Soil Acidity Correction by Employing Silicate Agrominerals Through Short-Term Incubation", "Assessment of Soil Acidity Correction Potential Using Metabasalt", and "Optimizing Research Protocols for Evaluating the Agronomic Efficiency of Soil Remineralizers".

The authors wish to express their sincere gratitude to the VALE-Embrapa Cerrados partnership for the financial support provided through the project "*Tecnologia de manejo de estéreis de mina como insumos minerais para a fertilidade do solo agrícola (AgroGeol)*", which enabled the development of Chapter "Assessment of Soil Acidity Correction Potential Using Metabasalt" of this book.

The authors gratefully acknowledge the financial support by Center for Carbon Research in Tropical Agriculture (CCARBON) at the University of São Paulo (USP), which is sponsored by the São Paulo Research Foundation (FAPESP) under grant 2021/10573–4, which made possible the development of Chapters "Agronomic Characterization of Magnesium Soil Remineralizers and Silicate Fertilizers in Tropical Conditions" and "Agronomic Characterization of Potassium Soil Remineralizers and Silicate Fertilizers in Tropical Conditions" of this book. The first author of these chapters also wishes to thank the National Council for Scientific and Technological Development (CNPq) for the research excellence award.

The authors wish to express their sincere gratitude to the Grantham Foundation for the Protection of the Environment (UnB/Finatec—7358) and to the National Council for Scientific and Technological Development (CNPq—409283/2022-3) for their financial support and for providing scholarships to members of the research team. The authors also acknowledge the Santa Catarina Research and Innovation Foundation (FAPESC—Edital 43/2024), whose support was essential to the development of Chapters "The Potential of Remineralizers and Bioinputs as a Strategy for Sovereignty in Tropical Agriculture" and "An Overview of Research and Innovation in Stonemeal Technology in Brazil" of this book.

The authors gratefully acknowledge the financial support provided by Embrapa, and Ministry of Science, Technology and Innovation, Contract 1AAFMI/2021 "*Adequações em protocolo de eficiência agronômica em função das particularidades do remineralizadores de solo*", which made possible the development of this book Chapter "Optimizing Research Protocols for Evaluating the Agronomic Efficiency of Soil Remineralizers".

Contents

Agrogeological Principles on Soil Remineralization and Brazilian Protagonism 1
Suzi Huff Theodoro and Éder de Souza Martins

Changing Time Scales: From Silicate Rocks to Agrominerals Through Bioweathering 23
Luise Lottici Krahl, Giuliano Marchi, Elis Marina de Freitas, Luiz Fernando dos Santos, and Éder de Souza Martins

Chemical and Mineralogical Classification of Silicate Agrominerals 49
Éder de Souza Martins, Elzino Rodrigues Júnior, Giuliano Marchi, Gisele Freitas Vilela, Philipp Swoboda, Carlos Augusto Posser Silveira, Eduardo de Souza Martins, Pablo Hardoim, and Suzi Huff Theodoro

Impact of Silicate Agrominerals on Charge Generation in Cerrado Soil 69
Giuliano Marchi, João Paulo Guimarães Soares, Éder de Souza Martins, Luise Lottici Krahl, Arminda Moreira de Carvalho, Luiz Fernando dos Santos, Priscila Brelaz da Silva, and Elis Marina de Freitas

Potential of Soil Remineralization to Carbon Dioxide Removal in Agricultural Soils 97
Antonio Carlos de Azevedo

Agronomic Characterization of Calcium Magnesium Remineralizers in Tropical Conditions 109
Augusto Vaghetti Luchese, Robson Fernando Missio, Antonio Carlos Saraiva da Costa, Ivan Granemann de Souza Junior, and Alessandra Algeri

A Protocol for Assessing the Potential of Soil Acidity Correction by Employing Silicate Agrominerals Through Short-Term Incubation 125
Giuliano Marchi, Luise Lottici Krahl, Ewerton Gonçalves de Abrantes, Paula Caroline Ferreira Rodrigues, Elis Marina de Freitas, João Paulo Guimarães Soares, and Éder de Souza Martins

Assessment of Soil Acidity Correction Potential Using Metabasalt 153
Luise Lottici Krahl, Giuliano Marchi, Elis Marina de Freitas, and Éder de Souza Martins

Agronomic Characterization of Magnesium Soil Remineralizers and Silicate Fertilizers in Tropical Conditions 177
Carlos Alexandre Costa Crusciol, Sirlene Lopes Oliveira, Luiz Gustavo Moretti, and João William Bossolani

Agronomic Characterization of Potassium Soil Remineralizers and Silicate Fertilizers in Tropical Conditions 199
Carlos Alexandre Costa Crusciol, Sirlene Lopes Oliveira, José Roberto Portugal, and Éder de Souza Martins

The Potential of Remineralizers and Bioinputs as a Strategy for Sovereignty in Tropical Agriculture 235
Caroline Siqueira Gomide, Fernanda de Paula Medeiros, Andreia Matheus, Viviane Sousa Oliveira, Luiza Coimbra Oliveira, Luiz Henrique Gomes de Moura, Claudete Gindri Ramos, and Suzi Huff Theodoro

An Overview of Research and Innovation in Stonemeal Technology in Brazil 259
Manoel Batista Viana, Caroline Siqueira Gomide, Fernanda de Paula Medeiros, Claudete Gindri Ramos, and Suzi Huff Theodoro

Optimizing Research Protocols for Evaluating the Agronomic Efficiency of Soil Remineralizers 279
Carlos Augusto Posser Silveira, Adilson Luis Bamberg, Rosane Martinazzo, Éder de Souza Martins, Magda Bergmann, Andréa Sander, Luís Henrique Gularte Ferreira, Ivan dos Santos Pereira, Giuliano Marchi, and Gisele Freitas Vilela

Agrogeological Principles on Soil Remineralization and Brazilian Protagonism

Suzi Huff Theodoro and Éder de Souza Martins

Abstract The application of silicate rock dust to enhance soil fertility and quality represents a vital strategy for countries heavily dependent on imported soluble fertilizers, particularly those localized in tropical regions where agricultural soils are highly weathered. Brazil faces this scenario and, as a result, must explore innovative technological approaches to sustain agricultural productivity, economic viability, and food and environmental security. With its rich geodiversity and extensive agricultural sector, Brazil is well positioned to capitalize on technologies that harness the diverse silicate-rich rocks located near its agricultural regions. Additionally, the country benefits from a regulatory framework that sets standards to ensure the agronomic efficiency and environmental safety of these materials, defined as soil remineralizers (REM). In light of these considerations, this chapter aims to outline the key aspects of Law 12,890/2013 and Normative Instruction 05/2016, highlighting the requirements set forth by these regulations, as well as the main advantages and limitations associated with the utilization of REM. These legal measures of Brazil integrate geochemical and mineralogical parameters for rock characterization and are based on methodological principles that adhere to recognized scientific practices in agrogeology. By doing so, they illustrate the connection between regulatory criteria and existing scientific methodologies in the field. In conclusion, the combination of Brazil's vast natural resources, regulatory framework, and agricultural potential creates a favorable environment for the adoption of REM. However, to fully constrain this potential, it is essential to develop new public policies focused on financing and capacitating technicians in the use and benefits of remineralizers. Additionally, holding training courses and providing information on the benefits of applying REM for rural consultants and extension agents would significantly increase the chances of

S. H. Theodoro
Postgraduate Program in Environment and Rural Development, Universidade de Brasília, Vila Nossa Sra de Fátima, Brasília 73345010, Federal District, Brazil

É. de Souza Martins (✉)
Agrogeological Laboratory, Empresa Brasileira de Pesquisa Agropecuária - Embrapa Cerrados, BR 020, km 18, Planaltina 73310970, Federal District, Brazil
e-mail: eder.martins@embrapa.br

É. de Souza Martins and S. Huff Theodoro (eds.), *Soil Remineralizers and Silicate Fertilizers*, Frontier Studies in Soil Science,
https://doi.org/10.1007/978-3-032-14656-4_1

success. This will ensure broader implementation and maximize the positive impact of this technology on Brazilian agriculture.

Keywords Nature-based technology · Soil renewal · Regional agrominerals · Enhanced weathering

1 Introduction

For over two centuries, the use of ground silicate rocks to enhance soil fertility and quality has been studied, with significant contributions from pioneers like Hutton in the 17th century (Perman 2022), Missoux (1853), Hensel (1894), Vernadsky (1945), Graham (1941), Keller (1948), Ilchenko (1955), Ilchenko and Guimarães (1953), Barshad (1969), and Leonardos et al. (1976, 1987). However, it was only in the late 1990s and early 2000s that this approach gained widespread attention. New research findings from around the globe, including those by Kronberg et al. (1987), Leonardos et al. (2000), van Straaten (2007), Manning (2010), Ramos et al. (2022), and Theodoro et al. (2000), have highlighted its potential. The increased interest was fueled by concerns over dependence on external agricultural inputs, rising prices of soluble fertilizers in the global market, and the positive environmental and productivity impacts identified in various agroecosystems. This scenario, together with observed tangible benefits, has significantly increased awareness and driven the adoption and visibility of silicate rock applications in modern agriculture.

Brazil has emerged as a leader in research on the use of silicate rocks, producing significant findings from studies by Leonardos et al. (2000), Theodoro (2000), Theodoro and Leonardos (2006, 2015, and 2021), Martins et al. (2014), Carvalho et al. (2018), and Ramos et al. (2019, 2020, 2022), among others. Additionally, research efforts in countries such as England (Manning 2010, 2013), Australia (Coroneos et al. 1995), Indonesia (Setiawati and Mutmainnah 2016), China (Yang et al. 2021; Zhang et al. 2020), Uganda and Kenya (van Straaten 2002), and Cameroon (Tchouankoue et al. 2014) have also contributed to the understanding of this technology. These studies propose that the application of crushed silicate rocks serves as a mechanism for restoring soil fertility by introducing essential macronutrients (such as K, P, Mg, Ca, and S), micronutrients, and beneficial elements (including Si, Zn, Fe, Li, Mo, B, V, Cu, Co, and others) derived from the weathering of minerals present in silicate rocks (Burbano et al. 2022). This body of research highlights the potential of silicate rocks to enrich soils and enhance agricultural productivity.

The application of ground silicate rocks to the soil serves as a mechanism for rejuvenating and remineralizing degraded soils (Leonardos et al. 2000; Theodoro 2000; Theodoro et al. 2013, van Straaten 2006). It is well established that many high-performing agricultural soils originate from naturally occurring silicate rocks, which have been shaped by physicochemical weathering processes, volcanic activity, glacial movement, and river erosion (Keller 1948). This evidence supports the hypothesis that utilizing these materials can enhance soil fertility and quality. In Brazil, soil remineralizers (REM) are defined as mineral products derived exclusively from comminution processes for finely ground silicate rocks that are rich in basic elements and fresh primary minerals, which can weather in agricultural soils over time. These materials improve soil quality and provide essential nutrients for both microbiota and cultivated plants.

REMs are managed through a practice known as "rock meal," which involves applying ground rocks to the soil to broadly improve fertility (chemical, physical, and biological). The term "rock meal" (stonemeal, cf. Fyfe and Leonardos 2006, or stone meal, cf. Swanback 1950) was coined as "rochagem" in Portuguese (something as "rockage") by Leonardos et al. (1976) to describe the general concept. These authors defined rock meal for specific uses of each type of rock: carbonates, phosphates, sulfates, and silicates. Leonardos et al. (1976) highlight that the idea of rock meal is quite ancient, possibly originating with agriculture itself.

In Brazil, there is an increasing recognition of the value of findings from scientific literature as tools to enhance agricultural capacity, optimize production profiles, reduce costs, and ensure environmental, production safety, food, and nutrition security. This recognition is particularly strong as the studies have been conducted using established methodologies from geosciences and agronomy, lending credibility to their results.

However, no matter how good the research results were, it became necessary to establish legal norms and rules that would guarantee the proper use of silicate rock powders. It was from the enactment of Law No. 12,890 of 2013 that it was established that certain types of silicate rock powders, known as REM, are classified as a category of agricultural input, with use and trade permitted in Brazil. The Law determined that REM are materials of mineral origin that have undergone only size reduction and classification by mechanical processes, that change soil fertility indices through the addition of macro- and micronutrients to plants, and promote the improvement of physicochemical properties or biological activity of the soil (Brazil 2013). This law amended Fertilizer Law No. 6894/1980 (Brazil 1980) by introducing remineralizers as a new category of inputs, which must comply with specific rules established through subsequent regulations (Brazil 2015). This general definition of REM needed a more specific definition to differentiate it from other inputs, such as amendments and soil conditioners.

The regulation process was intensely debated for more than two years and included contributions from researchers at various research and teaching institutions, as well as federal government technicians with expertise in the topic. The outcome of these debates ultimately enabled the Brazilian Ministry of Agriculture and Livestock (MAPA) to issue Normative Instruction (NI) 05/2016 (Brazil, 2016). This NI defined the main conditions and guarantees that soil remineralizers must present to obtain registration for use, trade, and inspection in the country (Theodoro et al. 2021). Additionally, since 2020, this input has been included in the National Fertilizer Plan 2022–2050 (Emerging Chains), as provided for in Decree No. 10,991 of March 11, 2022, amended by Decree No. 11,518/2023 of May 4, 2023 (Brazil 2023). This Decree stipulated that REM production should reach 18 million annual tons by 2050.

With these considerations in mind, the primary objective of this chapter is to discuss the agrogeological foundations that underpin the establishment of minimum rules and guarantees for soil remineralizers in Brazilian regulations, serving as a reference for initiatives in other countries. In addition, we will present a current diagnosis of remineralizers that have already been registered in Brazil, as well as the impact and benefits of this new category of input on ensuring agricultural production.

2 Global and Regional Agrominerals

Agrominerals refer to all mineral-origin sources used for soil fertility management and plant nutrition (van Straaten 2007). These key agrominerals can be categorized based on their principal anions, such as chloride, sulfate, carbonate, phosphate, and silicate (Table 1). The primary associated cations include calcium (Ca), magnesium (Mg), and potassium (K).

Salts associated with chlorides and sulfates are highly soluble in water, while carbonates, phosphates, and borates exhibit low solubility. Silicates, despite having

Table 1 Types of agrominerals classified according to the principal anion and cation association, with examples of existing products for agricultural use

Main anion		Rock source	Main cations	Crust abundance		Water solubility	Agricultural products examples
				area %	mass %		
Chloride	Cl^-	Evaporitic deposits (sedimentary)	K	0.00	< 1	Very high	KCl (potash)
Sulphate	SO_4^{2-}	Evaporitic deposits (sedimentary)	Ca, Mg, (K)	0.00	< 1	Very high	Gypsum; Polyhalite
Carbonate	CO_3^{2-}	Limestone (sedimentary) Carbonatite (igneous) Marble (metamorphic)	Ca, Mg, (K)	15.0	2	Low	Lime
Phosphate	PO_4^{3-}	Phosphorite (sedimentary) Phoscorite (igneous)	Ca	0.00	< 1	Low	Phosphate rock
Borate	BO_3^{3-}	Borate (sedimentary)	Ca	0.00	< 1	Low	Ulexite
Silicate	SiO_4^{4-}	Sedimentary Igneous Metamorphic	Ca, Mg, K	85.0	93.0	Very low	Glauconite, basalt, mica schist

very low water solubility, are the most abundant, covering approximately 85% of the Earth's surface (Goldscheider et al. 2020). In second place are carbonates, which cover 15% of the Earth's surface. Conversely, salts (including chlorides and sulfates) and phosphates are relatively rare, covering less than 0.01% of the Earth's surface.

Agrominerals can be categorized into two major groups: global and regional. *Global agrominerals* are raw materials to supply fertilizer commodities, capable of being transported over distances of tens of thousands of kilometers between continents. In contrast, *regional agrominerals* are raw materials used to develop regional products. These products are typically viable for use within the same region where they are produced, usually within a few hundred kilometers of their origin, depending on logistical models (Lompo et al. 2018; van Straaten 2022).

The advantages of products derived from global agrominerals include their long-distance transportation capabilities, the need for application doses on the order of 10^2 kg per hectare, and high-water solubility, allowing for direct nutrient absorption by plant roots in the soil solution (Havlin and Heiniger 2020). However, these agrominerals represent finite resources (Walan et al. 2014; Al Rawashdeh 2020) and their derived inputs exhibit low nutrient utilization efficiency in agricultural systems (Barłóg 2023), as they are rapidly washed out of soil, along with high energy costs for production and transportation (Wu et al. 2021). Excessive use of these fertilizers also degrades soil quality (Newton et al. 2020).

On the other hand, products derived from regional agrominerals, such as silicate and carbonate sources, are characterized by their high abundance, low relative concentrations, and application doses on the order of 10^3 kg per hectare (Martins et al. 2023). Silicate rocks constitute the primary parent material of soils (Macias and Chesworth 1992) and are the main sources of mineral nutrients necessary for the development of plants and animals (van Straaten 2006, 2007). Despite their ubiquitous occurrence, silicates were not considered agrominerals during the Green Revolution, which began in the 1960s. This was primarily due to their relatively low nutrient content and very low water solubility, which made them the last option for developing inputs aimed at soil fertility management. (Martins et al. 2014). Nevertheless, it has long been known that they weather (Goldich 1938), and that silicate minerals that they contain are thermodynamically unstable at the Earth's surface (Curtis 1976; Manning 2022). Kinetic factors control the decomposition of silicate minerals (Manning 2022).

However, numerous studies with silicate minerals have shown that biological processes associated with the rhizosphere of plants, especially the action of microorganisms, promote reactions compatible with the agronomic time scale, a process defined as bioweathering (Hinsinger et al. 1992, Hinsinger and Gilkes 1997, Hinsinger 2001, Krahl et al. 2020, 2022a, 2022b, Paredes Filho et al. 2020; Jafari et al. 2021; Baptista et al. 2022), and so accelerate the reaction kinetics.

Another essential nutrient for plants is nitrogen (N), the main reservoir of which is the atmosphere. The industrial synthesis of nitrogen fertilizers from atmospheric N combined with fossil natural gas currently constitutes the primary source for agriculture, mainly in the form of urea (Milani et al. 2022). Brazil is a pioneer in adopting biological nitrogen fixation as a large-scale solution, particularly for soybean production, having developed the first successful bioinoculants since the 1970s and 1980s.

Among regional agrominerals, agricultural limestone began its intensive use in Brazil in the 1970s to reduce the availability of toxic aluminum and increase base saturation, allowing intensive agricultural use in acidic soils. Systematic research for the use of silicate rocks began in the 2000s, as will be seen later, complementing existing soil fertility management practices.

The strategies for using regional agrominerals allow for soil improvement (Barker et al. 1998; Swoboda et al. 2022; Martins et al. 2023), higher food quality (Adamović et al. 2011; Reis et al. 2021), and a greater potential for soil carbon sequestration (Beerling et al. 2020). Economically, it enables the development of regional production chains, thereby reducing the flow of resources to foreign entities. The social impacts are also significant, as they create several direct and indirect jobs related to production chains in the same regions where agriculture is developed (Liverpool-Tasie et al. 2020).

The adoption and consolidation of regional agrominerals also have the potential to increase nutrient use efficiency in agricultural production systems, reducing nutrient waste from sources with high concentration and solubility. The expectation is to further extend the useful lifespan of fertilizer commodities with the more significant incorporation of regional sources. This is the proposal of the National Fertilizer Plan 2050, which aims to develop emerging chains of remineralizers, bioproducts, and nanotechnology to help increase the efficiency of nutrient use and, consequently, that of fertilizer commodities themselves (Brazil 2023). Silicate agrominerals (ASi) are derived directly from natural materials or industrial processes, composed of more than 50% silicates, and demonstrate agronomic efficiency as fertilizers, conditioners, and/or soil remineralizers (REM). ASi are not well-known in terms of their agronomic behavior, despite their greater abundance in the Earth's crust.

3 Soil Remineralization as Nature-Based Technology

Natural rock comminution processes, such as those caused by glacial and volcanic activity, can also generate sources of products for soil fertility management (Gunnarsen et al. 2019; Minasny et al. 2021). Various studies on bioweathering, especially those related to the interaction of plant rhizospheres with aluminosilicates, show that weathering rates can be very high, transforming the primary mineral present in ground rock on a timescale of 10^{-1}–10^{0} years while forming new functional mineral phases (Hinsinger et al. 1992; Hinsinger 2001; Burford et al. 2003; Bonneville et al. 2011; Gunnarsen et al. 2019; Krahl et al. 2020, 2022a, Minasny et al. 2021).

The transformation of primary silicate minerals in soil is a well-established process and constitutes the same process as soil remineralization. It can be understood through the general chemical weathering reaction of base-rich silicates by hydrolysis (Eq. 1):

$$\text{AlSi(Ca, Mg, K, Na}\ldots)\text{s} + (\text{H}_2\text{O})\text{aq} + \left(\text{H}^+\right)\text{aq} + \text{bio activity} \rightarrow (\text{new phases})\text{s} + (\text{Si, Ca, Mg, K, Na}\ldots)\text{aq} + \left(\text{OH}^-\right)\text{aq} \quad (1)$$

In this equation, "s" represents the rock-forming aluminosilicate minerals rich in bases and the mineral phases as reaction products, and "aq" represents the elements released in solution through weathering.

The foundations of soil remineralization are based on natural bioweathering processes, which regulate the rates and forms of dissolution (congruent and incongruent) in the soil, the action of roots in mineral dissolution, interaction with microorganisms, and the formation of charges on mineral particles (Pinheiro 2018; Krahl et al. 2022a). Congruent dissolution occurs when all constituents of the mineral are solubilized at the same rate, while incongruent dissolution happens when certain elements are released into the solution at different rates. Additionally, other beneficial effects co-occur, such as the stabilization of organic matter promoted by REMs, carbon sequestration through soil alkalinization, and carbon stabilization on mineral surfaces, soil reactions, mineral transformation in the soil, residual effects, and effects on soil and plant growth in relation to the mineralogical composition of each material, among others.

These direct benefits are associated with the chemical weathering process of aluminosilicates, a natural process that transforms rocks into soils. For this reason, soil remineralization is a nature-based technology that is accelerated by the comminution process. One issue that has complicated the understanding of the potential use of REMs is the concept that aluminosilicates have very low solubility in water and would not weather on an agronomic timescale. While this transformation process occurs in nature on scales of 10^3–10^6 years, in soil remineralization, it takes place over 10^0–10^1 years (Soares et al. 2022; Swoboda et al. 2022).

Some silicate minerals present in rocks contain nutrients required for plant growth. These nutrients are primarily released by the action of the plant root system. The main nutrients released by the rocks are Ca, Mg, K, and micronutrients such as B, Cu, Mn, and Zn (Marchi et al. 2020). Additionally, they mitigate soil acidity by releasing base cations during mineral weathering of Ca and Mg-bearing silicates (Krahl et al. 2020). They also have the potential to enhance agricultural products (Tessaro et al. 2013; Dias et al. 2018; Miranda et al. 2018; Rinaldi et al. 2022). At the same time, they promote increased cation exchange capacity (CEC) and pH in the soil, enhance nutrient supply, reduce nutrient loss, and stimulate biological activity (Medeiros et al. 2024). These general characteristics of REMs are very similar to those of soil conditioners. However, REMs are the only products that generate new mineral phases of both low and high crystallinity, which remain in the soil over the long term and modify its physical and chemical characteristics with management, thereby increasing specific surface area, cation exchange capacity, and water retention (Krahl et al. 2020, 2022a).

The new mineral phases can be of short-range order minerals (Hinsinger 2001; Krahl et al. 2020) or high crystallinity minerals (Hinsinger et al. 1991, Hinsinger and Gilkes 1997; Krahl et al. 2022a, b). The presence of these new mineral phases alters the physical and chemical characteristics of the soil (Santos et al. 2021). The elements released into solution contribute to nutrient supply and increase pH through the dissociation of silicate bases (Conceição et al. 2022). Thus, Eq. (1) highlights the three direct effects of soil remineralization: providing nutrients, increasing pH, and forming new functional mineral phases.

4 Brazilian Normative Instruction 05 Criteria

Throughout the process of discussing and elaborating on NI 05/2016, it has been consistently understood that, on average, the geochemical and mineralogical composition of the Earth's crust (both continental and oceanic) meets the main criteria defined in the Normative (Martins et al. 2023).

Nevertheless, in addition, it was necessary to establish some conditions, including: (i) the total sum of bases in oxide forms—SBr (K_2O, MgO and CaO) must be equal to or greater than 9% (w/w), with K_2O must be greater than 1%; (ii) the proportion of free silica (SiO_2 in the form of quartz) must be equal to or less than 25% of the total; (iii) the maximum allowable levels for four potentially toxic elements - PTE (Hg, Pb, As, Cd). In addition to geochemical and mineralogical aspects, NI 05/2016 also established that remineralizers must have a granulometric range that allows for different reactivity times in the short, medium, and long term (Fig. 1). The same NI 05/2016 defined that the presence of phosphorus and micronutrients can be declared on the label of registered products, as additional attributes.

Confirming this outline, NI 05 also established the need to carry out agronomic tests (art.9 of NI 05/2016), to verify the agronomic effectiveness and benefits to soils and plants. Such studies must be coordinated by official institutions or MAPA-accredited institutions, conducted with plants, in a greenhouse or in the field, optionally complemented with incubation tests or in leaching columns that demonstrate the agronomic efficiency of the different REM.

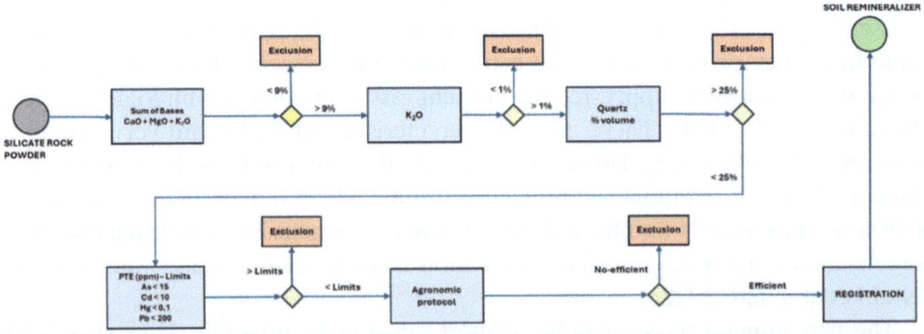

Fig. 1 Diagram showing the conditions that crushed silicate rocks must present in order to be registered as soil remineralizers

Table 2 Average geochemical composition in the form of oxides of the Earth's crust (Wedepohl 1995) and soil (Winogradow 1954)

Oxides (weight%)	Earth crust	Average soil
SiO_2	61.5	70.6
Al_2O_3	15.1	13.5
Fe_2O_3	6.28	5.4
K_2O	2.4	1.6
Na_2O	3.2	0.85
CaO	5.5	1.9
MgO	3.7	1.0
P_2O_5	0.18	0.18
MnO	0.10	0.11
TiO_2	0.68	0.77
Carbon	0.2	3.8
Total	98.84	99.11

Note that the average composition of the Earth's crust and the average composition of soils are directly related (Table 2), where the main elements (oxygen, silicon, and aluminum) are present in high proportions. The bases are present at lower levels in the soil, while organic matter occurs at high levels in the soil. Oxygen makes up around 50% of the Earth's crust and average soil. Silicon is present at higher levels in average soil than in the Earth's crust.

These general similarities in the average composition of the Earth's crust and the composition of soils are directly related to the soil formation process, where rocks serve as their source materials. The increase in silicon is associated with the accumulation of clay minerals and quartz, a mineral composed of SiO_2 that is highly resistant to weathering. At the same time, the more mobile elements, consisting of bases such as calcium, magnesium, potassium, and sodium, tend to decrease in soils due to leaching processes (Leonardos et al. 1987). Organic matter constitutes one of the systematic responses of soils to increase the residence time of nutrient elements in their interaction with biota (Kleber et al. 2021).

The total bases in the form of oxides ($SBr = CaO + MgO + K_2O$) in the Earth's crust amount to 11.6%. The proportion of quartz in the average mineralogical composition of the crust is 12% (Clarke and Washington 1924), 18.4% in sedimentary rocks, and 22.5% in continental crust rocks (Yaroshevsky and Bulakh 1994). In other words, the average composition of the Earth's crust meets the geochemical and mineralogical criteria established by NI 05/2016.

These facts are consistent with nature itself, as remineralization fundamentally involves the geochemical and mineralogical rejuvenation of soils. The timescale for soil formation ranges from 10^3 to 10^6 years, but the application of ground rocks considerably accelerates this process. The comminution of rocks represents a stage of physical weathering processed industrially. The weathering speed of remineralizers (REM) in soils can occur on the agronomic timescale, depending on factors such as REM granulometry, climate, and soil management practices. It ranges from 10^2 to

10^0 years, as it increases the specific surface area and the contact with the action of solutions and biota on the minerals.

5 Soil Remineralizers as Plentiful Regional Agrominerals

The REMs can have different chemical, physical, and mineralogical compositions, derived from various igneous, sedimentary, and metamorphic silicate rocks, which can exhibit basic, ultrabasic, alkaline, ultramafic, and calcsilicate compositions. Basic rocks, such as basalts, diabases, and gabbros, as well as their metamorphic derivatives, including amphibolite and basic granulite, are the most abundant REM sources (Beerling et al., 2020; Martins et al., 2024). These rocks are sources of calcium, magnesium, and silicon in addition to several micronutrients, such as Fe, Mn, Cu, Zn, Mo, and V, among others. At the same time, new mineral phases of low crystallinity are formed through the bioweathering process in the rhizosphere interaction, characterized by high specific surface area and high surface electric charge (Hinsinger 2001). Due to their high abundance on all continents (Fig. 2) and relatively high base content, they are considered excellent sources for promoting carbon sequestration in soil through the processes of enhanced weathering and carbonation in agricultural soils (Beerling et al. 2018, 2020; Renforth 2012, 2019).

Ultrabasic rocks, such as dunites and pyroxenites, and their metamorphic derivatives, including serpentinites, are sources of silicon and magnesium (Crusciol et al. 2019; Moretti et al. 2019, 2021). Despite being much less abundant than the basic rocks (Fig. 3), they also have potential for use as an acidity amendment (Moretti et al. 2021), carbon sequestration (Kelemen et al. 2020), and micronutrient supply, such as nickel (Beerling et al. 2018; Guo et al. 2023).

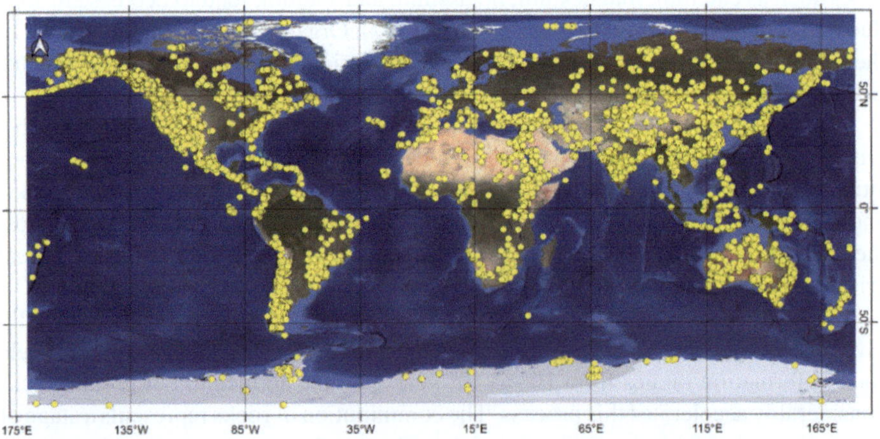

Fig. 2 Major worldwide occurrences of basalt spills (data extracted from EARTHCHEM, https://www.earthchem.org/)

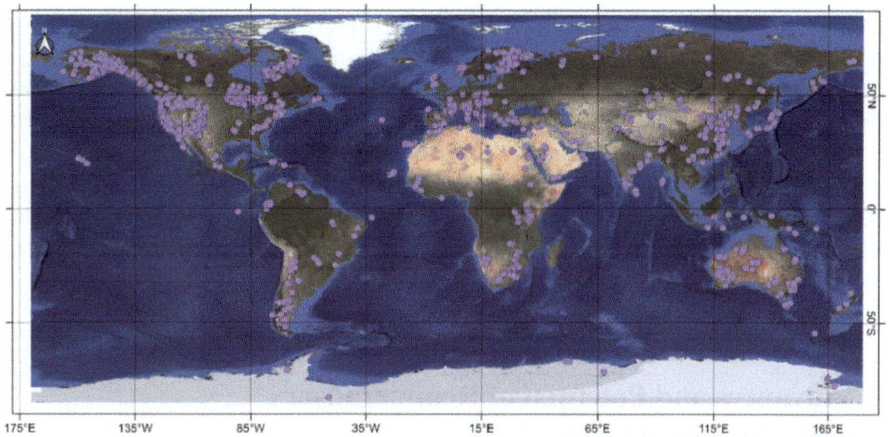

Fig. 3 Main world occurrences of ultramafic rocks (data extracted from EARTHCHEM, https://www.earthchem.org/)

Alkaline ultramafic rocks are the REM with the highest existing nutrient content, reaching base sums above 25%, and significant phosphorus contents above 1% P_2O_5 (Guo et al. 2014). Like ultrabasic rocks, alkaline ultramafic rocks (Fig. 4) also promote correction of soil acidity, increased base saturation, and availability of silicon and micronutrients to the soil and plants (Kelemen et al. 2020; Medeiros et al. 2025).

Silicate rocks with potential as potassium sources must present minerals with high chemical and biological reactivity. K-feldspar (microcline and orthoclase) and muscovite tend to have lower weathering potential and may release potassium in the

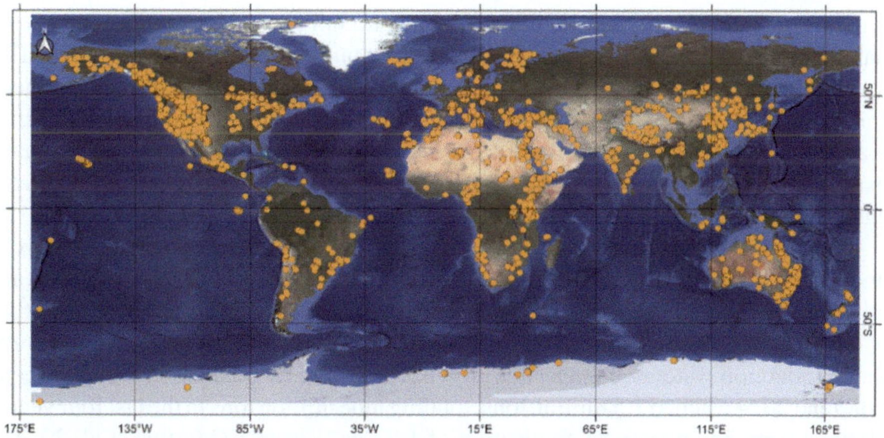

Fig. 4 Main world occurrences of alkaline rocks (data extracted from EARTHCHEM, https://www.earthchem.org/)

medium and long term (Manning and Theodoro 2018). All other silicate potassium minerals, such as feldspathoids and biotite, have a high potential to make potassium available from the first cycle (Aisha and Taalab 2008; Badr 2006; Setiawati and Mutmainnah 2016). The most common silicate potassium minerals that have potential as nutrient sources are K-feldspar, muscovite and biotite (Barral-Silva et al. 2005; Mohammed et al. 2014), which occur in igneous, metamorphic and sedimentary rocks (Shirale et al. 2019). Glauconite is a clay mineral that occurs in certain marine pelitic sedimentary rocks in environments with redox oscillations (Rudmin et al. 2019).

6 Advances and Potential of Remineralizers after Regulatory Standards

Since the regulation of Law No. 12,890/2013, complemented by Normative Instruction No. 05/2016 and Decree 8384/2014 (Brazil 2016), the Brazilian market for soil remineralizers (REM) has experienced exponential growth. The registration of these inputs has been intensifying at the Ministry of Agriculture, Livestock, and Supply (MAPA),[1] with a particularly sharp increase in the last three years, driven, in part, by the scarcity crisis and high prices of conventional soluble fertilizers and the high demand from various profiles of farmers (family and business).

This rapid growth is reflected in the number of registrations at MAPA: between 2019 and 2023, there was a 500% increase, jumping from 17 to 68 approved products (Brazil, 2023), as can be seen in Fig. 5. The potential for this number to grow even further is quite vigorous, considering the country's mineral profile and the trend of increased use, indicating that the PNF projections (18 million annual tons of remineralizers by 2050) could be reached very quickly. This expansion potential remains robust, considering: (i) the abundance of suitable mineral resources in the national territory; (ii) the increasing adoption by farmers, due to costs significantly lower than the values of soluble fertilizers (currently, a ton of REM can be obtained for values between U$ 28 and 82) and (iii) the prospect that the projections of the National Fertilizer Plan (18 million tons/year by 2050) can be reached earlier than expected. This scenario consolidates REM as a strategic alternative for Brazilian agriculture, aligning productivity with sustainability.

In this context, Brazil has consolidated itself as a global leader in both the research and practical application of ground silicate rocks in agriculture, also boasting an advanced regulatory framework for these inputs (Theodoro et al. 2022). Currently, about 10% of Brazilian agricultural properties already incorporate these inputs into their production systems. To meet this growth, the average annual production between 2020 and 2024 reached 1.5 million tons, totaling nine million tons in the last four years alone. Assuming an average application of 1.5 t ha^{-1} $year^{-1}$ (Martins et al. 2024), the benefited area reaches approximately four million hectares, which represents

[1] https://mapa-indicadores.agricultura.gov.br/publico/extensions/Fertilizantes/Fertilizantes.html.

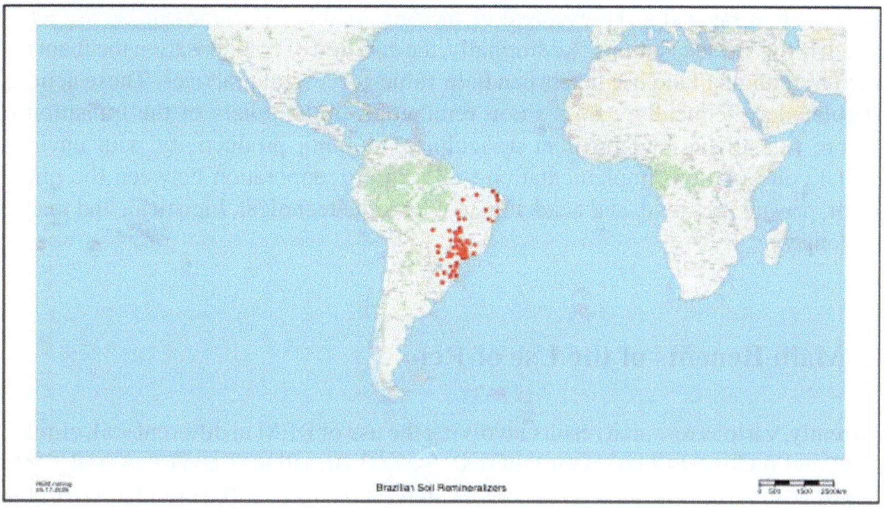

Fig. 5 Brazilian production units of soil remineralizers. In December 2024 there were 75 units (data from Martins et al. 2024)

about 6.25% of the country's arable land, which according to Embrapa totals 64 million hectares. This exponential growth positions REM as increasingly relevant alternatives in the Brazilian agricultural input matrix.

Although the data indicate vigorous and rapid growth in the use of REM as an agricultural input, much still needs to advance to implement the necessary structuring actions that can transform this technology into a vector of sustainable development for the country. According to Theodoro (2021), among the main strategic measures, the following stand out: (i) Integrated Public Policies, with the development of a complementary regulatory framework and financial and technical incentive programs for the adoption of REM by farmers; (ii) diversification of the supply of different products (diversity of macro and micronutrients) in order to expand the variety of REM to meet specific nutritional demands, different soil profiles, and regional realities; (iii) implement financing and technical support mechanisms, through specific rural credit lines and specialized technical assistance for the efficient use of inputs, considering the various Brazilian agroecosystems and the needs of agricultural crops; (iv) promotion of research and development through investments in applied studies to optimize formulations and application techniques; (v) professional training of different technician profiles, considering the interdisciplinary training of geologists, agronomists, and agricultural technicians with a focus on integrating geodiversity and agroecosystem needs; (vi) recognition of REM as a strategic input through its inclusion in new national food security and biodiversity policies; (vii) development of regional production chains, which value the creation of commercial routes that value territorial specificities and integrate mineral and agricultural production.

Beyond these, other measures need to be implemented, such as valuing the gains related to the sustainability of the agricultural-mineral sector and its adherence to practices that favor the achievement of the Sustainable Development Goals, established by the United Nations. Additionally, the creation of compensation mechanisms (such as reducing land use taxes) can help value successful practices. These actions, articulated systemically, can position remineralizers as pillars of the transition to a more sustainable and resilient agriculture, aligning productivity with environmental conservation. Implementation will require cooperation between the public sector, private initiative, and academia to overcome technical, logistical, and market challenges.

7 Main Benefits of the Use of Rem

Currently, various research results involving the use of REM in different soil, climate, plant, and rock types have been reported. Several Brazilian (Leonardos et al. 2000; Ramos et al. 2022; Theodoro et al. 2022; Cunha and Almeida 2021) and international (Swoboda et al. 2022; Manning 2018; van Straaten 2022) authors have demonstrated the benefits derived from the application of this type of input. Among them, the authors highlight the direct and indirect benefits.

Direct benefits:

Rejuvenation and Remineralization of Weathered Soils: This process enhances soil health (Carvalho et al. 2018; Hinsinger and Gilkes 1997; Kronberg et al. 1987; Leonardos et al. 2000; Theodoro and Leonardos 2015; van Straaten 2007).

Nutrient Efficiency: Plants selectively absorb nutrients necessary for their growth and development. The surplus nutrients remain within the system as secondary minerals, creating a nutrient bank that is available for an extended period (Theodoro and Leonardos 2006, 2015).

Enhanced Soil Fertility: This is achieved through increased pH, cation exchange capacity (CEC), and the availability of essential nutrients such as calcium (Ca), magnesium (Mg), potassium (K), phosphorus (P), silicon (Si), iron (Fe), zinc (Zn), copper (Cu), boron (B), and molybdenum (Mo) (Boniao et al. 2002; Marchi et al. 2020; Medeiros et al. 2024, 2025; Momesso et al. 2022; Swoboda et al. 2022).

Reduced Acidity: The process lowers aluminum (Al) toxicity and potential acidity, making the soil more conducive to plant growth (Conceição et al. 2022).

Increased Productivity: There has been a notable rise in the yield of various crops, including soybeans, corn, sugarcane, rice, and coffee (with improved quality), as well as fruit crops (Burbano et al. 2022; Ramos et al. 2021; Chiomento et al. 2021).

Indirect benefits:

Improved soil health: more stable structure, increased water infiltration, and intensified microbial activity (Curtis et al. 2023; Fang et al. 2023; Azeem et al. 2022; Busato et al. 2022; Basak et al. 2021).

Carbon sequestration: Contribution to mitigating climate change through atmospheric CO_2 capture (Beerling et al. 2018, 2020; Manning et al. 2024; Medeiros et al. 2025; Renforth et al., 2015).

Low costs: These products are less expensive than soluble fertilizers (Manning and Theodoro 2018; Theodoro et al. 2022).

Regional input chain: Diversification of economic gains by considering new business opportunities in the mineral sector, such as the use of waste and/or by-products from the production of other mineral goods. As examples, ornamental rocks or materials used in civil construction and infrastructure (Martins et al. 2024).

Beyond being a sustainable alternative, remineralizers are emerging as a pivotal solution for global food security and climate resilience. Their ability to function as a 'multi-nutrient bank' offers a path to transition the current agricultural model—which is overly dependent on inputs and land—towards a regenerative system that is crucial for decarbonizing agriculture on a global scale, reducing the dependence on synthetic fertilizers, promoting circular economies and increasing soil resilience, with positive effects throughout the agricultural production chain. These results reinforce the strategic role of remineralizers for a more sustainable and productive agriculture, aligning agronomic, environmental, and economic benefits.

8 Conclusion

The use of soil remineralizers represents a transformative approach in agriculture, providing a sustainable method to rejuvenate the fertility of depleted soils across various agroecosystems, with up-and-coming applications in tropical regions. Unlike conventional fertilizers, the gradual release of nutrients from soil remineralizers enhances resource efficiency, mimicking the natural nutrient cycles that have sustained ecosystems for thousands of years.

In Brazil, the exploration of soil remineralizers began over 50 years ago. However, the method only gained substantial traction in the early 2000s, primarily due to the increasing reliance on costly imported soluble fertilizers. This shift was driven by a pressing need for indigenous strategies that promote both environmental sustainability and agricultural productivity.

Extensive research and favorable outcomes, documented in both national and international studies, have led to the establishment of a comprehensive legal framework in Brazil, guiding the appropriate use of soil remineralizers. This strategic move underscores Brazil's pioneering role in the field, as the nation stands as the first to develop legal and methodological guidelines essential for the widespread implementation of this innovative technology.

Acknowledgments We express our deep gratitude to the Brazilian Agricultural Research Corporation (Embrapa) and the University of Brasília (UnB) for their continuous institutional support, which has sustained research into the use of rock dust (remineralizers) for over two decades. Our recognition also extends to all colleagues from various public institutions who make up the Interinstitutional Group of Remineralizers. Their effort is crucial in reconciling the distinct, and sometimes complex, interests of the agricultural and mineral sectors in Brazil.

Declaration of Competing Interest The authors declare that they have no known competing financial interests or personal relationships that could have appeared to influence the work reported in this paper.

References

Adamović M, Stojanović M, Grubišić M, Ileš D, Milojković J (2011) Importance of aluminosilicate minerals in safe food production. Maced J Anim Sci 1(1):175–180

Aisha HA, Taalab AS (2008) Effect of natural and/or chemical potassium fertilizers on growth, bulbs yield and some physical and chemical constituents of onion (*Allium cepa L.*). Res J Agric Biol Sci 4(3):228–237

Al Rawashdeh R (2020) World peak potash: an analytical study. Resour Policy 69:101834. https://doi.org/10.1016/j.resourpol.2020.101834

Azeem M, Haider MZ, Javed S, Saleem MH, Alatawi A (2022) Drought stress amelioration in maize (Zea mays L.) by inoculation of bacillus spp. strains under sterile soil conditions. Agriculture 12(1):50. https://doi.org/10.3390/agriculture12010050

Badr MA (2006) Efficiency of K-feldspar combined with organic materials and silicate dissolving bacteria on tomato yield. J Appl Sci Res 2:1191–1198

Baptista JDC, Gray ND, Tarumoto MB, Singleton I, McCann CM, Manning DAC (2022) Bacterial communities in soils as indicators of the potential of syenite as an agromineral. Pesq Agrop Brasileira 57:e01414

Barker WW, Welch SA, Chu S, Banfield JF (1998) Experimental observations of the effects of bacteria on aluminosilicate weathering. Am Mineral 83(11–12 Part 2):1551–1563. https://doi.org/10.2138/am-1998-11-1243

Barłóg P (2023) Improving fertilizer use efficiency—methods and strategies for the future. Plants 12(20):3658. https://doi.org/10.3390/plants12203658

Barral-Silva M, Hermo BS, García-Rodeja E, Freire NV (2005) Reutilization of granite powder as an amendment and fertilizer for acid soils. Chemosphere 61(7):993–1002

Barshad I (1969) Chemistry of soil development. In: Bear F (ed) Chemistry of the soil, 2nd ed. Reinhold, New York, pp 160:01–70

Basak B, Sarkar B, Naidu R (2021) Environmentally safe release of plant available potassium and micronutrients from organically amended rock mineral powder. Environ Geochem Health 43:3273–3286. https://doi.org/10.1007/s10653-020-00677-1

Beerling DJ, Leake JR, Long SP, Scoles JD, Ton J, Nelson PN, Bird M, Kantzas E, Taylor LL, Sarkar B, Kelland M, DeLucia E, Kantola I, Muller C, Rau G, Hansen J (2018) Farming with crops and rocks to address global climate, food and soil security. Nat Plants 4:138–147. https://doi.org/10.1038/s41477-018-0108-y

Beerling DJ et al (2020) Potential for large-scale CO2 removal via enhanced rock weathering with croplands. Nature 583:242–248. https://doi.org/10.1038/s41586-020-2448-9

Boniao RD, Shamshuddin J, Van Ranst E, Zauyah S, Omar SRS (2002) Changes in chemical properties and growth of corn in volcanic soils treated with peat, ground basalt pyroclastics, and calcium silicate. Commun Soil Sci Plant Anal 33(7–8):1219–1233. https://doi.org/10.1081/CSS-120003883

Bonneville S, Morgan DJ, Schmalenberger A, Bray A, Brown A, Banwart SA, Benning LG (2011) Tree-mycorrhiza symbiosis accelerate mineral weathering: evidences from nanometer-scale elemental fluxes at the hypha–mineral interface. Geochim Cosmochim Acta 75(22):6988–7005

Brazil (1980) Lei 6,894/1980. Available: https://www.planalto.gov.br/ccivil_03/leis/1980-1988/l6894.htm
Brazil (2013) Lei 12.890/2013 de 10 de dezembro de 2013—Altera a Lei no 6.894, de 16 de dezembro de 1980. Available: https://www.planalto.gov.br/ccivil_03/_ato2011-2014/2013/lei/l12890.htm
Brazil (2015) Decreto No. 8, 384—Regulamenta a Lei No 6.894, de 16 de dezembro de 1980. http://www.planalto.gov.br/ccivil_03/_ato2011-2014/2014/decreto/D8384.htm
Brazil (2016) Ministério da Agricultura, Pecuária e Abastecimento—Brasil Instrução Normativa N° 05/ 2016. https://www.in.gov.br/materia/-/asset_publisher/Kujrw0TZC2Mb/content/id/21393137/do1-2016-03-14-instrucao-normativa-n-5-de-10-de-marco-de-2016-21393106
Brazil (2023) Ministério da Indústria, Comércio Exterior e Serviços. Secretaria de Desenvolvimento Industrial, Inovação, Comércio e Serviços. Plano nacional de fertilizantes 2050: uma estratégia para os fertilizantes no Brasil/Ministério da Indústria, Comércio Exterior e Serviços, Secretaria de Desenvolvimento Industrial, Inovação, Comércio e Serviços. – Brasília: SDIC/MGI, 197p. https://www.gov.br/mdic/pt-br/assuntos/sdic/confert/pnf/pnf-v-08-06-12-23.pdf, Accessed Nov 24, 2025
Burbano DFM, Theodoro SH, Carvalho AXM, Ramos CG (2022) Crushed vulcanic rock as soil remineralizer: a strategy to overcome the global fertilizer crisis. Nat Resour Res 31:2197–2210. https://doi.org/10.1007/s11053-022-10107-x
Burford EP, Kierans M, Gadd GM (2003) Geomycology: fungi in mineral substrata. Mycologist 17(3):98–107
Busato J, Santos L, Paula A, Sodré F, de Oliveira A, Dobbss L, Martins E, Jindo K (2022) Can co-application of silicate rock powder and humic-like acids increase nutrient uptake and plant growth in weathered tropical soil? Acta Agric Scand Sect B: Soil Plant Sci, Sect B—Soil Plant Sci 72(1):761–774
Carvalho AMX, Cardoso IM, Souza MEP, Theodoro SH (2018) Rochagem: o que se sabe sobre essa técnica? In: Cardoso I, Fávero C. Solos e Agroecologia. 1st edn. vol 4. Embrapa, Brasília, pp 103–128, ISBN: 978–85–7035-774-8
Chiomento JLT, Korchagin J, Fornari M, Brugnera LP, dos Santos Trentin T, Dornelles AG, Zanin E, Borges LM (2021) Agronomic performance of strawberry cultivated in substrate with rock powder. Braz J Dev 7(2):15646–15657. https://doi.org/10.34117/bjdv7n2-267
Clarke FW, Washington HS (1924) The composition of the earth's crust. US government printing office, USGS, professional paper 127, p 117
Conceição LT, Silva GN, Holsback HMS, Oliveira CF, Marcante NC, Martins ÉS et al (2022) Potential of basalt dust to improve soil fertility and crop nutrition. J Agric Food Res 10:100443
Coroneos C, Hinsinger P, Gilkes RJ (1995) Granite powder as a source of potassium for plants: a glasshouse bioassay comparing two pasture species. Fertil Res 45:143–152
Crusciol CAC, Moretti LG, Bossolani JW, Moreira A, Micheri PH, Rossi R (2019) Can dunite promote physiological changes, magnesium nutrition and increased corn grain yield? Commun Soil Sci Plant Anal 50(18):2343–2353
Cunha GO, Almeida JA (2021) Agronomic potential of four rock powders, pure or mixed, as soil remineralizers. Res, Soc Dev 10(17). https://doi.org/10.33448/rsd-v10i17.24828
Curtis CD (1976) Stability of minerals in surface weathering reactions: a general thermochemical approach. Earth Surf Process 1:63–70
Curtis J, Luchese A, Missio R (2023) Application of soil remineralizer to poultry liter as an efficient and sustainable alternative for fertilizing maize crop. J Plant Nutr:423–438. https://doi.org/10.1080/01904167.2022.2071728
Dias KGDL, Guimarães PTG, Carmo DL, Reis THP, Lacerda JJDJ (2018) Alternative sources of potassium in coffee plants for better soil fertility, productivity, and beverage quality. Pesq Agrop Brasileira 53:1355–1362
Fang Q, Lu A, Hong H et al (2023) Mineral weathering is linked to microbial priming in the critical zone. Nat Commun 14:345. https://doi.org/10.1038/s41467-022-35671-x

Fyfe WS, Leonardos OH (2006) Theodoro SH (2006) sustainable farming with native rocks: the transition without revolution. An Acad Bras Ciências 78:715–720
Goldscheider N, Chen Z, Auler AS, Bakalowicz M, Broda S, Drew D et al (2020) Global distribution of carbonate rocks and karst water resources. Hydrogeol J 28:1661–1677
Goldich SS (1938) A study in rock-weathering. J Geol 46:17–58
Graham RR (1941) Colloidal organic acids as factors in the weathering of anorthite. Soil Sci 52:291–295
Gunnarsen KC, Jensen LS, Gómez-Muñoz B, Rosing MT, de Neergaard A (2019) Glacially abraded rock flour from Greenland: potential for macronutrient supply to plants. J Plant Nutr Soil Sci 182(5):846–856
Guo F, Sun H, Yang J, Zhang L, Mu Y, Wang Y, Wu F (2023) Improving food security and farmland carbon sequestration in China through enhanced rock weathering: field evidence and potential assessment in different humid regions. Sci Total Environ 903:166118
Guo P, Niu Y, Yu X (2014) A synthesis and new perspective on the petrogenesis of kamafugites from west Qinling, China, in a global context. J Asian Earth Sci 79:86–96
Havlin J, Heiniger R (2020) Soil fertility management for better crop production. Agronomy 10(9):1349
Hensel J (1894) Bread from stones: a new and rational system of land fertilization and physical regeneration. AJ Tafel
Hinsinger P, Jaillard B, Dufey JE (1992) Rapid weathering of a Trioctahedral mica by the roots of ryegrass. Soil Sci Soc Am J 56(3):977–982. https://doi.org/10.2136/sssaj1992.03615995005600030049x
Hinsinger P, Gilkes RJ (1997) Dissolution of phosphate rock in the rhizosphere of five plant species grown in an acid, P-fixing mineral substrate. Geoderma 75(3–4):231–249. https://doi.org/10.1016/S0016-7061(96)00094-8
Hinsinger P (2001) Bioavailability of inorganic P in the rhizosphere as affected by root-induced chemical changes: a review. Plant Soil 237:173–195. https://doi.org/10.1023/A:1013351617532
Hinsinger PH, Dufey JE, Jaillard B (1991) Biological weathering of micas in the rhizosphere as related to potassium absorption by plant roots. In: Developments in agricultural and managed forest ecology, vol 24. Elsevier, pp 98–105
Ilchenko W (1955) Os tufos da Mata da Corda e seu emprego na Agricultura. Dep Prod Vegetal Belo Horizonte Boletim Agricultura 9-10:39–71
Ilchenko W, Guimarães D (1953) Sobre a possibilidade de utilização agrícola dos sienitos nefelinicos do Planalto de Poços de Caldas. MG Inst Teen Avulso 15:16
Jafari F, Khademi H, Shahrokh V, Angel FAZ, Acosta JA, Khormali F (2021) Biological weathering of phlogopite during enriched vermicomposting. Pedosphere 31(3):440–451
Kelemen PB, McQueen N, Wilcox J, Renforth P, Dipple G, Vankeuren AP (2020) Engineered carbon mineralization in ultramafic rocks for CO2 removal from air: review and new insights. Chem Geol 550:119628. https://doi.org/10.1016/j.chemgeo.2020.119628
Keller WD (1948) Native rocks and minerals as fertilizers. Sci Month 66:122–130
Kleber M, Bourg IC, Coward EK, Hansel CM, Myneni SC, Nunan N (2021) Dynamic interactions at the mineral–organic matter interface. Nat Rev Earth Environ 2(6):402–421. https://doi.org/10.1038/s43017-021-00162-y
Krahl LL, da Paz SPA, Valadares LF, Marchi G, Valadares LF, Sousa-Silva JC, Martins ÉS (2020) Successive offtake of elements by maize grown in pure basalt powder. Afr J Agric Res 15(2):229–239
Krahl LL, Marchi G, Paz SPA, Angélica RS, Sousa-Silva JC, Valadares LF, Martins ÉS (2022a) Increase in cation exchange capacity by the action of maize rhizosphere on Mg or Fe biotite-rich rocks. Pesquisa Agropecuária Tropical 52:e72376
Krahl LL, Valadares LF, Sousa-Silva JC, Marchi G, Martins ÉS (2022b) Dissolution of silicate minerals and nutrient availability for corn grown successively. Pesq Agrop Brasileira 57.(Regional Agrominerals):e01467

Kronberg BI, Leonardos OH, Fyfe WS (1987) The use of ground rocks in laterite systems—an improvement to the use of conventional soluble fertilizers. Chem Geol 60(1–4):361–370. https://doi.org/10.1016/0009-2541(87)90143-4

Leonardos OH, Fyfe WS, Kronberg BI (1987) The use of ground rocks in laterite systems: an improvement to the use of conventional soluble fertilizers? Chem Geol 60(1–4):361–370. https://doi.org/10.1016/0009-2541(87)90143-4

Leonardos OH, Fyfe WS, Kronberg BI (1976) Rochagem: o método de aumento da fertilidade em solos lixiviados e arenosos. Anais 29th Congresso Brasileiro de Geologia, Brasil, p 137–145

Leonardos OH, Theodoro SH, Assad ML (2000) Remineralization for sustainable agriculture: a tropical perspective from a Brazilian viewpoint. Nutr Cycl Agroecosyst 56:3–9. https://doi.org/10.1023/A:1009855409700

Liverpool-Tasie LSO, Wineman A, Young S, Tambo J, Vargas C, Reardon T et al (2020) A scoping review of market links between value chain actors and small-scale producers in developing regions. Nat Sustainability 3(10):799–808

Lompo F, Bationo A, Sedogo MP, Bado VB, Hien V, Ouattara B (2018) Role of local agro-minerals in mineral fertilizer Recommandations for crops: examples of some West Africa phosphate rocks. In: Bationo A, Ngaradoum D, Youl S, Lompo F, Fening J (eds) Improving the profitability, sustainability and efficiency of nutrients through site specific fertilizer recommendations in West Africa agroecosystems. Springer, Cham. https://doi.org/10.1007/978-3-319-58789-9_9

Macias F, Chesworth W (1992) Weathering in humid regions, with emphasis on igneous rocks and their metamorphic equivalents. In: Developments in earth surface processes (vol. 2, cap. 12, pp. 283–306). Elsevier

Manning DAC (2010) Mineral sources of potassium for plant nutrition. A Rev Agron Sustain Dev 30:281–294. https://doi.org/10.1051/agro/2009023

Manning DAC (2013) Resolving the conflict between mining and sustainability applied. Environ Res. http://www.tci-thaijo.org/index.php/jer 36(1):3–12

Manning DAC (2018) Innovation in resourcing geological materials as crop nutrients. Nat Resour Res 27(2):217–227. https://doi.org/10.1007/s11053-017-9347-2

Manning DAC (2022) Mineral stabilities in soils: how minerals can feed the world and mitigate climate change. Clay Miner 57:31–40. https://doi.org/10.1180/clm.2022.17

Manning DAC, Theodoro SH (2018) Enabling food security through use of local rocks and minerals. Extr Ind Soc 7(2):480–487. https://doi.org/10.1016/j.exis.2018.11.002

Manning DA, de Azevedo, AC, Zani CF, Barneze AS (2024) Soil carbon management and enhanced rock weathering: the separate fates of organic and inorganic carbon. Eur J Soil Sci 75(4):e13534. https://doi.org/10.1111/ejss.13534

Marchi G, Guelfi-Silva DR, Malaquias J, Guilherme LRG, Spehar CR, Martins ÉS (2020) Solubility and availability of micronutrients extracted from silicate agrominerals. Pesq Agrop Brasileira 55:e00807

Martins ÉS, Silveira CAP, Bamberg AL, Martinazzo R, Bergmann M, Angélica RS (2014) Silicate agrominerals as nutrient sources and as soil conditioners for tropical agriculture. In: International scientific centre of fertilizers (ed), world fertilizer congress of CIEC, 16, Technological innovation for sustainable tropical agriculture: proceedings, 2014, Rio de Janeiro

Martins ÉS, Martins ES, Hardoim PR (2023) Princípios geoquímicos, mineralógicos e biológicos do manejo de remineralizadores de solos. Informe Agropecuário, EPAMIG 44(321):26–39. Available: https://www.alice.cnptia.embrapa.br/alice/bitstream/doc/1156783/1/Eder-principios-geoquimicos-mineralogicos.pdf

Martins ÉS, Theodoro SH, Bernardez F, Siqueira DS, Caixeta AP, Curtis JCD, Luchese AV (2024) Produção Brasileira de Remineralizadores e Fertilizantes Naturais: 2019 a 2022. Revista Novo Solo 3:10–19. Available: https://abrefen.org.br/2023/06/29/producao-brasileirade-remineralizadorese-fertilizantes-naturais-2019-a-2022/

Medeiros FP, Theodoro SH, Carvalho AMX, Oliveira VS, Oliveira LV, Almeida RAP, Viana MB, Gomide CS (2025) The combination of crushed rock and organic matter enhances the capture of inorganic carbon in tropical soils. J S Am Earth Sci 152. https://doi.org/10.1016/j.jsames.2024.105254

Medeiros FP, Carvalho AMX, Gindri Ramos C, Dotto GL, Cardoso IM, Theodoro SH (2024) Rock powder enhances soil nutrition and coffee quality in agroforestry systems. Sustainability 16:354. https://doi.org/10.3390/su16010354

Milani D, Kiani A, Haque N, Giddey S, Feron P (2022) Green pathways for urea synthesis: a review from Australia's perspective. Sustain Chem Clim Action 1:100008. https://doi.org/10.1016/j.scca.2022.100008

Minasny B, Fiantis D, Hairiah K, Van Noordwijk M (2021) Applying volcanic ash to croplands–the untapped natural solution. Soil Secur 3:100006

Miranda CC, Florentino LA, Rezende AV, Nogueira DA, Leite RF, Naves LP (2018) Desenvolvimento de Urochloa brizantha adubada com fonolito e inoculada com bactérias diazotróficas solubilizadoras de potássio. Revista de Ciências Agrárias 41(3):625–632

Missoux M (1853) Sur l'emploi de la poudre des roches granitiques comme excitant de la vegetation. Compt Rend Acad Sci (Paris) 36:1136

Mohammed SMO, Brandt K, Gray ND, White ML, Manning DAC (2014) Comparison of silicate minerals as sources of potassium for plant nutrition in sandy soil. Eur J Soil Sci 65(5):653–662

Momesso L, Crusciol CAC, Nascimento CAC, Soratto RP, Canisares PL, Moretti LG, Rosolem CA, Trivelin PCO, Kuramae EE, Cantarella H (2022) Feasibility of early fertilization of maize with 15 N application to preceding cover crop. Eur J Agron 135. https://doi.org/10.1016/j.eja.2022.126485

Moretti LG, Bossolani JW, Crusciol CAC, Moreira A, Micheri PH, Rossi R, Imaizumi C (2019) Dunite in agriculture: physiological changes, nutritional status and soybean yield. Commun Soil Sci Plant Anal 50(14):1775–1784. https://doi.org/10.1080/00103624.2019.1635143

Moretti LG, Crusciol CAC, Bossolani JW, Rossi R, Moreira A (2021) Dunite solubilization kinetics in silicon-magnesium fertilization. J Plant Nutr 44(16):2435–2446

Newton P, Civita N, Frankel-Goldwater L, Bartel K, Johns C (2020) What is regenerative agriculture? A review of scholar and practitioner definitions based on processes and outcomes. Front Sustain Food Syst 4:577723

Paredes Filho MV, Mesquita AC, Mantovani JR, Rezende AV, Florentino LA (2020) Bacterial strains decrease solubility of potassium in the soil. Biosci J 36(5). https://doi.org/10.14393/BJ-v36n5a2020-42326

Perman R (2022) James Hutton: The Genius of Time. Birlinn Ltd.

Pinheiro S (2018) Agroecologia 7.0. Juquira Candiru Satyagraha, Brazil, Porto Alegre, p 663

Ramos CG, Dalmora AC, Kautzmann RM, Hower J, Dotto GL, Oliveira LFS (2021) Sustainable release of macronutrients to black oat and maize crops from organically-altered Dacite rock powder. Nat Resour Res 30:1941–1953. https://doi.org/10.1007/s11053-021-09862-0

Ramos CG, Hower JC, Blanc E, Oliveira MLS, Theodoro SH (2022) Possibilities of using silicate rock powder: an overview. Geosci Front 13(1). https://doi.org/10.1016/j.gsf.2021.101185

Reis JC, Rodrigues GS, Barros I, Rodrigues RDAR, Garrett RD, Valentim JF et al (2021) Integrated crop-livestock systems: a sustainable land-use alternative for food production in the Brazilian Cerrado and Amazon. J Clean Prod 283:124580

Renforth P (2019) The negative emission potential of alkaline materials. Nat Commun 10(1). https://doi.org/10.1038/s41467-019-09475-5

Renforth P, von Strandmann PAE, Henderson GM (2015) The dissolution of olivine added to soil: implications for enhanced weathering. Appl Geochem 61:109–118. https://doi.org/10.1016/j.apgeochem.2015.05.016

Renforth P (2012) The potential of enhanced weathering in the UK. Int J Greenh Gas Con 10:229–243. https://doi.org/10.1016/j.ijggc.2012.06.011

Rinaldi MM, Costa AM, Malaquias JV, Martins ÉS (2022) Qualidade pós-colheita e vida útil de frutos de Passiflora cincinnata BRS Sertão Forte em função do tipo de adubação e armazenamento. Rev Bras Frutic 44:e-051

Rudmin M, Banerjee S, Makarov B, Mazurov A, Ruban A, Oskina Y, Tolkachev O, Buyakov A, Shaldybin M (2019) An investigation of plant growth by the addition of glauconitic fertilizer. Appl Clay Sci 180:105178. https://doi.org/10.1016/j.clay.2019.105178

Santos LFD, Sodré FF, Martins ÉS, Figueiredo CCD, Busato JG (2021) Effects of biotite syenite on the nutrient levels and electrical charges in a Brazilian savanna Ferralsol. Pesquisa Agropecuária Tropical 51:e66691

Setiawati TC, Mutmainnah L (2016) Solubilization of potassium containing mineral by microorganisms from sugarcane rhizosphere. Agric Agric Sci Procedia 9:108–117. https://doi.org/10.1016/j.aaspro.2016.02.134

Shirale AO, Meena BP, Gurav PP, Srivastava S, Biswas AK, Thakur JK et al (2019) Prospects and challenges in utilization of indigenous rocks and minerals as source of potassium in farming. J Plant Nutr 42(19):2682–2701. https://doi.org/10.1080/01904167.2019.1659353

Soares GJ, Theodoro SH, Carvalho AXM, Burbano DFM, Ramos CG (2022) Remineralizadores de suelos y sistemas agroforestales: una opción para la captura de CO_2. Revista Agroecología 15(1). https://doi.org/10.59187/revistaagroecologia.v15i1

Swanback TR (1950) Granite stone meal as a source of potash for tobacco. Connecticut agricultural experiment station, New Haven, Bulletin, vol 536, 14p

Swoboda P, Döring TF, Hamer M (2022) Remineralizing soils? The agricultural usage of silicate rock powders: a review. In: Science of the Total environment, vol 807. Elsevier B.V. https://doi.org/10.1016/j.scitotenv.2021.150976

Tchouankoue JP, Tchekambou ANT, Angue MA, Ngansop C, Theodoro SH (2014) Rock fertilizers as an alternative to conventional fertilizers: the use of basalt from the Cameroon volcanic line for maize farming on ferralitic soils. In: Goreau TJ, Larson RW, Campe J (eds) Geotherapy: innovative methods of soil fertility restoration, carbon sequestration, and reversing CO2 increase, 1st edn. CRC Press. https://doi.org/10.1201/b13788

Tessaro D, Matter JM, Kuczman O, Furtado LDF, Costa LADM, Costa MSSDM (2013) Produção agroecológica de mudas e desenvolvimento a campo de couve-chinesa. Ciência Rural 43:831–837

Theodoro SH, Manning DAC, Carvalho AXM, Ferrão FR, Almeida GR (2022) Soil remineralizer: a new rote to sustentability for Brazil, a giant exporting agro-mineral commoditites. In: Yakovleva, N, Nickless E (eds) Routledge handbook of the extractive industries and sustainable development, 1st edn. pp 261–281. https://doi.org/10.4324/9781003001317

Theodoro SH, Medeiros FP, Ianniruberto M, TKB J (2021) Soil remineralization and recovery of degraded areas: an experience in the tropical region. J S Am Earth Sci 103014. https://doi.org/10.1016/j.jsames.2020.103014

Theodoro SH, Leonardos OH (2015) Stonemeal: principles, potential and perspective from Brazil. In: Goreau TJ, Larson RW, Campe J (eds) Geotherapy: innovative methods of soil fertility restoration, carbon sequestration and reversing CO_2 increase. CRC Press, pp 403–418. ISBN: 978-1-4665-9539-2

Theodoro SH, Leonardos OH, Rocha EL, Macedo I, Rego K (2013) Stonemeal of amazon soils with sediments from reservoirs: a case study of remineralization of the Tucuruí degraded land for agroforest reclamation. Acad Brasil Ciências 85(1):23–34. https://doi.org/10.1590/S0001-37652013000100003

Theodoro SH (2000) A Fertilização da terra pela Terra: uma alternativa para a sustentabilidade do pequeno produtor rural. Ph.D. Universidade de Brasília, Brasília, DF, p 221. Available: https://repositorio.unb.br/handle/10482/20881

Theodoro SH, Leonardos OH (2006) The use of rocks to improve family agriculture in Brazil. An Acad Bras Cienc 78(4):721–730. https://doi.org/10.1590/S0001-37652006000400007

Van Straaten P (2002) Rocks for crops: Agrominerals of sub-Saharan Africa. International Centre for Research in Agroforestry, Nairobi. isbn:0-88955-512-5

Van Straaten P (2006) Farming with rocks and minerals: challenges and opportunities. 78(4):731–747. https://doi.org/10.1590/S0001-37652006000400009

Van Straaten P (2007) Agrogeology: the use of rocks for crops. Enviroquest Ltd., Canada, p 440

Van Straaten P (2022) Distribution of agromineral resources in space and time—a global geological perspective. Pesq Agrop Brasileira 57. https://doi.org/10.1590/S1678-3921
Vernadsky WI (1945) The biosphere and the noösphere. Am Sci 33(1):1–12
Walan P, Davidsson S, Johansson S, Höök M (2014) Phosphate rock production and depletion: regional disaggregated modeling and global implications. Resour Conserv Recycl 93:178–187
Wedepohl KH (1995) The composition of the continental crust. Geochim Cosmochim Acta 59(7):1217–1232
Winogradow AP (1954) Geochemieseltener und nur in spuren vorhanden er chemischer elementeim boden. The Gryter, Berlin
Wu H, MacDonald GK, Galloway JN, Zhang L, Gao L, Yang L et al (2021) The influence of crop and chemical fertilizer combinations on greenhouse gas emissions: a partial life-cycle assessment of fertilizer production and use in China. Resour Conserv Recycl 168:105303
Yang X, Long Y, Sarkar B, Li Y, Lü G, Ali A, Yang J, Cao YE (2021) Influence of soil microorganisms and physicochemical properties on plant diversity in an arid desert of Western China. J For Res. https://doi.org/10.1007/s11676-021-01292-1
Yaroshevsky AA, Bulakh AG (1994) The mineral composition of the earth's crust, mantle, meteorites, moon, and planets. In: Advanced mineralogy. Springer, Berlin Heidelberg, pp 27–36. https://doi.org/10.1007/978-3-642-78523-8_3
Zhang L, Gadd GM, Li Z (2020) Microbial biomodification of clay minerals. Academic. ISSN 0065-2164, Advances in Applied Microbiology. https://doi.org/10.1016/bs.aambs.2020.07.002

Changing Time Scales: From Silicate Rocks to Agrominerals Through Bioweathering

Luise Lottici Krahl, Giuliano Marchi, Elis Marina de Freitas, Luiz Fernando dos Santos, and Éder de Souza Martins

Abstract Bioweathering, the biological transformation of Earth's lithosphere, represents a critical intersection between geological and biological processes. This chapter examines the mechanisms, evidence, and applications of bioweathering, highlighting its potential to accelerate mineral dissolution on shorter timescales compared to traditional abiotic weathering. Central to this process is the rhizosphere — a hotspot of biogeochemical activity where plant root interactions and microbial dynamics significantly enhance mineral solubility and nutrient release. By analyzing experimental findings and advanced methodologies such as synchrotron-based spectroscopy and isotopic tracing, this work highlights the transformative implications of bioweathering for sustainable agriculture, nutrient cycling, and soil health. Special attention is given to the challenges and opportunities of integrating bioweathering into agricultural systems, including optimizing silicate mineral dissolution rates and leveraging secondary mineral phases to improve cation exchange capacity and long-term soil fertility. This chapter aims to bridge the gap between geological processes and practical agricultural applications, offering insights into how bioweathering can address the pressing demands of a changing world.

Keywords Biological dissolution · Rhizosphere microbiome · Biological-mineral interactions

L. L. Krahl (✉) · G. Marchi · E. M. de Freitas · L. F. dos Santos · É. de Souza Martins
Agrogeological Laboratory, Empresa Brasileira de Pesquisa Agropecuária - Embrapa Cerrados, BR 020, km 18, Planaltina 73310970, Federal District, Brazil
e-mail: luisekrahl82@gmail.com; luisekrahl@yahoo.com.br

G. Marchi
e-mail: giuliano.marchi@embrapa.br

E. M. de Freitas
e-mail: elismarina1.freitas@gmail.com

L. F. dos Santos
e-mail: luiz.santos1@educa.go.gov.br

É. de Souza Martins
e-mail: eder.martins@embrapa.br

É. de Souza Martins and S. Huff Theodoro (eds.), *Soil Remineralizers and Silicate Fertilizers*, Frontier Studies in Soil Science,
https://doi.org/10.1007/978-3-032-14656-4_2

1 Introduction

The transformation of Earth's lithosphere through weathering processes plays a pivotal role in shaping ecosystems, cycling nutrients, and supporting life. Among these processes, bioweathering — facilitated by biological agents such as plant roots, microorganisms, and their interactions — emerges as a fascinating mechanism with implications extending far beyond geochemistry. While traditional weathering is often measured across geological timescales, modern research reveals that biological influences can significantly accelerate mineral dissolution, creating pathways for nutrient availability at shorter time frames. This paradigm shift, connecting geological phenomena with the immediacy of biological activity, underscores the role of bioweathering in nutrient cycling, sustainable agriculture, and even carbon sequestration.

At the heart of bioweathering lies the rhizosphere, an intricate microenvironment influenced by plant roots and teeming with microbial activity. These biogeochemical hotspots revolutionized our understanding of soil-mineral interactions, from the release of essential nutrients to the formation of secondary minerals. Unlike abiotic processes, the rhizosphere introduces a dynamic interplay of organic acids, enzymes, and microbial metabolites that modulate mineral solubility and dissolution rates. Studies, such as those conducted by Hinsinger et al. (2006) and Jones (1998), have shown that root exudates can alter pH and redox conditions, thereby enhancing the dissolution of silicate minerals and promoting the bioavailability of essential nutrients for plant growth.

The potential of bioweathering extends beyond mere academic curiosity, bearing practical implications in addressing global agricultural challenges. As global food demand increases, innovative approaches to enhance soil fertility and improve nutrient cycling are crucial. Weathering of silicate minerals under biotic influence provides an opportunity not only to supply essential nutrients, such as potassium and phosphorus, but also to improve soil structure and water retention. Recent advancements in experimental methodologies, such as synchrotron-based spectroscopy and isotopic tracing, have shed light on the precise mechanisms by which bioweathering contributes to soil health and plant productivity (Yu 2018).

This chapter examines the transformative impact of bioweathering, investigating its mechanisms, experimental evidence, and implications for agricultural practices. From the dissolution kinetics of aluminosilicates to the role of the rhizosphere in accelerating mineral transformations, the discussion integrates cutting-edge research with applied perspectives. By bridging the gap between geological processes and agricultural applications, this work aims to expand our understanding of how weathering processes can be harnessed to meet the demands of a changing world.

2 Bioweathering Time Frame

In natural systems, the dissolution of rock and soil-forming processes occur over geological timescales (Hartmann et al. 2013; White and Brantley 1995). Dissolution kinetics in diluted aqueous solutions are extremely slow, typically ranging from 104 to 107 years for the transformation of primary minerals to secondary minerals (Lasaga 1995; Palandri and Kharaka 2004). Experimental weathering rates in bioreactors for silicate minerals are commonly two to four orders of magnitude faster than natural field-derived rates (White and Brantley 2003). Therefore, rock dissolution and soil formation in natural systems occur on a geological timescale, influenced by the intensity of soil formation factors (Hartmann et al. 2013; White and Brantley 1995). Consequently, at these experimental and natural rates, aluminosilicates would have no significant impact on agricultural processes.

Mineral dissolution in natural environments involves a complex interplay of factors, including specific surface area and chemical processes like acid-base reactions, redox transformations, and adsorption–desorption–precipitation equilibria. These processes are likely to have critical roles in the biogeochemical cycling of elements (Van Schöll et al. 2008; Hemingway et al. 2018).

No accelerated process can provide the necessary nutrients to plants within a production cycle if the source material dissolves at an insufficient rate. In agricultural systems, the grinding process for agrominerals increases the surface area that comes into contact with the soil. Biological processes enhance the rates of dissolution, thereby accelerating weathering and promoting the formation of secondary mineral phases such as short-range-order minerals (e.g., allophane, ferrihydrite) and high-activity clays (e.g., smectite, vermiculite). Notably, these newly formed phases exhibit an elevated cation exchange capacity (CEC > 100 $cmol_c$ kg^{-1}), thereby enhancing soil nutrient retention and long-term fertility (Basak et al. 2022; Yu et al. 2019).

Recent evidence suggests that silicate agrominerals may exhibit relatively rapid dissolution rates (Basak et al. 2018; Burghelea et al. 2015; Li et al. 2015b; Manning et al. 2017; Wang et al. 2000). The weathering of silicate minerals by root activity and its associated organisms, known as biological weathering or bioweathering, not only releases nutrients but also has the potential to form 2:1 clay minerals and low-crystallinity minerals within the annual crop cycle time scale. These periods range from 1 to 10 years. (Finlay et al. 2020; Hinsinger et al. 1991; Krahl et al. 2022a, b; Mohammed et al. 2014; Naderizadeh et al. 2010; Norouzi and Khademi 2010).

Associations of organic matter with pedogenic minerals, known as mineral–organic associations, in the soil are one of the factors that determine the long-term storage of SOC. Such mineral organic associations limit microbial and enzymatic access and are quantitatively the most important mechanism protecting SOC from microbial use for centuries or millennia (Keiluweit et al. 2015). The conditions needed for the formation of mineral–organic associations will be most abundant in regions with soil moisture supply to produce high plant biomass, that is, under conditions of intensive weathering, low pH, and soil formation, such as the tropical

Table 1 Particle size distribution of biotite schist utilized in the experiment

	Particle size distribution (μm)			
	< 53	53–300	300–1000	> 1000–2000
	%			
Biotite schist	21.1	17.4	2.4	59.1

environment (Kleber et al. 2015; Sarkar et al. 2018). However, unlike other climate zones, the rate of natural rock weathering in the tropics is limited by the supply of fresh mineral surfaces.

In these tropical environments, applications of silicate rocks in agricultural soils can add raw minerals to the system (Fig. 2). Besides improving soil quality, this is a promising alternative to stabilize the SOM by the bioweathering process and newly formed minerals (Edwards et al. 2017; Beerling et al. 2018).

2.1 *Experimental Evidence*

A preliminary experiment on greenhouse bioweathering at Embrapa Cerrados, a public research center funded by the Brazilian government,[1] aimed to assess the role of the maize rhizosphere in the bioweathering of biotite schist (as characterized in Tables 1 and 2). The findings indicated that the transformation of biotite to hydrobiotite, a precursor of vermiculite (Coleman et al. 1963; Farmer and Wilson 1970; Fordham 1990), occurred within a 45-day period. The study entailed successive maize cultivation using pure biotite schist fractions of varying particle sizes. Control treatments were established using pots without plants. Maize was cultivated over seven consecutive cropping cycles, each lasting 45 days, culminating in a total experimental duration of 315 days.

The X-ray diffraction (XRD) analysis revealed that the biotite schist consisted of phyllosilicates, including biotite, muscovite, and chlorite, in conjunction with quartz, plagioclase, and pyroxene (Fig. 1). No differences were detected in the X-ray diffractograms between the original biotite schist and control treatments. However, treatments involving maize plants resulted in the emergence of a new phase, hydrobiotite (Fig. 2).

[1] The Embrapa Cerrados National Research Center, based in Planaltina (Federal District, Brazil), is a unit of Embrapa focused on developing scientific and technological solutions for the sustainable use of the Cerrado biome. It conducts research and innovation in areas such as agricultural and livestock production systems adapted to savanna and seasonally dry conditions, soil and water resource management and conservation, genetic improvement of cultivated species, integration of crop-livestock-forest systems (ILPF), and climate change mitigation and adaptation. In addition to research, the center promotes technology transfer, technical training, and partnerships with farmers, cooperatives, and public and private institutions to enhance sustainability, productivity, and the conservation of regional biodiversity (https://www.embrapa.br/en/cerrados)

Table 2 Chemical composition of biotite schist determined by X-ray fluorescence (XRF) spectroscopy

	SiO_2	Al_2O_3	Fe_2O_3	CaO	MgO	TiO_2	P_2O_5	Na_2O	K_2O	MnO	BaO	Cr_2O_3	LOI[a]
Fraction (μm)	%												
< 53	59.6	16.1	8.06	2.31	3.84	1.08	0.29	2.57	2.61	0.06	0.06	0.02	2.18
53–300	59.7	19	7.74	0.87	3.68	0.57	0.15	1.84	4.14	0.05	0.11	0.02	2.72
Bulk sample[b]	62.6	16.8	7.8	1.44	3.2	0.86	0.2	1.93	3.19	0.12	0.07	0.02	2.17

[a]Loss on ignition
[b]Original sample. Adapted from Krahl et al. (2022a)

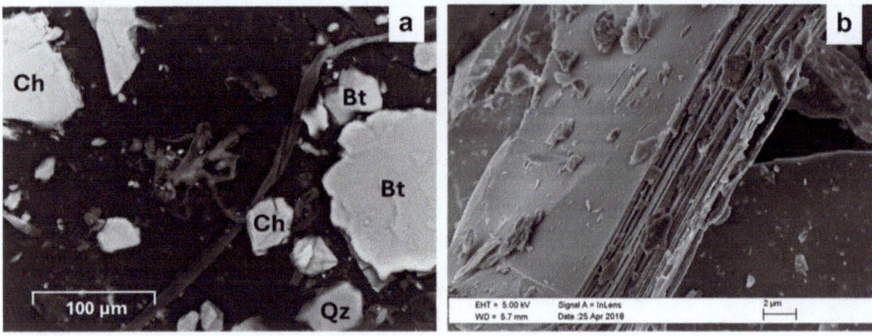

Fig. 1 **a** Backscattered electron (BSE) image of biotite schist minerals around the maize root, showing quartz (Qz), biotite (Bt), and chlorite (Ch). **b** Secondary electron (SE) images of biotite grains. Author: Luise Lottici Krahl

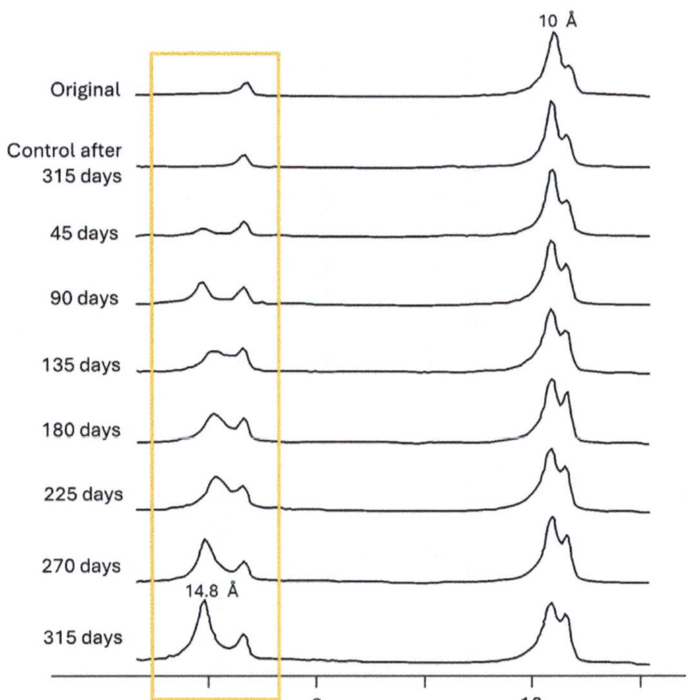

Fig. 2 X-ray diffraction patterns of particles smaller than 53 µm in biotite schist, following interaction with maize rhizosphere, indicate the formation of hydrobiotite. Adapted from Krahl et al. (2022a)

Many studies have found evidence of bioweathering in primary minerals. A study of the weathering of trioctahedral mica (phlogopite, with a particle size range of 2–100 μm) by the rhizosphere of Italian ryegrass (Lolium multiflorum Lam.) resulted in the release of potassium and an expansion of the mica lattice after three days of continuous exposure to the root mat. This phenomenon is attributed to the plant's uptake of potassium (Hinsinger et al. 1992). By the fourth day, vermiculitization of phlogopite was observable up to 1.5 mm from the root surface, as detected by X-ray diffractometry. Similarly, in a study conducted by Norouzi and Khademi (2010), the emergence of a 14.5 Å peak in phlogopite and biotite weathered by the rhizosphere of alfalfa (Medicago sativa L.) over a period of 90 days was interpreted as a reflection from the 001 planes of typical vermiculite. In a study conducted by Naderizadeh et al. (2010) regarding the transformation of phlogopite in the presence of added organic matter within the rhizosphere, it was found that increased acidity in the rhizosphere enhanced the release of potassium from phlogopite. This process facilitated the formation of vermiculite, with smaller quantities of smectite and hydroxy-interlayered vermiculite (HIV) also being produced.

Alterations in primary minerals were also detected during chemical extraction processes. These procedures facilitate the release of non-exchangeable potassium from micas by promoting both the exchange of interlayer K^+ with hydrated cations and the subsequent dissolution of the mineral, leading to the formation of weathering products (Andrist-Rangel et al. 2013; Li et al. 2015a).

3 Agromineral Dissolution and Its Products

Various strategies have been developed to enhance the permanent cation exchange capacity (CEC) in tropical agricultural soils (Table 3). These strategies involve the use of rock (Van Straaten 2007), 2:1 clay minerals (Noble et al. 2001; Qafoku et al. 2004), and biochar (Gaskin et al. 2010; Mukherjee et al. 2011), aiming to enhance long-term agricultural productivity.

Table 3 Permanent charge of materials from temperate and tropical climate soils

Material	CEC ($cmol_c\ kg^{-1}$)
Soil organic matter[a]	100–300
Vermiculite[a]	80–150
Smectite[a]	60-100
Illite[a]	15-25
Kaolinite[a]	2–8
Biochar[b, c]	10–70

[a] van Straaten (2007)
[b] Gaskin et al. (2008)
[c] Mukherjee et al. (2011)

Natural smectite formation from basalt weathering can reach up to 30% (Sherman 1952), with the potential to form up to 50%. Conversely, biotite schist can produce 10% vermiculite. Given that the cation exchange capacity (CEC) of smectite is approximately 100 $cmol_c\ kg^{-1}$, and that of vermiculite is around 150 $cmol_c\ kg^{-1}$, it is necessary to apply 40 tons of basalt and 140 tons of biotite schist to increase the soil's CEC by 1 $cmol_c\ kg^{-1}$ within the 0–20 cm layer. As the accumulation increases with each application, the resulting permanent charges within these ground rocks hold significant potential for generating soils with a surface horizon characterized by high permanent charge over the long term.

4 The Bioweathering Process

The *rhizosphere* is regarded as an area of significant biogeochemical activity, marked by increased microbial activity, soil organic matter production, accelerated mineral weathering, and the formation of secondary mineral phases. Traditionally, the rhizosphere refers to the few millimeters of soil that adhere to plant roots. However, when considering processes such as water uptake and mobile nutrient absorption, its functional extent can be significantly greater (Jones 1998; Uroz et al. 2015; Van Dam and Bouwmeester 2016; White et al. 2017). Microorganisms in the *rhizosphere*, including fungi and bacteria, facilitate biogeochemical processes that break down rocks through biomechanical forces, biochemical reactions, and the secretion of metabolites.

Roots release various organic acids with different chain lengths. The main anionic components include lactate, acetate, oxalate, succinate, fumarate, malate, citrate, isocitrate, and aconitate, which dissolve minerals and serve as nutrient sources for microorganisms. The concentrations of organic acids in the soil solution typically range from 0.1 to 0.1 mmol L^{-1}, but can exceed 50 mmol L^{-1} at the root surface (Jones et al. 1996). Mycorrhizal fungal mycelia secrete a variety of molecules, including low-molecular-weight (LMW) organic acids, amino acids, polyols, peptides, siderophores, glycoproteins, and diverse enzymes such as proteases, phosphatases, lignin peroxidases, and laccases (Finlay et al. 2020). This process releases essential nutrients (e.g., K, Ca, Fe, Si) while sequestering CO_2 through enhanced mineral dissolution, thereby contributing to soil fertility regeneration and agricultural practices such as rock powder amendment (Reis et al. 2024; Yu et al. 2019). Additionally, the release of phytosiderophores by roots, particularly in particular grass species, is important for the uptake of poorly mobile micronutrients such as iron and various other metals (Basak et al. 2022; Burghelea et al. 2015).

The efficacy of silicate rock powders (e.g., phonolite, basalts, granite) as soil remineralizers, therefore, hinges on microbe-mineral interactions (Fig. 3). Microbial colonization can occur in all mineral and rock environments, influenced by their physical and chemical properties (Uroz et al. 2015; Dong et al. 2022). To describe the ecological significance of minerals, Uroz et al. (2015) introduced the term *mineralosphere*, which refers to the interface and habitat that includes both the rocks (or mineral surfaces) and the surrounding soil, influenced by physical, chemical,

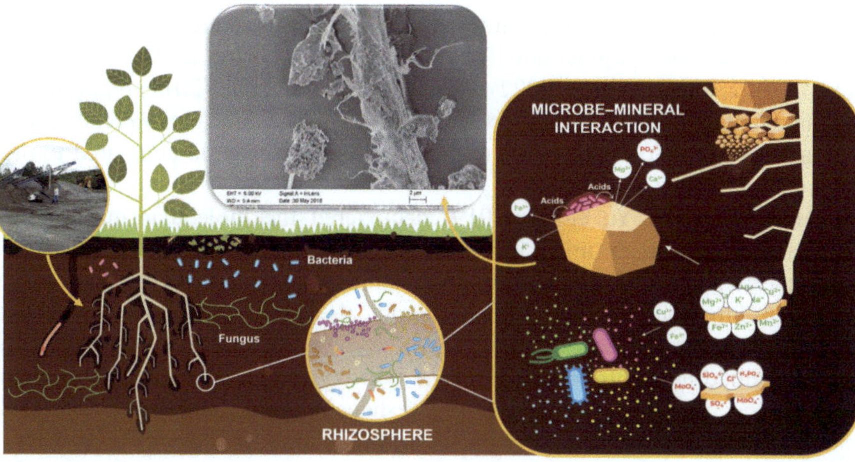

Fig. 3 The bioweathering process. Adapted from André Vicente by Tropica and Krahl et al. (2020)

and biological interactions with minerals. The mineralosphere possesses properties comparable to those of the rhizosphere. Minerals, due to their physicochemical characteristics and the nutrients they release upon dissolution, modify their environment both chemically and physically.

In an 8-month *Brachiaria* cultivation trial in Ferralsol, Reis et al. (2024) demonstrated that rock mineralogy—particularly reactive minerals like zeolites in phonolite and Si-Ca-Fe-rich mafic minerals in basalts—modulated soil bacterial community structure (SBC) and microbial biomass carbon (MBC). Methodologically, T-RFLP (Terminal Restriction Fragment Length Polymorphism), a molecular biology technique for analyzing microbial community diversity, and NMDS (Non-metric Multidimensional Scaling), a statistical ordering technique for visualizing patterns of similarity or dissimilarity between samples based on multivariate data, revealed that 15% of bacterial populations specifically responded to mineral composition after 8 months. Additionally, MBC dynamics indicated metabolic adaptation to progressive mineral dissolution. These findings underscore that mineral-driven microbial recruitment enhances nutrient cycling (e.g., K^+, Ca^{2+} mobilization) and long-term soil health (Reis et al. 2024).

Minerals also protect microorganisms by shielding them from environmental stressors, such as UV radiation, mechanical wear, and temperature changes. Structural features, such as fractures, fissures, and pores within the mineral matrix, offer this protection (Uroz et al. 2015) as well as the nutrients necessary for microbial metabolism (Dong et al. 2022). The importance of minerals as nutritional sources is evident in their impact on the structural and functional dynamics of microbial communities. Both laboratory and field studies have demonstrated that microbial assemblages vary according to the elemental composition of mineral substrates, as summarized by Dong et al. (2022). The nutrient content of minerals influences both the abundance and diversity of surface-adherent microbial communities. This phenomenon may facilitate access to limiting nutrients from minerals, thereby creating a niche for specific microbial populations (Basak et al. 2022; Ghosh et al. 2023; Mauck and Roberts 2007; Tshering et al. 2022). Research by Colin et al. (2017) highlighted an enrichment of particular bacterial taxa on minerals, suggesting that minerals resistant to weathering tend to attract more mineral-weathering bacteria compared to those that weather easily.

Complementarily, Yu et al. (2019) deciphered the molecular machinery of bio-weathering at fungal-mineral interfaces. Using synchrotron radiation spectro-microscopy and hydroxyl radical quantification, they observed that *Trichoderma guizhouense* hyphae produced superoxide radicals at their tips, triggering catalytic Fenton-like reactions. These reactions dissolved hematite into bioavailable Fe^{2+} and formed ferrihydrite within 66 h (> 1000 nM hydroxyl radicals detected), highlighting fungi as biogeochemical "engineers" that reshape mineral reactivity through redox reactions (Yu et al. 2019).

Oxidation, often paired with hydrolysis, is a chemical weathering process where an element in a crystal loses an electron. In the process of weathering primary soil minerals, iron (Fe^{2+}) is the most frequently oxidized element. Manganese (Mn^{2+}) is also subject to oxidation, albeit to a lesser extent. In primary minerals, these elements are consistently found in their reduced states. Oxidation occurs through two mechanisms: (1) oxidation of structural Fe^{2+}, which starts mineral breakdown, and (2) oxidation of Fe^{2+} is released into the soil solution via hydrolysis. The oxidation of Fe^{2+} to Fe^{3+} within the mineral structure compromises the electrostatic neutrality of the mineral. Consequently, cations (excluding Fe^{3+}) leave the crystal structure to preserve electroneutrality. This leaves vacancies (holes) in the mineral structure. Subsequently, the mineral may undergo structural collapse or become increasingly vulnerable to other chemical weathering processes such as hydrolysis, chelation, and solutioning—simple dissolution and solvation (Essington 2015). Oxidation is the primary dissolution process affecting minerals like olivine and biotite that can dissolve easily in the short term.

Generally, monocotyledons, such as grasses, utilize soil potassium reserves more effectively than dicotyledons. Grasses may have developed an efficient mechanism to mobilize iron (Broadley et al. 2012). Graminaceous plants employ a chelation-based mechanism, known as strategy II, to acquire essential nutrients like ferric iron from the soil as well as from minerals. Through this process, they release phytosiderophores—specialized molecules from the mugineic acid family—into the rhizosphere. These molecules chelate Fe^{3+}, effectively solubilizing it for plant uptake. Once chelated, the Fe^{3+} phytosiderophore complex is transported into root cells via Yellow Stripe1 (YS1) proteins, ensuring efficient iron acquisition and supporting plant growth and development (Kobayashi et al. 2019; Walker and Connolly 2008).

However, the rates at which silicate rocks undergo weathering induced by plants are contingent upon the rock's mineral composition and granulometry (Deng et al. 2023). Minerals in soils and other environments also play a role in shaping specific microbial communities, influenced by their mineral composition, nutrient content, and weatherability (Uroz et al. 2015). The communities within the rhizosphere enhance mineral dissolution rates. This occurs primarily through their capacity to generate acidity, absorb and release elements, and produce low-molecular-weight organic acids (Hinsinger et al. 1993; Jones 1998; Kong et al. 2014; Uroz et al. 2009; White et al. 2017; Yu 2018). These root activities lead to either an accumulation or a reduction of ions present in the soil solution within the rhizosphere. The nature and extent of these changes in ionic concentrations depend on the plant's requirements and the soil's supply.

Furthermore, root exudates of organic acids and enzymes play a substantial role in providing major nutrients for plants. These exudates are critical agents driving silicate weathering (Basak et al. 2022; Bray et al. 2015; Ghosh et al. 2023; Kong et al. 2014; Li et al. 2014; Machado et al. 2016), attacking the surfaces of all minerals present in a silicate agromineral. The secretion of organic acids by plants and microorganisms can modify the structure of certain minerals, including potassium feldspar and biotite (Fig. 3). This process releases elemental nutrients and changes the environmental chemistry. These acids possess carboxyl and hydroxyl groups that interact with soil elements, significantly contributing to mineral weathering, nutrient mobilization, and aluminum detoxification (Kong et al. 2014).

A study performed by Bray et al. (2015) quantified the impact of organic ligands on biotite weathering rates. The research demonstrated that at low concentration levels—such as those exuded by microorganisms —the presence of organic ligands enhanced the release of di-and trivalent metals from both octahedral and tetrahedral layers. The authors observed that the formation of metal–organic ligand complexes likely facilitated an equilibrium shift, thereby promoting the further release of aluminum, iron, and magnesium from biotite.

5 Potassium Mobilization

The ability of various plant types to uptake nutrients, particularly potassium (K), from silicate minerals has been investigated through field and laboratory experiments. Studies have examined nutrient uptake from minerals such as muscovite and/or biotite (Basak et al. 2018, 2022, Burghelea et al. 2015; Ghosh et al. 2023; Hinsinger et al. 1992; Krahl et al. 2022b; Mohammed et al. 2014; Naderizadeh et al. 2010; Norouzi and Khademi 2010), glauconite (Franzosi et al. 2014; Rudmin et al. 2018), K-feldspar (Ciceri et al. 2019; Manning et al. 2017; Ramos et al. 2019), waste micas (Madaras et al. 2013; Basak et al. 2018; Basak 2019), basalt powder (Akter and Akagi 2005; Anda et al. 2015; Hinsinger et al. 2001; Shafar et al. 2017) and gneiss (Wang et al. 2000). Research indicates that plants can extract a significant amount of potassium (K) from K-rich minerals in silicate rocks, such as micas and K-feldspars, along with their weathering products (Basak et al. 2022; Burghelea et al. 2015; Ciceri et al. 2019; Ghosh et al. 2023; Mohammed et al. 2014; Naderizadeh et al. 2010; Zorb et al. 2014).

Certain soil microorganisms possess the capability to mobilize insoluble or fixed forms of potassium from minerals such as micas, illite, and orthoclase. These organisms, as pointed out by Ghosh et al. (2023), are classified as potassium-mobilizing microorganisms, encompassing both bacteria and fungi that significantly influence the availability of soil minerals. As such, they play a critical role in ion cycling and the maintenance of soil fertility (Ghosh et al. 2023). Bacterial species such as *Bacillus mucilaginosus*, *B. edaphicus*, *B. circulans*, *Paenibacillus* spp., *Acidithiobacillus ferrooxidans*, *Pseudomonas*, and *Burkholderia* are known as effective potassium-mobilizing bacteria. Fungal species such as *Aspergillus terreus* and *A. niger* were isolated from potassium-rich soil and demonstrated the ability to mobilize insoluble potassium in a liquid medium when insoluble sources of potassium, like feldspar and potassium aluminium silicate, were present (Ghosh et al. 2023; Mauck and Roberts 2007; Wang et al. 2015).

By secreting organic acids and other bioactive compounds, these microorganisms enhance nutrient cycling and contribute to soil fertility, making them vital agents in sustainable nutrient management (Ghosh et al. 2023). In addition to their role in potassium solubilization, they exhibit diverse functionalities, including the production of plant growth-regulating substances, antibiotics, and enzymes that biodegrade organic matter, thereby facilitating nutrient recycling and supporting overall soil health.

The release rates of K during mica weathering are influenced by its mineralogy (Krahl et al. 2022a). Trioctahedral micas, such as biotite and phlogopite, have a less stable structure compared to dioctahedral micas, like muscovite. As a result, potassium can be more efficiently extracted by plants from trioctahedral micas (Li et al. 2015b). The release of cations from these trioctahedral micas during incongruent dissolution may facilitate the formation of secondary minerals with 2:1 layers, such as illite or hydrous mica, vermiculite, chlorite, and interstratified minerals (Mohammed et al. 2014; Naderizadeh et al. 2010; Norouzi and Khademi 2010).

At the onset of the natural weathering process of biotite minerals, an expanded interstratified biotite-vermiculite phase, known as hydrobiotite, is formed. Hydrobiotite exhibits a significant peak in X-ray diffraction analysis, ranging from 10 Å to approximately 24 Å, which results from the interlayer expansion at the edges of biotite layers. Over time, a peak is anticipated to form at 14 Å, indicating the presence of vermiculite minerals (Fanning et al. 1989; Murakami et al. 2003; Pozzuoli et al. 1992). Hydrobiotite and vermiculite have the potential to enhance soil cation exchange capacity, particularly in soils that are deficient in 2:1 minerals and possess low organic matter content (Churchman and Lowe 2012). The chemical composition, particularly the Fe and Mg content of biotite, along with the substitution of elements in the octahedral and tetrahedral silicate sheets (for example, Si^{4+} by Al^{3+} and Fe^{3+} in tetrahedron sites or Al^{3+} by Mg^{2+} in octahedron sites), results in varying weathering rates (Bonneville et al. 2016; Gilkes et al. 1973; Murakami et al. 2003). Under oxidizing conditions, reduced iron released from the mineral matrix, if not absorbed by plant roots or lost through leaching or runoff, quickly precipitates as amorphous Fe^{3+} hydroxides. The precipitation of Fe on the surface of transforming minerals may contribute to delaying the weathering process and blocking charged sites formed on mineral surfaces (Bray et al. 2015; Tamrat et al. 2018; Tombolini et al. 2002; Zavarzina et al. 2016).

Li et al. (2025) found that biotite had the highest K-supplying capacity among phyllosilicate minerals, followed by vermiculite, illite, smectite, muscovite, and phlogopite. However, Norouzi and Khademi (2010) reported a different order for alfalfa K uptake: phlogopite, biotite, then muscovite. These differences may be due to variations in experimental conditions, data interpretation, and mineral characteristics, such as structure, particle size, and weathering state.

The significance of silicate sources goes beyond nutrient availability alone (Anda et al. 2015). Plant-induced weathering rates are influenced by the mineralogy of the rocks, soil characteristics, and crop species (Burghelea et al. 2015; Harley and Gilkes 2000; Mohammed et al. 2014; Wang et al. 2000). These factors determine the type and proportion of newly formed minerals. Additionally, rocks are heterogeneous and possess complex compositions, which are not fully understood, especially regarding their behavior in agricultural systems (Basak et al. 2023; Harley and Gilkes 2000; Krahl et al. 2020; Li et al. 2015a; Nunes et al. 2014; Ramos et al. 2015; Zhang et al. 2018).

6 Nutrient Availability

The prediction of nutrient availability from rocks begins with mineralogical identification, as certain minerals are known to dissolve rapidly in the soil under rhizosphere activity, providing significant amounts of nutrients for plant uptake. Conversely, others may exhibit very slow dissolution. The rapid dissolution in the rhizosphere can result in adequate nutrient levels for plant growth.

To accelerate the dissolution of minerals and release nutrients to plants, several factors must be met. Practices that promote soil health, biodiversity, carbon sequestration, and the resilience of agricultural systems, such as regenerative practices, should be employed. These practices enhance biological activity in the soil and at the soil-rock interface, speeding up the decomposition of minerals and the release of nutrients.

Some regenerative practices that promote the dissolution of silicate agrominerals include:

- Addition of organic matter (compost, manure, plant waste) to feed microorganisms;
- Inoculation of mineral-solubilizing microorganisms (e.g., bacteria and mycorrhizal fungi);
- Maintenance of soil moisture, essential for microbial survival;
- Species with deep, aggressive roots, such as grasses and legumes;
- Plants release root exudates, such as organic acids and enzymes, which facilitate the dissolution of minerals;
- Implement crop rotation and utilize cover crops to ensure the soil remains biologically active throughout the year;
- Implement no-tillage with permanent soil cover;

- The integration of crops, livestock, and forests, which enhances biological diversity and nutrient cycling;
- Implementing intercropping strategies to enhance the diversity of biological interactions.

Multiple experiments have demonstrated that the use of silicate minerals can lead to an increase in root volume across various crops (Martins et al. 2023). The increase in roots results in a larger volume of soil being explored, which enhances nutrient exploitation and access to water. Additionally, it increases the supply of root exudates and promotes microbial interactions.

A diversified and permanent vegetation cover is a crucial strategy for enhancing soil biological activity. This approach increases the bioavailability of various nutrients complexed in clay minerals, soil organic matter, or present in rhizosphere-environment matrices (REM) through bioweathering processes (Martins et al. 2023). Consequently, it releases a portion of these nutrients for subsequent crops (Fig. 4). Cover crops have roots that differ from those of main crops and produce various exudates, which attract a wider range of microorganisms. They also contribute to the cycling of nutrients from deeper soil layers to the surface. Cover crops maintain and enhance the soil with living vegetation during the off-season. They minimize surface runoff and erosion by intercepting precipitation with green leaves and reducing its impact, while their roots improve soil structure and increase water infiltration. Additional benefits include: harnessing solar energy and atmospheric CO_2 to enrich soil organic matter, boosting soil nitrogen levels, reducing compaction, creating habitats for beneficial organisms, and encouraging the growth of mycorrhizal fungi for future crops.

Various chemical extractants have been employed to assess nutrient availability from ground rocks for the purpose of recommending the use of silicate agrominerals to supply nutrients to crops (Li et al. 2014; Basak et al. 2018). However, these methods remain in development due to the mineralogical, chemical, and physical heterogeneity of rocks, which currently precludes the use of a single extractant for

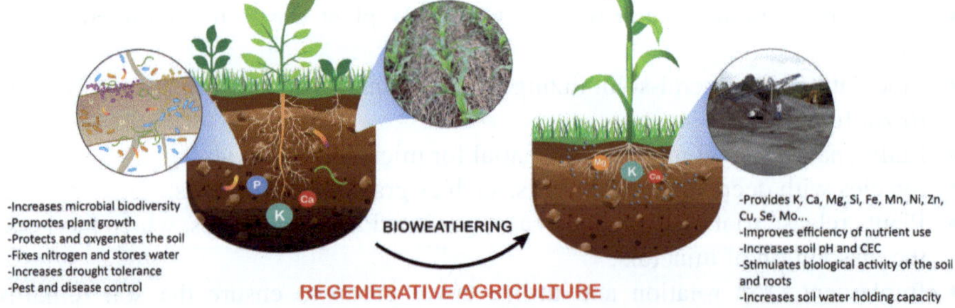

Fig. 4 Regenerative management via the implementation of cover crops, bioinputs, and silicate agrominerals

determining nutrient availability and release rates. It is essential to evaluate silicate agrominerals through plant-based nutrient extraction methods.

7 Advances in Bioweathering Technology

To assess biological weathering at the nanoscale to microscale (Finlay et al. 2020), several studies have employed atomic force microscopy (AFM) and scanning transmission electron microscopy coupled with energy-dispersive X-ray spectroscopy (STEM–EDX). They intend to reveal the physical and chemical alterations of mineral surfaces driven by fungal activity (Bonneville et al. 2009; McMaster 2012). These include the biomechanical penetration of minerals, secretion of organic acids and siderophores, and the formation of microtunnels that enhance mineral dissolution.

Additionally, stable isotope techniques (Uhlig et al. 2017) have been utilized to monitor nutrient release and uptake pathways, offering quantitative insights into biologically mediated weathering, as emphasized by Finlay et al. (2020). The authors argue that existing weathering models often underestimate or overlook the importance of biological processes.

Research using synchrotron radiation-based spectromicroscopy at fungus-mineral interfaces has shown that fungal hyphae can alter the local redox state of iron, indicating that they are redox-active. These findings suggest that fungus-initiated catalytic reactions occur at hyphal-mineral interfaces, as superoxide does not diffuse far from its site of formation. Additionally, the results imply that these catalytic reactions may represent a new strategy for microbial iron uptake. Collectively, these observations enhance the understanding of how fungi make minerals available to biological systems (Yu et al. 2019).

8 Dissolution and Regulating Soil Carbon Storage

The role of newly formed minerals and low-crystallinity minerals, which arise from the congruent or incongruent dissolution of silicate agrominerals, is a topic of ongoing research. This investigation is particularly focused on their potential to sequester atmospheric carbon in tropical environments. Despite their significance, the effects of these minerals in such settings have not been thoroughly examined. In these regions, the application of silicate rocks in agricultural soils introduces raw minerals into the ecosystem, which not only enhances soil quality but also stabilizes soil organic matter (SOM) through bioweathering processes and the formation of low-crystallinity minerals. These minerals, such as vermiculites, smectites, amorphous phases, and others with high cation exchange capacities (CEC), contribute to the adsorption or occlusion of SOM, thereby playing a critical role in the long-term stabilization of soil organic carbon (SOC) (Edwards et al. 2017; Beerling et al. 2018).

Carbon stabilization in soil can be achieved through two primary mechanisms: as soil organic carbon (SOC), predominantly in soil organic matter (SOM), and as soil inorganic carbon (SIC). Silicate applications have the potential to affect SOC, influenced by soil mineral phases (Kleber et al. 2015; Sarkar et al. 2018; Singh et al. 2017; Wang et al. 2018), and SIC (Beerling et al. 2018; Hartmann et al. 2013; Kantola et al. 2017; Moosdorf et al. 2014; Strefler et al. 2018). The discussion on soil carbon sequestration has traditionally centered on SOC, due to its ability to rapidly respond to land management practices (Singh et al. 2018, Fig. 2).

Research has shown that the adsorption of soil organic matter (SOM) onto mineral surfaces, as well as its occlusion within soil aggregates, can enhance its stabilization by reducing its accessibility to microbial decomposition (Angst et al. 2018; Kleber et al. 2015; Sarkar et al. 2018; Singh et al. 2017). The type and amount of clay minerals have a significant influence on SOM stabilization in soils due to their structure and physicochemical characteristics (Angst et al. 2018; Kleber et al. 2015; Singh et al. 2017; Wang et al. 2018). Soils with higher clay contents generally protect more organic carbon from mineralization through physical, biological, and chemical mechanisms (Angst et al. 2018; Han et al. 2016; Sarkar et al. 2018). The large specific surface area and higher charge (CEC) in smectites and vermiculites (2:1 expansive layers type) favor a higher sorption capacity and greater ability to protect SOM. Additionally, clay minerals such as allophane and imogolite can strongly adsorb SOM through these mechanisms (Fig. 5).

The association of organic matter with pedogenic minerals, referred to as mineral-organic associations, is a crucial factor in the long-term retention of soil organic carbon (SOC). These mineral-organic associations restrict microbial and enzymatic

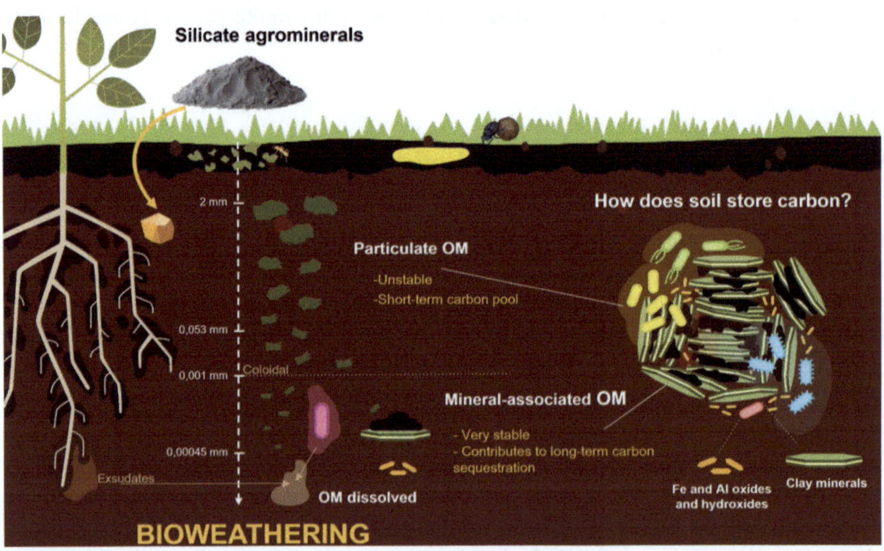

Fig. 5 Potential effects of weathering of crushed silicate rocks as agrominerals. Adapted from André Vicente by Tropica

access, representing the primary mechanism for protecting SOC from microbial utilization over periods extending from centuries to millennia (Keiluweit et al. 2015). The conditions favorable for the formation of mineral-organic associations are most prevalent in regions where soil moisture supports high plant biomass production, characterized by intensive weathering, low pH, and soil formation, such as tropical environments (Kleber et al. 2015; Sarkar et al. 2018). However, unlike other climatic zones, the rate of natural rock weathering in tropical regions is constrained by the availability of fresh mineral surfaces.

The hydrolysis of silicate minerals is an endothermic process; therefore, higher temperatures enhance mineral dissolution rates, contributing to more effective Enhanced Silicate Weathering (ESW) in tropical environments (Vandeginste et al. 2024). The best-suited locations are warm and humid areas, particularly in India, Brazil, Southeast Asia, and China, where almost 75% of the global potential can be exploited (Strefler et al. 2018).

ESW is a climate mitigation strategy designed to accelerate the natural breakdown of silicate minerals, thereby capturing atmospheric CO_2 and improving soil health. This process mimics natural geologic mechanisms but on a timescale relevant to human activities. ESW generally involves the release of cations such as Ca^{2+} and Mg^{2+} into solution, which consumes CO_2 and helps regulate atmospheric levels. Amending soils with crushed silicate rocks, such as basalt, olivine, or wollastonite, can capture and store dissolved inorganic carbon (DIC) within soil minerals through weathering processes (Beerling et al. 2018; Deng et al. 2023; Hartmann et al. 2013; Kantola et al. 2017; Lefebvre et al. 2019; Li et al. 2025; Moosdorf et al. 2014; Renforth 2012; Strefler et al. 2018).

Theoretical studies highlight the significant potential of ESW for CO_2 removal and its multiple agronomic benefits when applied to agricultural soils. As a promising strategy, ESW has gained attention for its ability to enhance CO_2 removal while simultaneously improving soil quality.

These minerals react with CO_2 and water to form bicarbonates. These bicarbonates can either be leached into groundwater and eventually reach the ocean, where they contribute to long-term carbon storage, or precipitate as solid carbonates, thereby locking carbon in a mineral form. In tropical soils with a pH below 6 (acidic soil), however, carbon sequestration is impeded due to the predominance of dissolved CO_2 and reduced bicarbonate formation (Holden et al. 2024). For the H_2CO_3/HCO_3^- system, the pKa value is approximately 6.35. At a pH of 5.5, the ratio of bicarbonate (HCO_3^-) to carbonic acid (H_2CO_3) is roughly 0.16:1, indicating that there is about six times more H_2CO_3 than HCO_3^-.

Atmospheric carbon sequestration is most efficient in environments with a moderate pH range, between 6 and 9. In this condition, bicarbonate (HCO_3^-) is the predominant chemical species, allowing carbon dioxide to be captured in a dissolved form. The oceans play a crucial role in this process, with approximately 90% of dissolved atmospheric CO_2 present in the form of bicarbonate.

Carbon sequestration, however, is influenced by numerous factors. For example, the presence of divalent cations such as Ca^{2+} and Mg^{2+} is crucial for the efficacy of ESW, as these cations are integral to promoting the dissolution of silicate minerals

(Abdalqadir et al. 2024). The CO_2 removal potential of silicate rocks can be determined based on their chemical composition. This is typically expressed as the theoretical maximum carbon dioxide removal capacity per metric ton of material. The estimation takes into account the molecular masses of CO_2 and the relevant basic oxides—CaO, MgO, Na_2O, and K_2O—as well as their concentrations, reported as weight percentages (% CaO, % MgO, % Na_2O, and % K_2O) (Li et al. 2025). Direct methods measure bicarbonate ions (HCO_3^-) in leachates and pedogenic carbonates within soils (te Pas et al. 2023; Larkin et al. 2022). Alternative methods assess the mass loss of base cations (Ca^{2+}, Mg^{2+}, K^+, Na^+) (Abdalqadir et al. 2024; Beerling et al. 2024; Buckingham 2022) and use soil electrical conductivity as an indirect parameter (Rieder et al. 2024).

The highest potential for CO_2 sequestration is anticipated from ultramafic rocks with low silica content. These igneous rocks are distinguished by elevated concentrations of MgO and FeO. They contain significant amounts of olivine $(Mg,Fe)_2SiO_4$ and pyroxene $Cax(Mg,Fe)_{2-x}Si_2O_6$, both of which are relatively fast-reacting minerals that contribute to their high weathering potential (Vandeginste et al. 2024). Basalts, the most common mafic volcanic rock, are rich in Ca, Mg, and Fe and are primarily composed of plagioclase, pyroxene, and olivine. This composition makes basalts particularly suitable for ESW, as these elements play a crucial role in promoting effective CO_2 sequestration through mineral carbonation (Beerling et al. 2024; Kelemen et al. 2020).

The hierarchy of CO_2 efficiency among silicate minerals suitable for ESW is ranked as follows: wollastonite $\geqslant$ olivine > basalt > albite $\geqslant$ anorthite (Kelemen et al. 2020; Li et al. 2025; Vandeginste et al. 2024). Wollastonite ($CaSiO_3$), rich in calcium and silicon, is considered suitable for ESW due to its high reactivity and potential agronomic benefits, such as contributing to soil pH regulation (Haque et al. 2020a, b).

Studies have shown that the efficiency of rocks in removing CO_2 depends on their mineralogy and particle size, based on mesocosm and field experiments (Beerling et al. 2024; Kelemen et al. 2020; Vandeginste et al. 2024).

In a comprehensive review, Li et al. (2025) summarized Carbon Dioxide Removal (CDR) rates for basalt, olivine, and wollastonite. For basalt, CO_2 removal efficiencies reported in mesocosm experiments range from 0.03 to 0.12 t CO_2 t^{-1} silicate ha^{-1} y^{-1} (Reershemius et al. 2023, te Pas et al. 2023; Buckingham 2022; Kelland et al., 2020). In contrast, field trials have shown higher efficiencies, ranging from 0.05 to 0.17 t CO_2 t^{-1} silicate ha^{-1} y^{-1} (Beerling et al., 2024; Kantola et al., 2023). Similarly, for wollastonite, efficiencies in field trials reported range from 0.19 to 1.90 t CO_2 t^{-1} silicate ha^{-1} y^{-1} (Haque et al. 2020a), which is approximately two orders of magnitude higher than those observed in mesocosm experiments, where values range from 6.7×10^{-4} to 0.05 t CO_2 t^{-1} silicate ha^{-1} y^{-1} (te Pas et al. 2023; Wood et al. 2023; Haque et al. 2020b).

Field-based estimates of potential carbon dioxide removal have reached approximately 3.4 t CO_2 ha^{-1} y^{-1} following the application of crushed metabasalt (Beerling et al. 2024). These findings indicate that increasing the global application of silicates in soils may contribute to climate change mitigation efforts.

8.1 Future Research Directions

Chemical, physical, and biological modification processes within agroecosystems could enhance the efficacy of silicate agrominerals. The active metabolism of plants, coupled with soil microbial activity, has the potential to accelerate the weathering process significantly. These various mechanisms can facilitate the mobilization of elements from silicate minerals, which may have substantial implications for nutrient availability.

The mechanisms underlying silicate dissolution in the rhizosphere, as well as the nature of their bioweathering products, remain only partially understood (Basak 2019; Li et al., 2014). Limited information is available on the chemical alterations and generation of negative charges resulting from weathering, particularly from the perspective of plant nutrition. Likewise, the influence of different particle-size fractions on the release patterns of elements to meet plant demands is not yet fully clarified (Harley and Gilkes 2000). Considering that the formation of new mineral phases can increase negative charges (CEC), it is proposed that the successive application of silicate agrominerals, in combination with rhizosphere-driven weathering by cultivated plants, may generate emergent properties. Such processes could progressively enhance the physical and chemical attributes of soils, especially in highly weathered and nutrient-depleted environments such as Ferralsols. Over the medium to long term, this practice holds the potential to foster the development of resilient, sustainable, and highly productive anthropogenic soils.

On a field scale, various factors can influence the dissolution process and rates of silicate agrominerals. The impact of these silicates when applied in agricultural soils as fertilizers or remineralizers is contingent on soil properties, crop type, climatic conditions (such as temperature and precipitation), and management practices. Further laboratory studies, along with greenhouse and field experiments, are required to investigate the dissolution kinetics of these agrominerals within such complex natural systems.

Acknowledgements Author's Contribution All authors made significant contributions to the conception, development, and refinement of this chapter. Luise Lottici Krahl was primarily responsible for the literature search, data organization, and drafting of the initial sections and discussion. Giuliano Marchi and Éder de Souza Martins contributed to the critical analysis, data integration, and writing of the discussion. Elis Marina de Freitas and Luiz Fernando dos Santos revised the manuscript, providing intellectual input and ensuring overall consistency and clarity. All authors read and approved the final version of the chapter and agree to be accountable for its content.

Acknowledgments

The authors gratefully acknowledge the financial support provided by Embrapa and by FINEP—Fundo Setorial Mineral through the project 'Rede de PD&I sobre remineralizadores e agrominerais (REMAg)' (Project No. 1541/2022), which made possible the development of this book chapter.

Declaration of Competing Interest The authors declare that they have no known competing financial interests or personal relationships that could have appeared to influence the work reported in this paper.

References

Abdalqadir M, Hughes D, Rezaei Gomari S, Rafiq U (2024) A state of the art of review on factors affecting the enhanced weathering in agricultural soil: strategies for carbon sequestration and climate mitigation. Environ Sci Pollut Res 31(13):19047–19070

Akter M, Akagi T (2005) Effect of fine root contact on plant-induced weathering of basalt. Soil Sci Plant Nutr 51:861–871. https://doi.org/10.1111/j.1747-0765.2005.tb00121.x

Anda M, Shamshuddin J, Fauziah CI (2015) Improving chemical properties of a highly weathered soil using finely ground basalt rocks. Catena 124:147–161. https://doi.org/10.1016/j.catena.2014.09.012

Andrist-Rangel Y, Simonsson M, Öborn I, Hillier S (2013) Acid-extractable potassium in agricultural soils: source minerals assessed by differential and quantitative X-ray diffraction. J Plant Nutr Soil Sc 176:407–419

Angst G, Messinger J, Greiner M, Häusler W, Hertel D, Kirfel K, Kögel-Knabnera I, Leuschner C, Rethemeyer J, Mueller CW (2018) Soil organic carbon stocks in topsoil and subsoil controlled by parent material, carbon input in the rhizosphere, and microbial-derived compounds. Soil Biol Biochem 122:19–30. https://doi.org/10.1016/j.soilbio.2018.03.026

Basak BB, Maity A, Ray P, Biswas DR, Roy S (2022) Potassium supply in agriculture through biological potassium fertilizer: a promising and sustainable option for developing countries. Arch Agron Soil Sci 68(1):101–114. https://doi.org/10.1080/03650340.2020.1821191

Basak BB (2019) Waste mica as alternative source of plant-available potassium: evaluation of agronomic potential through chemical and biological methods. Nat Resour J 27:1–13. https://doi.org/10.1007/s11053-018-9430-3

Basak BB, Sarkar B, Sanderson P, Naidu R (2018) Waste mineral powder supplies plant available potassium: evaluation of chemical and biological interventions. J Geochem Explor 186:114–120. https://doi.org/10.1016/j.gexplo.2017.11.023

Beerling DJ, Leake JR, Long SP, Scholes JD, Ton J, Nelson PN, Bird M, Kantzas E, Taylor LL, Sarkar B, Kelland M, DeLucia E, Kantola I, Müller C, Rau G, Hansen J (2018) Farming with crops and rocks to address global climate, food and soil security. Nat Plants 4:138–147. https://doi.org/10.1038/s41477-018-0108-y

Beerling DJ, Epihov DZ, Kantola IB, Masters MD, Reershemius T, Planavsky NJ et al (2024) Enhanced weathering in the US Corn Belt delivers carbon removal with agronomic benefits. Proc Natl Acad Sci 121(9):e2319436121. https://doi.org/10.1073/pnas.2319436121

Bonneville S, Bray AW, Benning LG (2016) Structural Fe (II) oxidation in biotite by an ectomycorrhizal fungi drives mechanical forcing. Environ Sci Technol 50(11):5589–5596

Bonneville S, Smits MM, Brown A, Harrington J, Leake JR, Brydson R, Benning LG (2009) Plant-driven fungal weathering: early stages of mineral alteration at the nanometer scale. Geology 37:615–618

Bray AW, Oelkers EH, Bonneville S, Wolff-Boenisch D, Potts NJ, Fones G, Benning LG (2015) The effect of pH, grain size, and organic ligands on biotite weathering rates. Geochim Cosmochim Acta 164:127–145

Broadley M, Brown P, Cakmak I, Rengel Z, Zhao F (2012) Function of nutrients: micronutrients. In: Marschner P (ed) Marschner's mineral nutrition of higher plants. Academic, London, pp 191–248. https://doi.org/10.1016/B978-0-12-384905-2.00007-8

Buckingham F (2022) The potential of enhanced weathering as a CO_2 removal technique in UK agricultural soils. (Doctoral dissertation,. University of Oxford

Burghelea C, Zaharescu DG, Dontsova K, Maier R, Huxman T, Chorover J (2015) Mineral nutrient mobilization by plants from rock: influence of rock type and arbuscular mycorrhiza. Biogeochemistry 124:187–203. https://doi.org/10.1007/s10533-015-0092-5

Churchman GJ, Lowe DJ (2012) Alteration, formation, and occurrence of minerals in soils. In: Huang PM, Li Y, Sumner ME (eds) Handbook of soil sciences, vol 1. CRC Press, Boca Raton, pp 20.1–20.72. https://doi.org/10.1201/b11267

Ciceri D, Close TC, Barker AV, Allanore A (2019) Fertilizing properties of potassium feldspar altered hydrothermally. Commun Soil Sci Plant Anal 50:482–491. https://doi.org/10.1080/00103624.2019.1566922

Coleman NT, LeRoux FH, Cady JG (1963) Biotite-hydrobiotite-vermiculite in soils. Nature 198(4878):409–410. https://doi.org/10.1038/198409c0

Colin Y, Nicolitch O, Turpault MP, Uroz S (2017) Mineral types and tree species determine the functional and taxonomic structures of forest soil bacterial communities. Appl Environ Microbiol 83:e02684–e02616

Deng Z, Wang G, Jin W, Tang N, Ren H, Chen X (2023) Characteristics and quantification of fine particle loss in internally unstable sandy gravels induced by seepage flow. Eng Geol 321:107150

Dong H, Huang LQ, Zhao L, Zeng Q, Liu X, Sheng Y, Shi L, Wu G, Jiang H, Li F, Zhang L, Guo D, Li G, Hou W, Chen HY (2022) A critical review of mineral–microbe interaction and co-evolution: mechanisms and applications. Natl Sci Rev 9(10):nwac128. https://doi.org/10.1093/nsr/nwac128

Edwards DP, Lim F, James RH, Pearce CR, Scholes J, Freckleton RP, Beerling DJ (2017) Climate change mitigation: potential benefits and pitfalls of enhanced rock weathering in tropical agriculture. Biol Lett 13:1–7. https://doi.org/10.1098/rsbl.2016.0715

Essington ME (2015) Soil and water chemistry: an integrative approach, 2nd edn. CRC Press, Boca Raton, FL. https://doi.org/10.1201/b18385

Fanning DS, Keramidas VZ, El-Desoky MA (1989) Micas. In: Dixon JB, Weed SB (eds) Madison. Wisconsin, SSSA, pp 551–634

Farmer VC, Wilson MJ (1970) Experimental conversion of biotite to hydrobiotite. Nat Lond 226:841–842

Finlay RD, Mahmood S, Rosenstock N, Bolou-Bi EB, Köhler SJ, Fahad Z, Rosling A, Wallander H, Belyazid S, Bishop K, Lian B (2020) Reviews and syntheses: biological weathering and its consequences at different spatial levels–from nanoscale to global scale. Biogeosciences 17(6):1507–1533

Fordham AW (1990) Formation of trioctahedral illite from biotite in a soil profile over granite gneiss. Clay Clay Miner 38:187–195

Franzosi C, Castro LN, Celeda AM (2014) Technical evaluation of glauconies as alternative potassium fertilizer from the Salamanca formation, Patagonia, Southwest Argentina. Nat Resour Res 23:311–320. https://doi.org/10.1007/s11053-014-9232-1

Gaskin JW, Speir RA, Harris K, Das KC, Lee RD, Morris LA, Fisher DS (2010) Effect of peanut hull and pine chip biochar on soil nutrients, corn nutrient status, and yield. Agron J 102:623–633. https://doi.org/10.2134/agronj2009.0083

Ghosh S, Mondal S, Banerjee S, Mukherjee A, Bhattacharyya P (2023) Temporal dynamics of potassium release from waste mica as influenced by potassium mobilizing bacteria. J Pure Appl Microbiol 17(1):273–288

Gilkes RJ, Young RC, Quirk JP (1973) Artificial weathering of oxidized biotite: I. Potassium removal by sodium chloride and sodium tetraphenylboron solutions. Soil Sci Soc Am Proc 37:25–28. https://doi.org/10.2136/sssaj1973.03615995003700010013x

Han L, Sun K, Jin J, Xing B (2016) Some concepts of soil organic carbon characteristics and mineral interaction from a review of literature. Soil Biol Biochem 94:107–121. https://doi.org/10.1016/j.soilbio.2015.11.023

Haque F, Santos RM, Chiang YW (2020a) CO_2 sequestration by wollastonite-amended agricultural soils–an Ontario field study. Int J Greenh Gas Con 97:103017

Haque F, Santos RM, Chiang YW (2020b) Optimizing inorganic carbon sequestration and crop yield with wollastonite soil amendment in a microplot study. Front Plant Sci 11:1012

Harley AD, Gilkes RJ (2000) Factors influencing the release of plant nutrient elements from silicate rock powders: a geochemical overview. Nutr Cycl Agroecosyst 56:11–36. https://doi.org/10.1023/A:100985930

Hartmann J, West AJ, Renforth P, Köhler P, De La Rocha CL, Wolf-Gladrow DA, Dürr HH, Scheffran J (2013) Enhanced chemical weathering as a geoengineering strategy to reduce atmospheric carbon dioxide, supply nutrients, and mitigate ocean acidification. Rev Geophys 51(2):113–149. https://doi.org/10.1002/rog.20004

Hemingway JD, Hilton RG, Hovius N, Eglinton TI, Haghipour N, Wacker L, Chen M-C, Galy VV (2018) Microbial oxidation of lithospheric organic carbon in rapidly eroding tropical mountain soils. Science 360(6385):209–212. https://doi.org/10.1126/science.aao6463

Hinsinger P, Barros ONF, Benedetti MF, Noack Y, Callot G (2001) Plant-induced weathering of a basaltic rock: experimental evidence. Geochim Cosmochim Acta 65:137–152. https://doi.org/10.1016/S0016-7037(00)00524-X

Hinsinger P, Dufey JE, Jaillard B (1991) Biological weathering of micas in the rhizosphere as related to potassium absorption by plant roots. In: Developments in agricultural and managed Forest ecology, vol 24. Elsevier, pp 98–105

Hinsinger P, Elsass F, Jaillard B, Robert M (1993) Root induced irreversible transformation of a trioctahedral mica in the rhizosphere of rape. J Soil Sci 44(3):535–545

Hinsinger P, Jaillard B, Dufey JE (1992) Rapid weathering of a trioctahedral mica by the roots of ryegrass. Soil Sci Soc Am J 56:977–982. https://doi.org/10.2136/sssaj1992.03615995005600030049x

Hinsinger P, Plassard C, Jaillard B (2006) Rhizosphere: a new frontier for soil biogeochemistry. J Geochem Explor 88(1–3):210–213

Holden FJ, Davies K, Bird MI, Hume R, Green H, Beerling DJ, Nelson PN (2024) In-field carbon dioxide removal via weathering of crushed basalt applied to acidic tropical agricultural soil. Sci Total Environ 955:176568

Jones DL (1998) Organic acids in the rhizosphere—a critical review. Plant Soil 205:25–44. https://doi.org/10.1023/A:1004356007312

Jones DL, Darah PR, Kochian LV (1996) Critical evaluation of organic acid mediated iron dissolution in the rhizosphere and its potential role in root iron uptake. Plant Soil 180:57–66. https://doi.org/10.1007/BF00015411

Kantola IB, Blanc-Betes E, Masters MD, Chang E, Marklein A, Moore CE et al (2023) Improved net carbon budgets in the US Midwest through direct measured impacts of enhanced weathering. Glob Chang Biol 29(24):7012–7028

Kantola IB, Masters MD, Beerling DJ, Long SP, DeLucia EH (2017) Potential of global croplands and bioenergy crops for climate change mitigation through dep. Biol Lett 13:1–7. https://doi.org/10.1098/rsbl.2016.0714

Keiluweit M, Bougoure JJ, Nico PS, Pett-Ridge J, Weber PK, Kleber M (2015) Mineral protection of soil carbon counteracted by root exudates. Nat Clim Chang 5:588–595. https://doi.org/10.1038/NCLIMATE2580

Kelemen PB, McQueen N, Wilcox J, Renforth P, Dipple G, Vankeuren AP (2020) Engineered carbon mineralization in ultramafic rocks for CO_2 removal from air: review and new insights. Chem Geol 550:119628

Kelland ME, Wade PW, Lewis AL, Taylor LL, Sarkar B, Andrews MG, Lomas MR, Cotton TEA, Kemp SJ, James RH, Pearce CR, Hartley SE, Hodson ME, Leake JR, Banwart SA, Beerling DJ (2020) Increased yield and CO_2 sequestration potential with the C4 cereal Sorghum bicolor cultivated in basaltic rock dust-amended agricultural soil. Glob Change Biology 26:3658–3676. https://doi.org/10.1111/gcb.15089

Kleber M, Eusterhues K, Keiluweit M, Mikutta C, Mikutta R, Nico PS (2015) Mineral-organic associations: formation, properties, and relevance in soil environments. Adv Agron 130:1–140. https://doi.org/10.1016/bs.agron.2014.10.005

Kobayashi T, Nozoye T, Nishizawa NK (2019) Iron transport and its regulation in plants. Free Radic Biol Med 133:11–20

Kong M, Huang L, Li L, Zhang Z, Zheng S, Wang MK (2014) Effects of oxalic and citric acids on three clay minerals after incubation. Appl Clay Sci 99:207–214. https://doi.org/10.1016/j.clay.2014.06.035

Krahl LL, da Paz SPA, Valadares LF, Marchi G, Valadares LF, Sousa-Silva JC, Martins ÉS (2020) Successive offtake of elements by maize grown in pure basalt powder. Afr J Agric Res 15(2):229–239

Krahl LL, Marchi G, Paz SPA, Angélica RS, Sousa-Silva JC, Valadares LF, Martins ÉS (2022a) Increase in cation exchange capacity by the action of maize rhizosphere on Mg or Fe biotite-rich rocks. Pesquisa Agropecuária Tropical 52:e72376

Krahl LL, Valadares LF, Sousa-Silva JC, Marchi G, Martins ÉS (2022b) Dissolution of silicate minerals and nutrient availability for corn grown successively. Pesq Agrop Brasileira 57:e01467

Larkin CS, Andrews MG, Pearce CR, Yeong KL, Beerling DJ, Bellamy J, James RH (2022) Quantification of CO_2 removal in a large-scale enhanced weathering field trial on an oil palm plantation in Sabah. Malays Front Clim 4:959229

Lasaga AC (1995) Fundamental approaches in describing mineral dissolution and precipitation rates. Rev Mineral Geochem 31:21–86

Lefebvre D, Goglio P, Williams A, Manning DAC, de Azevedo AC, Bergmann M, Meersmans J, Smith P (2019) Assessing the potential of soil carbonation and enhanced weathering through life cycle assessment: a case study for São Paulo State, Brazil. J Clean Prod 233:468–481. https://doi.org/10.1016/j.jclepro.2019.06.099

Li T, Wang H, Zhou Z, Chen X, Zhou J (2015a) A nano-scale study of the mechanisms of non-exchangeable potassium release from micas. Appl Clay Sci 118:131–137. https://doi.org/10.1016/j.clay.2015.09.013

Li J, Zhang W, Li S, Li X, Lu J (2014) Effects of citrate on the dissolution and transformation of biotite, analyzed by chemical and atomic force microscopy. Appl Geochem 51:101–108. https://doi.org/10.1016/j.apgeochem.2014.10.001

Li M, Zhang Z, Zeng X, He Z (2025) A systematic review on intelligent prediction of inorganic building materials performance. Int J Comput Intell Syst 18(1):1–44

Li T, Wang H, Wang J, Zhou Z, Zhou J (2015b) Exploring the potential of phyllosilicate minerals as potassium fertilizers using sodium tetraphenylboron and intensive cropping with perennial ryegrass. Sci Rep 5:1–7. https://doi.org/10.1038/srep09249

Machado RV, Andrade FV, Passos RR, Ribeiro RCC, Mendonça ES, Mesquita LF (2016) Characterization of ornamental rock residue and potassium liberation via organic acid application. Rev Bras Cienc Solo 40:1–13. https://doi.org/10.1590/18069657rbcs20150153

Madaras M, Mayerová M, Kulhánek M, Koubová M, Faltus M (2013) Waste silicate minerals as potassium sources: a greenhouse study on spring barley. Arch Agron Soil Sci 59:671–683. https://doi.org/10.1080/03650340.2012.667079

Manning DAC, Baptista J, Limon MS, Brandt K (2017) Testing the ability of plants to access potassium from framework silicate minerals. Sci Total Environ 574:476–481. https://doi.org/10.1016/j.scitotenv.2016.09.086

Martins ÉS, Martins ES, Hardoim PR (2023) Princípios geoquímicos, mineralógicos e biológicos do manejo de remineralizadores de solos. Informe Agropecuário 44(321):26–39

Mauck BS, Roberts JA (2007) Mineralogic control on abundance and diversity of surface-adherent microbial communities. Geomicrobiol J 24(3–4):167–177. https://doi.org/10.1080/01490450701457162

McMaster TJ (2012) Atomic force microscopy of the fungi-mineral interface: applications in mineral dissolution, weathering and biogeochemistry. Curr Opin Biotech 23:562–569

Mohammed SMO, Brandt K, Gray ND, White ML, Manning DAC (2014) Comparison of silicate minerals as sources of potassium for plant nutrition in sandy soil. Eur J Soil Sci 65(5):653–662

Moosdorf N, Renforth P, Hartmann J (2014) Carbon dioxide efficiency of terrestrial enhanced weathering. Environ Sci Technol 48(9):4809–4816

Murakami T, Utsunomiya S, Yokoyama T, Kasama T (2003) Biotite dissolution processes and mechanisms in the laboratory and in nature: early stage weathering environment and vermiculitization. Am Mineral 88(2–3):377–386

Mukherjee A, Zimmerman AR, Harris W (2011) Surface chemistry variations among a series of laboratory-produced biochars. Geoderma 163(3–4):247–255. https://doi.org/10.1016/j.geoderma.2011.04.021

Naderizadeh Z, Khademi H, Arocena JM (2010) Organic matter induced mineralogical changes in clay-sized phlogopite and muscovite in alfalfa rhizosphere. Geoderma 159(3–4):296–303. https://doi.org/10.1016/j.geoderma.2010.08.003

Noble AD, Gillman GP, Nath S, Srivastava RJ (2001) Changes in the surface charge characteristics of degraded soils in the wet tropics through the addition of beneficiated bentonite. Soil Res 39(5):991–1001

Norouzi S, Khademi H (2010) Ability of alfalfa (*Medicago sativa* L.) to take up potassium from different micaceous minerals and consequent vermiculitization. Plant Soil 328:83–93. https://doi.org/10.1007/s11104-009-0084-0

Nunes JMG, Kautzmann RM, Oliveira C (2014) Evaluation of the natural fertilizing potential of basalt dust wastes from the mining district of Nova Prata (Brazil). J Clean Prod 84:649–656. https://doi.org/10.1016/j.jclepro.2014.04.032

Palandri JL, Kharaka YK (2004) A complitation of rate parameters of water-mineral interaction kinetics for application to geochemical modelling. U.S. Geological Survey, Menlo Park, California

Pozzuoli A, Vila E, Franco E, Ruiz-Amil A (1992) Weathering of biotite to vermiculite in quaternary lahars from Monti Ernici. Cent Italy Clay Min 27:175–184. https://doi.org/10.1180/claymin.1992.027.2.03

Qafoku NP, Ranst EV, Noble A, Baert G (2004) Variable charge soils: their mineralogy, chemistry and management. Adv Agron 84(84):159–215. https://doi.org/10.1016/S0065-2113(04)84004-5

Ramos CG, Medeiros DS, Gomez L, Oliveira LFS, Schneider IAH, Kautzmann RM (2019) Evaluation of soil re-mineralizer from by-product of volcanic rock mining: experimental proof using black oats and maize crops. Nat Resour Res 1-18. https://doi.org/10.1007/s11053-019-09529-x

Ramos CG, Querol X, Oliveira MLS, Pires K, Kautzmann RM, Oliveira LFS (2015) A preliminary evaluation of volcanic rock powder for application in agriculture as soil a remineralizer. Sci Total Environ 512-513:371–380. https://doi.org/10.1016/j.scitotenv.2014.12.070

Reershemius T, Kelland ME, Davis IR, D'Ascanio R, Kalderon Asael B, Asael D, Epihov DE, Beerling DJ, Reinhard CT, Planavsky NJ (2023) A new soil-based approach for empirical monitoring of enhanced rock weathering rates. Environ Sci Technol 57(48). https://doi.org/10.1021/acs.est.3c03609

Reis BR, Vasconcelos ALS, Silva AMM, Andreote FD, Azevedo AC (2024) Changes in soil bacterial community structure in a short - term trial with different silicate rock powders. Chem Biol Technol Agric (S. l.):1–16. https://doi.org/10.1186/s40538-024-00586-w

Renforth P (2012) The potential of enhanced weathering in the UK. Int J Greenh Gas Con 10:229–243

Rieder L, Amann T, Hartmann J (2024) Soil electrical conductivity as a proxy for enhanced weathering in soils. Front Clim 5:1283107

Rudmin M, Oskina Y, Banerjee S, Mazurov A, Soktoev B, Shaldybin M (2018) Roasting-leaching experiments on glauconitic rocks of Bakchar ironstone deposit (Western Siberia) for evaluation their fertilizer potential. Appl Clay Sci 162:121–128. https://doi.org/10.1016/j.clay.2018.05.033

Sarkar B, Singh M, Mandal S, Churchman GJ, Bolan NS (2018) Clay minerals - organic matter interactions in relation to carbon stabilization in soils. In: Garcia C, Nannipieri P, Hernandez T (eds) The future of soil carbon. Academic Press Elservier, London, pp 71–86. https://doi.org/10.1016/B978-0-12-811687-6.00003-1

Shafar JM, Noordin WD, Zulkefly S, Shamshuddin J, Hanafi MM (2017) Improving soil chemical properties and growth performance of Hevea brasiliensis through basalt application. Proc Int Rubber Conf, Bali, Indonesia. https://doi.org/10.22302/ppk.procirc2017.v1i1.483

Sherman GD (1952) The genesis and morphology of the alumina-rich literite clays. In: Problems of clay and laterite genesis. American Institute of Mining and Metallurgical Engineers, New York, pp 154–161

Singh M, Sarkar B, Biswas B, Bolan NS, Churchman GJ (2017) Relationship between soil clay mineralogy and carbon protection capacity as influenced by temperature and moisture. Soil Biol Biochem 109:95–106. https://doi.org/10.1016/j.soilbio.2017.02.003

Singh M, Sarkar B, Sarkar S, Churchman GJ, Bolan NS, Mandal S, Menon M, Purakayastha TJ, Beerling DJ (2018) Stabilization of soil organic carbon as influenced by clay mineralogy. Adv Agron 148:33–84. https://doi.org/10.1016/bs.agron.2017.11.001

Strefler J, Amann T, Bauer N, Kriegler E, Hartmann J (2018) Potential and costs of carbon dioxide removal by enhanced weathering of rocks. Environ Res Lett 13:1–9. https://doi.org/10.1088/1748-9326/aaa9c4

Tamrat WZ, Rose J, Grauby O, Doelsch E, Levard C, Chaurand P, Basile-Doelsch I (2018) Composition and molecular scale structure of nanophases formed by precipitation of biotite weathering products. Geochim Cosmochim Acta 229:53–64. https://doi.org/10.1016/j.gca.2018.03.012

te Pas EEEM, Hagens M, Comans RNJ (2023) Assessment of the enhanced weathering potential of different silicate minerals to improve soil quality and sequester CO_2. Front Clim 4:954064. https://doi.org/10.3389/fclim.2022.954064

Tombolini F, Brigatti MF, Marcelli A, Cibin G, Mottana A, Giuli G (2002) Local and average Fe distribution in trioctahedral micas: analysis of Fe K-edge XANES spectra in the phlogopite–annite and phlogopite–tetraferriphlogopite joins on the basis of single-crystal XRD refinements. Eur J Mineral 14:1075–1085. https://doi.org/10.1127/0935-1221/2002/0014-1075

Tshering K, Rengel Z, Storer P, Solaiman ZM (2022) Microbial consortium inoculum with rock minerals increased wheat grain yield, nitrogen-use efficiency, and protein yield due to larger root growth and architecture. Agronomy 12(10):2481. https://doi.org/10.3390/agronomy12102481

Uhlig D, Schuessler JA, Bouchez J, Dixon JL, Von Blanckenburg F (2017) Quantifying nutrient uptake as driver of rockweathering in forest ecosystems by magnesium stable isotopes. Biogeosciences 14(12):3111–3128. https://doi.org/10.5194/bg-14-3111-2017

Uroz S, Calvaruso C, Turpault M-P, Frey-Klett P (2009) Mineral weathering by bacteria: ecology, actors and mechanisms. Trends Microbiol 17:378–387. https://doi.org/10.1016/j.tim.2009.05.004

Uroz S, Kelly LC, Turpault M, Lepleux C, Frey-Klett P (2015) The mineralosphere concept: mineralogical control of the distribution and function of mineral-associated bacterial communities. Trends Microbiol 23:751–762

Van Dam NM, Bouwmeester HJ (2016) Metabolomics in the rhizosphere: tapping into belowground chemical communication. Trends Plant Sci 21:256–265. https://doi.org/10.1016/j.tplants.2016.01.008

Van Schöll L, Kuyper TW, Smits MM, Landeweert R, Hoffland E, Breemen N (2008) Rock-eating mycorrhizas: their role in plant nutrition and biogeochemical cycles. Plant Soil 303(1–2):35–47. https://doi.org/10.1007/s11104-007-9491-2

Van Straaten P (2007) Agrogeology: the use of rocks for crops. Enviroquest Ltd., Canada, p 440

Vandeginste V, Lim C, Ji Y (2024) Exploratory review on environmental aspects of enhanced weathering as a carbon dioxide removal method. Minerals 14(1):75

Walker EL, Connolly EL (2008) Time to pump iron: iron-deficiency-signaling mechanisms of higher plants. Curr Opin Plant Biol 11:530–535

Wang JG, Zhang FS, Cao YP, Zhang XL (2000) Effect of plant types on release of mineral potassium from gneiss. Nutr Cycl Agroecosyst 56:37–44. https://doi.org/10.1023/A:1009826111929

Wang WY, Lian B, Pan L (2015) An RNA-sequencing study of the genes and metabolic pathways involved in aspergillus Niger weathering of potassium feldspar. Geomicrobiol J 32:689–700

Wang X, Yoo K, Wackett AA, Gutknecht J, Amundson R, Heimsath A (2018) Soil organic carbon and mineral interactions on climatically different hillslopes. Geoderma 322:71–80. https://doi.org/10.1016/j.geoderma.2018.02.021

White AF, Brantley SL (2003) The effect of time on the weathering of silicate minerals: why do weathering rates differ in the laboratory and field? Chem Geol 202(3/4):479–506. https://doi.org/10.1016/j.chemgeo.2003.03.001

White AF, Brantley SL (1995) Chemical weathering rates of silicate minerals: an overview. In: Chemical weathering rates of silicate minerals. De Gruyter, pp 1–22. https://doi.org/10.1515/9781501509650-003

White RA, Rivas-Ubach A, Borkum MI, Koberl M, Bilbao A, Colby SM, Hoyt DW, Bingol K, Kim Y-M, Wendler JP, Hixson KK, Jansson C (2017) The state of rhizospheric science in the era of multi-omics: a practical guide to omics technologies. Rhizosphere 3:212–221. https://doi.org/10.1016/j.rhisph.2017.05.003

Wood C, Harrison AL, Power IM (2023) Impacts of dissolved phosphorus and soil-mineral-fluid interactions on CO_2 removal through enhanced weathering of wollastonite in soils. Appl Geochem 148:105511

Yu GH, Chi ZL, Teng HH, Dong HL, Kappler A, Gillings MR et al (2019) Fungus-initiated catalytic reactions at hyphal-mineral interfaces drive iron redox cycling and biomineralization. Geochim Cosmochim Acta 260:192–203

Yu G (2018) Root exudates and microbial communities drive mineral dissolution and the formation of Nano-size minerals in soils: implications for soil carbon storage. In: Giri B, Prasad RAV (eds) Root biology, vol 52. Springer International Publishing, pp 143–166. https://doi.org/10.1007/978-3-319-75910-4_5

Zhang G, Kang J, Wang T, Zhu C (2018) Review and outlook for agromineral research in agriculture and climate mitigation. Soil Res 56:113–122. https://doi.org/10.1071/SR17157

Zavarzina DG, Chistyakova NI, Shapkin AV, Savenko AV, Zhilina TN, Kevbrin VV, Alekseeva TV, Mardanov AV, Gavrilov SN, Bychkov AY (2016) Oxidative biotransformation of biotite and glauconite by alkaliphilic anaerobes: the effect of Fe oxidation on the weathering of phyllosilicates. Chem Geol 439:98–109. https://doi.org/10.1016/j.chemgeo.2016.06.015

Zorb C, Senbayram M, Peiter E (2014) Potassium in agriculture–status and perspectives. J Plant Physiol 171:656–669. https://doi.org/10.1016/j.jplph.2013.08.008

Chemical and Mineralogical Classification of Silicate Agrominerals

Éder de Souza Martins, Elzino Rodrigues Júnior, Giuliano Marchi, Gisele Freitas Vilela, Philipp Swoboda, Carlos Augusto Posser Silveira, Eduardo de Souza Martins, Pablo Hardoim, and Suzi Huff Theodoro

Abstract This chapter aims to present a chemical and mineralogical classification of silicate agrominerals (ASi) and their potential for soil fertility management and plant nutrition. Derived from silicate rocks and some silicate materials from industrial processes, ASi are composed of more than 50% silicate minerals and are rich in bases, calcium, magnesium, and potassium. These mineral sources are designed for regional use and have direct effects on nutrient availability, increased soil pH, and the formation of functional neoformed minerals. Additionally, indirect effects include inorganic carbon sequestration, organic matter stabilization, and increased nutrient use efficiency compared to highly concentrated and soluble sources. This chapter proposes a chemical classification of silicate agrominerals into eight classes:

É. de Souza Martins (✉) · E. Rodrigues Júnior · G. Marchi
Agrogeological Laboratory, Empresa Brasileira de Pesquisa Agropecuária—Embrapa Cerrados, BR 020, km 18, Planaltina, Brasília, Distrito Federal 73310970, Federal District, Brazil
e-mail: eder.martins@embrapa.br

G. Marchi
e-mail: giuliano.marchi@embrapa.br

G. F. Vilela
Empresa Brasileira de Pesquisa Agropecuária—Embrapa Territorial, Campinas, São Paulo, Brazil
e-mail: gisele.vilela@embrapa.br

P. Swoboda
InPlanet GmbH, Heinrich-Geißler-Straße, München, Germany
e-mail: philipp.swoboda@inplanet.earth

C. A. P. Silveira
Empresa Brasileira de Pesquisa Agropecuária—Embrapa Clima Temperado, Pelotas, Rio Grande Do Sul, Brazil
e-mail: augusto.posser@embrapa.br

E. de Souza Martins · P. Hardoim
Grupo Associado de Agricultura Sustentável, Brasília, Distrito Federal, Brazil

S. H. Theodoro
Universidade de Brasília (UnB), Campus Planaltina, Vila Nossa Sra. de Fátima, Brasília, Distrito Federal, Brazil
e-mail: suzitheodoro@unb.br

É. de Souza Martins and S. Huff Theodoro (eds.), *Soil Remineralizers and Silicate Fertilizers*, Frontier Studies in Soil Science,
https://doi.org/10.1007/978-3-032-14656-4_3

magnesium-bearing (ASiMg), calcium-bearing (ASiCa), potassium-bearing (ASiK), calcium-magnesium-bearing (ASiCaMg), magnesium-calcium-potassium-bearing (ASiMgCaK), calcium-magnesium-potassium-bearing (ASiCaMgK), potassium-magnesium-calcium-bearing (ASiKMgCa), and potassium-calcium-magnesium-bearing (ASiKCaMg). This classification enables the differentiation of potential short-, medium-, and long-term effects, which are directly related to the reactivity of the minerals present in these sources. Although ASi represents one of the most abundant potential sources of agrominerals, systematic research on it has only gained momentum in the past 25 years. The development of these products has markedly influenced tropical agricultural soils, which tend to be more responsive to their application. Further research is required to establish practical application guidelines and management strategies tailored to these region-specific mineral products.

Keywords Regional fertilizers · Rock classification · Geodiversity

1 Introduction

Agriculture nowadays increasingly demands productive systems that are efficient in biological and energy terms, focusing on cultural management, soil fertility, and plant nutrition (Denny et al. 2023). These sustainable management strategies, often referred to as regenerative agriculture, advocate for a shift from the intensive use of globalized inputs and fossil fuels to a process-oriented agriculture rooted in nature-based, local, and regional solutions (Newton et al. 2020).

This transition from global to local and regional solutions enhances the One Health framework (Persano 2024), leading to improvements in soil quality, plant health, and agricultural products with increased nutritional density (Ramkumar et al. 2024). Brazil is at the forefront of this transition, extensively utilizing biofertilizers, biocontrollers, biostimulants (Parra 2023; Soares et al. 2023), and regional nutrient sources (Theodoro et al. 2022). Among these, silicates and carbonates stand out as the most abundant agrominerals in nature (Van Straaten 2022).

The complex chemical and mineralogical composition of silicate rocks and industrial silicate sources underscores the need for systematic classification to evaluate their primary effects on soil properties. Products derived from these agrominerals can influence nutrient availability, aid in correcting soil acidity, and promote the formation of new mineral phases. The extent of these effects is closely linked to the chemical and mineralogical character of the agrominerals. Additionally, these chemical and mineralogical characteristics inform the selection of appropriate beneficiation processes, ensuring that the resulting products effectively enhance soil fertility management.

The chapter presents a chemical and mineralogical classification of silicate agrominerals (ASi). This classification is designed to enhance the evaluation of the beneficiation process for these materials, aiding in the production of high-quality products and providing valuable insights into their relative reactivity in terms of agronomic performance. It also highlights the primary expected effects on soil for each ASi class.

2 Chemical Silicate Agromineral Classification

The proposed ASi classification is based on the sum of total bases in the form of oxides—SBr ($CaO + MgO + K_2O$)—and the concentrations of each of these oxides. The first division separates the ASi into two groups: one with SBr between 9 and 20%, and the other with SBr above 20% (Fig. 1). The 9% minimum threshold is based on the definition of soil remineralizers within the Brazilian legislation, as established in Normative Instruction No. 5 of 2016 (Brazil 2016). The upper limit of 20% SBr is associated with the typical maximum concentrations found in potassium-rich rocks and basic rocks. The other group, with SBr above 20%, is linked to ultramafic and ultramafic alkaline rocks, as well as silicate rocks with carbonate contributions, such as marls, certain marbles, and calc-silicate rocks.

When the sum of bases in the form of oxides is equal to or exceeds 20%, ASi is divided into four groups:

Magnesium-bearing silicate agrominerals (ASiMg)—containing more than or equal to 20% MgO. These are derived from typical ultramafic rocks (e.g., dunites, pyroxenites, serpentinites) and nickel ore slags from industrial processes.

Calcium-bearing silicate agrominerals (ASiCa)—containing more than or equal to 20% CaO, like certain calcsilicate rocks (e.g., certain marl, calcium paragranulites), and certain steel slags from industrial processes,

Calcium-magnesium-potassium-bearing silicate agrominerals (ASiCaMgK)—containing less than 20% MgO, less than 20% CaO, and MgO $<$ CaO, rich in calcium and derived from igneous rocks like anorthosites and calcsilicate metamorphic rocks (e.g., calcschists), as well as from steel slags.

Magnesium-calcium-potassium-bearing silicate agrominerals (ASiMgCaK)—containing less than 20% MgO, less than 20% CaO, and MgO $\geq$ CaO, primarily derived from alkaline ultramafic rocks (e.g., kamafugites, ugandites, kimberlites, olivine melilitites) and certain magnesium-rich metamorphic rocks.

When the sum of bases in the form of oxides is between 9 and 20%, ASi can be classified into four groups:

Potassium-bearing silicate agrominerals (ASiK)—with more than or equal to 8% K_2O derived from alkaline rocks, like high K syenites and phonolites, sedimentary and metamorphic rocks rich in glauconite, e.g., glauconitic siltstones and sandstones.

Potassium-magnesium-calcium-bearing silicate agrominerals (ASiK-MgCa)—between 3 and 8% K_2O, MgO $\geq$ CaO, derived from rocks rich in biotite or phlogopite (certain gneisses, schists, silicate portions of carbonatite bodies, and hydrothermal rocks).

Potassium-calcium-magnesium-bearing silicate agrominerals (ASiK-CaMg)—between 3 and 8% K_2O, MgO $<$ CaO, derived from determined alkaline

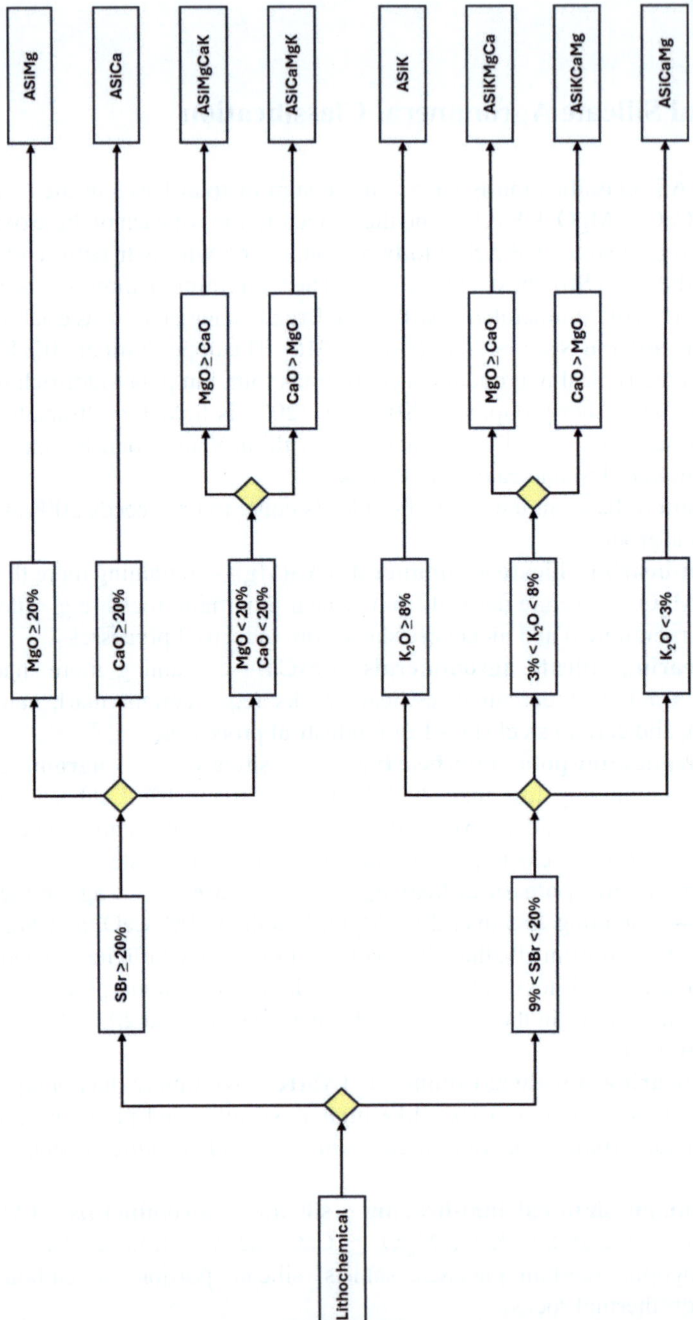

Fig. 1 Chemical classification of silicate agrominerals considering the sum of total bases in the form of oxides (SBr = CaO + MgO + K_2O) and the base contents

rocks with intermediate K levels, like some syenites and phonolites, and granitic rocks (e.g., granodiorite, dacite, rhyolite).

Calcium-magnesium-bearing silicate agrominerals (ASiCaMg)—with less than 3% K_2O. These come from basic rocks (e.g., basalts, diabases, gabbros, andesites, amphibolites, basic granulites).

Silicate agrominerals can generate two main types of products for managing soil fertility: soil remineralizers (REMs, according to criteria NI 05 from Brazilian legislation, Brazil 2016) and silicate fertilizers (FSi). REMs act as soil enhancers, providing three key benefits from the weathering of silicates: they supply essential nutrients and beneficial elements, help increase soil pH, and generate new mineral phases. Although FSi shares these beneficial effects of REMs, its primary function is to serve as a direct nutrient source. There are potential products that may fall outside this classification, such as potassic sources with an SBr below 9%.

Products derived from ASiMg, ASiCa, and ASiK are generally classified as FSi. In contrast, the other compositions—ASiCaMg, ASiKMgCa, ASiKCaMg, ASiCaMgK, and ASiMgCaK—are typically recognized as REMs.

Another type of silicate agrominerals originates from partially weathered or hydrothermally altered silicate rocks. For instance, the saprolite of nepheline syenite is characterized by an accumulation of K-feldspar (Chiwona et al. 2020). In contrast, the saprolite of kamafugite undergoes partial base depletion and accumulates 2:1 clay minerals (Corrêa et al. 2025; Ferreira et al. 2020). These types of ASi can be found in tropical regions, such as Brazil and Africa, where agrominerals from weathered rock, including both silicate (Chiwona et al. 2020) and phosphate-rich materials (Silva et al. 2019; Toledo et al. 2004), are present.

3 Classification of Silicate Agrominerals from Whole-Rock Database Compilation

The classification process utilized a global database comprising the chemical compositions of various rock types—sedimentary, igneous, and metamorphic—expressed as major elements in oxide form, trace elements, and additional relevant attributes. The database used is available in Gard et al. (2019), which is a compilation initiative of open data from various existing databases. The *complete.zip* file was extracted, and the database was processed using the Python programming language (Gupta and Bagchi 2024). Initially, samples from a specific continent were selected, reducing the database from 1,022,092 records to 780,587. The target attributes used are major oxides in weight %, such as SiO_2, TiO_2, Al_2O_3, FeO, MgO, CaO, MnO, K_2O, Na_2O, P_2O_5, contaminants As, Cd, Hg, Pb, and nutrients Co, Zn, Mn, Mo, Ni, Cl, and B in ppm. Other auxiliary attributes were also extracted, such as sample latitude and longitude, rock name, and the group to which it belongs (igneous, metamorphic, or sedimentary).

One of the applied treatments involved filtering for silicate rocks, which were required to contain at least 50% silicate minerals—typically corresponding to a total SiO_2 content greater than 20%. Negative oxide values in the database were handled using two procedures: records whose deviation exceeded the element's positive average were removed, while values slightly below the average (i.e., close to zero) were replaced by their absolute value, since many samples contained small negative numbers. High contents that are rarely seen in nature were also removed. Records without values were filled with zeros.

The processed database contained 459,965 records. The sum of bases method was employed by adding a field to the database that calculated the sum of K_2O, CaO, and MgO. Subsequently, a new field was created to store the ASi classes.

Inserting all records into the diagram at once made visualization and interpretation of the information impossible. Since it is a global database, 100 samples for each ASi class were randomly selected. The samples were normalized to have contents between 0 and 100 for elements K_2O, CaO, and MgO. After normalization, they were inserted into the ternary diagram for each ASi class.

The proposed classification in divisions for the ASi classes was tested with a K-means machine learning algorithm (Ikotun et al. 2023). K-means is an unsupervised learning method, meaning it does not require any predefined classes or labels for the samples. The algorithm utilizes a clustering principle, where K instances (cluster centers) are randomly initialized and tend to converge with closer classes. It was executed in two ways: one considering eight ASi classes and an unclassified category, and the other with only ASi classes. The input database for K-means was a vector with MgO, CaO, and K_2O for each sample, and the results of the K-means classes were compared with the proposed model.

It was observed that 63% of the database is classified as ASi (Fig. 2). Among the unclassified samples are carbonate rocks and silicate rocks with an SBr lower than 9%. The most abundant samples are those classified as ASiCaMg, particularly associated with basic rocks such as basalts and their intrusive and metamorphic equivalents.

All continents exhibit a high abundance of ASi occurrences, both on active continental margins and in the continental interior, in cratonic areas and tectonically active zones (Fig. 3). This finding supports the studies conducted by Van Straaten (2022), which highlight the environments where agrominerals are found worldwide.

A study conducted on ASi in Brazil by Embrapa and the Geological Survey of Brazil revealed that 97% of agricultural areas are located near a potential source of ASi (Brazil 2018). The results of this study in Brazil demonstrate a shorter distance between ASi occurrences and agricultural land (Fig. 4). Notably, 97.5% of agricultural land is within 300 km of a potential source of ASi, demonstrating that these inputs are used regionally.

The most abundant among all existing agrominerals are ASiCaMg samples, as indicated by the world map illustrating their distribution (Fig. 5). This is one of the reasons for selecting ASiCaMg as primary sources for inorganic carbon sequestration through enhanced weathering (Beerling et al. 2020). The development of REMs in

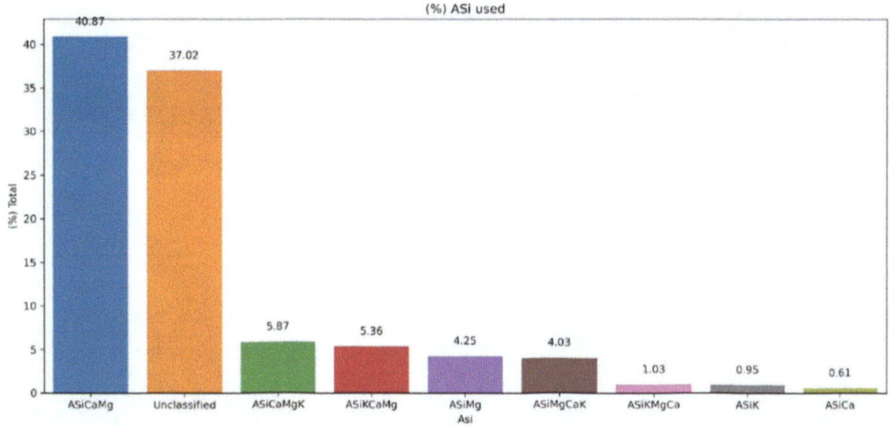

Fig. 2 Frequency distribution of rock samples according to the chemical classification of silicate agrominerals. ASiCaMg (calcium-magnesium), ASiCaMgK (calcium-magnesium-potassium), ASiKCaMg (potassium-calcium-magnesium), ASiMg (magnesium), ASiMgCaK (magnesium-calcium-potassium), ASiKMgCa (potassium-magnesium-calcium), ASiK (potassium), ASiCa (calcium). Data extracted from the database referenced in Gard et al. (2019)

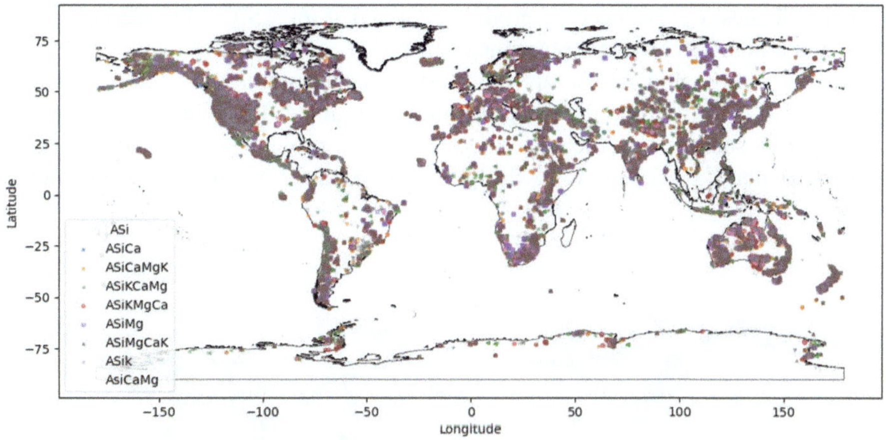

Fig. 3 World map showing the distribution of all silicate agrominerals. ASiCaMg (calcium-magnesium-bearing), ASiCaMgK (calcium-magnesium-potassium-bearing), ASiKCaMg (potassium-calcium-magnesium-bearing), ASiMg (magnesium-bearing), ASiMgCaK (magnesium-calcium-potassium-bearing), ASiKMgCa (potassium-magnesium-calcium-bearing), ASiK (potassium-bearing), ASiCa (calcium-bearing). Data extracted from the database referenced in Gard et al. (2019)

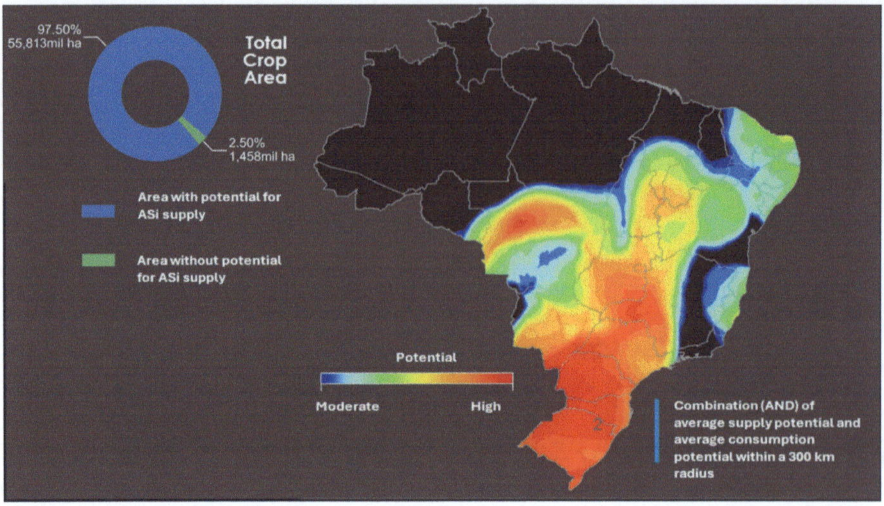

Fig. 4 Economic potential for silicate agrominerals in Brazil. Data from Brazil (2018)

Brazil indicates a prevalence of products derived from ASiCaMg, accounting for 32% of a total of 75 products (Martins et al. 2024).

ASiMg, ASiK, and ASiCa are the least abundant among the ASi, accounting for only 5.6% of the total database from Gard et al. (2019) (Fig. 2), yet they are important as sources for silicate fertilizers (FSi) to provide silicon and associated main nutrients (Ca, Mg, and K). ASiMg serve as a source of magnesium and nickel.

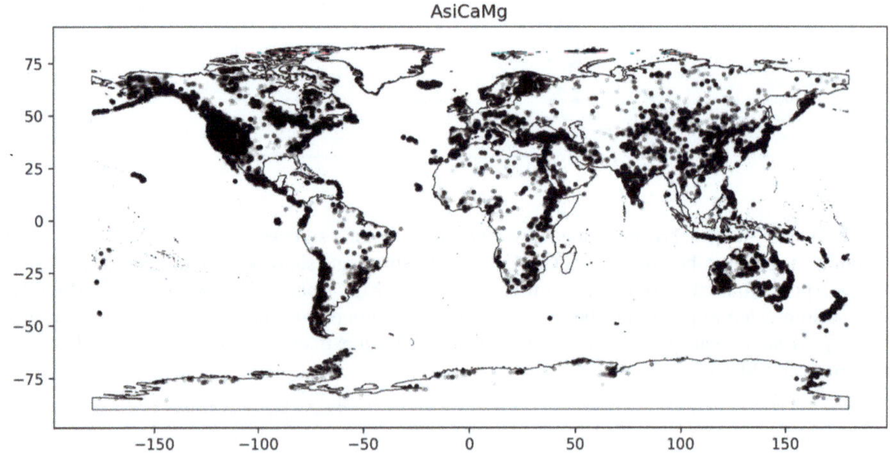

Fig. 5 Global distribution of calcium-magnesium-bearing silicate agromineral occurrences (ASiCaMg). Data extracted from the database referenced in Gard et al. (2019)

ASiCa are sources of calcium and silicon. ASiK are sources of potassium. The global distribution of these ASi is found on all continents (Fig. 6).

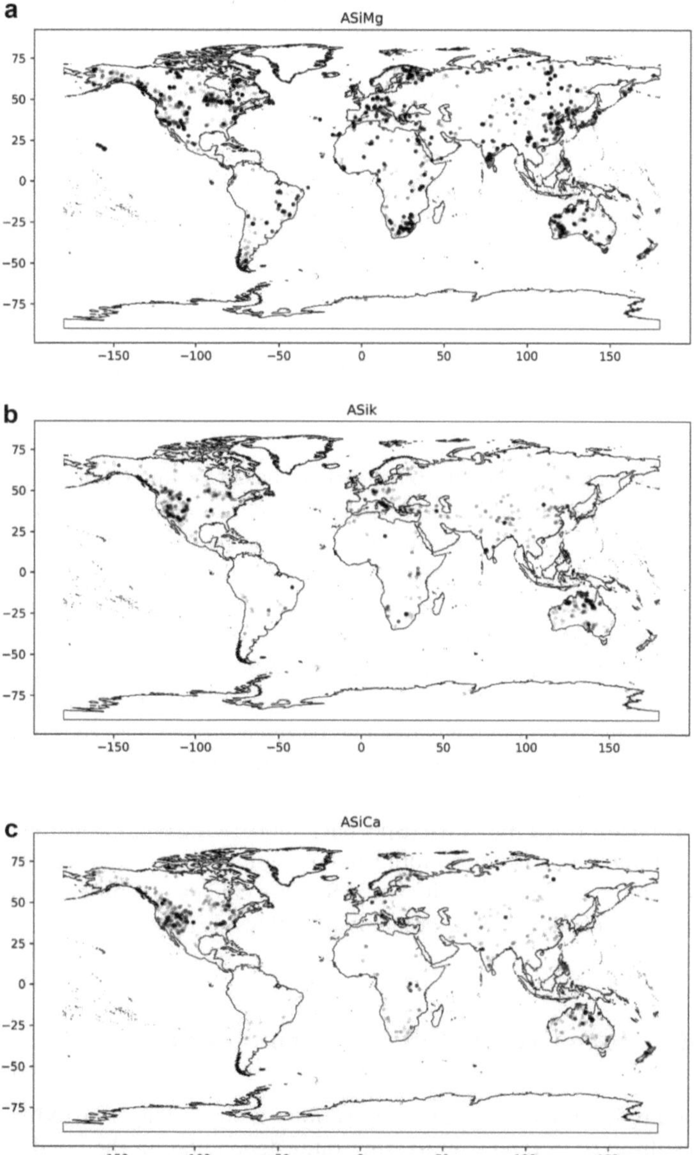

Fig. 6 Global distribution of magnesium-bearing silicate agrominerals (ASiMg), potassium-bearing silicate agrominerals (ASiK), and calcium-bearing silicate agrominerals (ASiCa), labeled A, B, and C, respectively. Data extracted from the database referenced in Gard et al. (2019)

The other ASi, ASiCaMgK, ASiKCaMg, ASiMgCaK, and ASiKMgCa, which account for 16.3% of the total database from Gard et al. (2019) (Fig. 2), are typical REMs that provide nutrients, contribute to soil acidity correction, and form new mineral phases. Although much less abundant than calcium-magnesium-bearing silicate agrominerals (ASiCaMg), these ASi classes are also found on all continents (Fig. 7).

The main rock types of igneous, metamorphic, and sedimentary origin related to the classes of ASi are presented in Table 1. There is a variety of common and abundant rocks that can be characterized as ASi. Another way to classify ASi is through the CaO–MgO–K_2O ternary diagram (Fig. 8). The distribution of classes suggests the potential to develop classification fields in the ternary diagram that may later serve as a basis for ASi characterization.

3.1 Classification of Silicate Agrominerals Regarding Their pH Neutralization Potential

A different approach to characterizing ASi is based on their potential to correct soil pH, such as their neutralization potential (NP), calculated as an equivalent in calcium carbonate (%$CaCO_3$ equivalent, calculated by the formula: $\%eqCaCO_3 = \%CaO * 1.79 + \%MgO * 2.48$). Figure 9 shows the distribution of NP for each ASi class. It is observed that there is a decreasing order of NP as follows: ASiMg, ASiCa, ASiMgCaK, ASiCaMgK, ASiCaMg, ASiKMgCa, and ASiK.

Similarly, the NP range for each ASi also indicates that the relative reactivity of weathering through hydrolysis reactions tends to have a direct relationship with the NP, such as the main elements that can be released and the types of neoformed minerals (Table 2). While at one end, ASiMg are the most reactive and have the highest NP, at the other end, ASiK are the least reactive and have the lowest NP. All ASi can form neoformed minerals, especially short-range order minerals, as well as high-crystallinity minerals when biotite and phlogopite micas are present in the ASi. This relationship between reactivity and ASi class can also be observed through the inverse correlation between K_2O content and NP levels (Fig. 10).

4 Relative Reactivity of Silicate Agrominerals

ASi have complex mineralogical compositions with varying reactivity potentials in soils, resulting in different weathering rates depending on the conditions of the agroecosystems. Generally, minerals richer in calcium and magnesium exhibit higher reactivity and a greater potential for weathering in agricultural soils, such as those with a relative dissolution rate of 10^2 or more, comparable to K-feldspar (Table 3). Incubation studies of basalt powder in acidic soils, where moisture is maintained

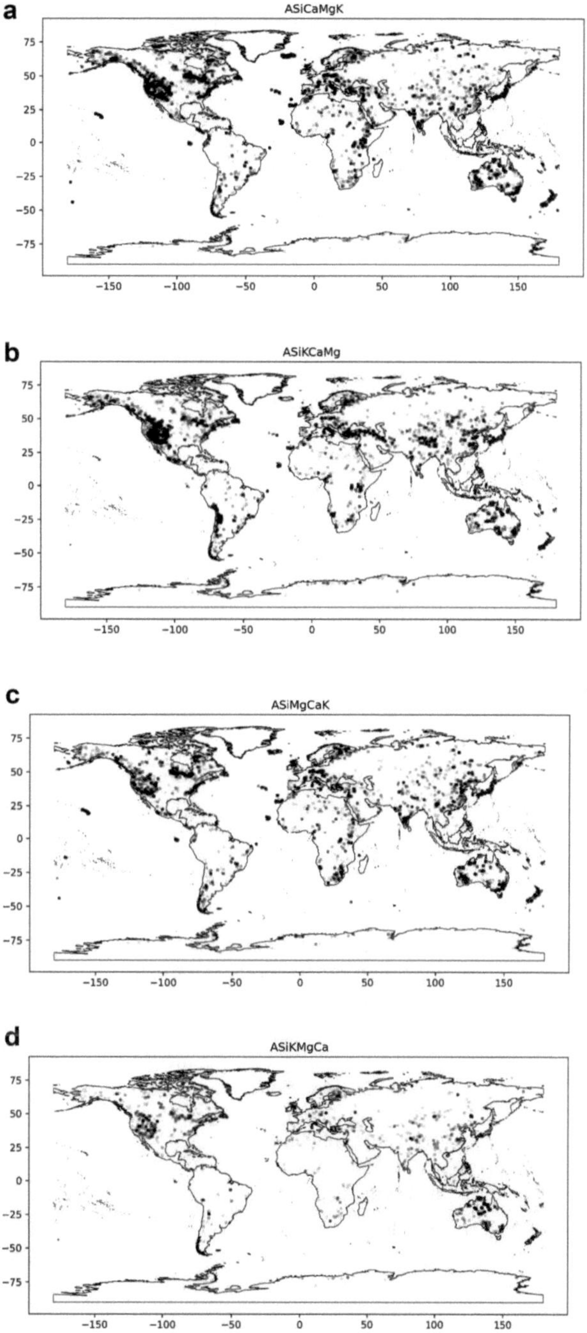

Fig. 7 Global distribution of calcium-magnesium-potassium-bearing silicate agrominerals (ASiCaMgK), potassium-calcium-magnesium-bearing (ASiKCaMg), magnesium-calcium-potassium-bearing (ASiMgCaK), and potassium-magnesium-calcium-bearing (ASiKMgCa), respectively, labeled as A, B, C, and D. Data extracted from the database referenced in Gard et al. (2019)

Table 1 Main rock types according to the class of silicate agromineral (ASi). Data extracted from the database referenced in Gard et al. (2019)

ASi class	Igneous rocks	Metamorphic rocks	Sedimentary rocks
ASiCaMg	Basalt, andesite, diabase, gabbro, diorite, dolerite, granodiorite, basanite, shoshonite	Amphibolite, granulite, schist, metabasalt, greenschist, metagabbro, metabasite, eclogite, hornblende gneiss, phyllite, greyschist, hornfels, orthogneiss, granofels, mica gneiss, migmatite, slate, charnockite, metadiorite	Black shale, siltstone, mudstone, turbidite, claystone, greywacke, argillite, shale, arkose
ASiMg	Dunite, serpentinite, pyroxenite, peridotite, komatiite, lherzolite, wehrlite, picrite	Serpentinite, soapstone, amphibolite, talc schist	Not reported
ASiK	Rhyolite, trachyte, syenite, phonolite, lamproite	Slate, gneiss	Siltstone, shale, mudstone
ASiCa	Carbonate-rich kimberlite	Skarn, marble, metasomatite	Sandstone, siltstone, mudstone, shale, argillite, conglomerate, claystone
ASiCaMgK	Basalt, gabbro, basanite, diabase, nephelinite, pyroxenite, gabbronorite, anorthosite	Amphibolite, granulite, gneiss, schist, eclogite, metabasalt, skarn, metagabbro, metabasite	Siltstone, black shale, mudstone, shale, turbidite, argillite, claystone
ASiKCaMg	Basalt, andesite, syenite, phonolite, granodiorite, dacite, rhyolite	Gneiss, skarn, granulite, amphibolite, phyllite	Not reported
ASiMgCaK	Basalt, basanite, gabbro, picrite, lamproite, komatiite, pyroxenite, kimberlite, lamprophyre, dolerite, nephelinite, kamafugite, olivine melilitite, ugandite	Granulite, amphibolite, eclogite, metabasalt, skarn, metabasite	Shale
ASiKMgCa	Basalt, andesite	Schist, gneiss, hornfels, mica schist, mica gneiss, slate, amphibolite	Shale

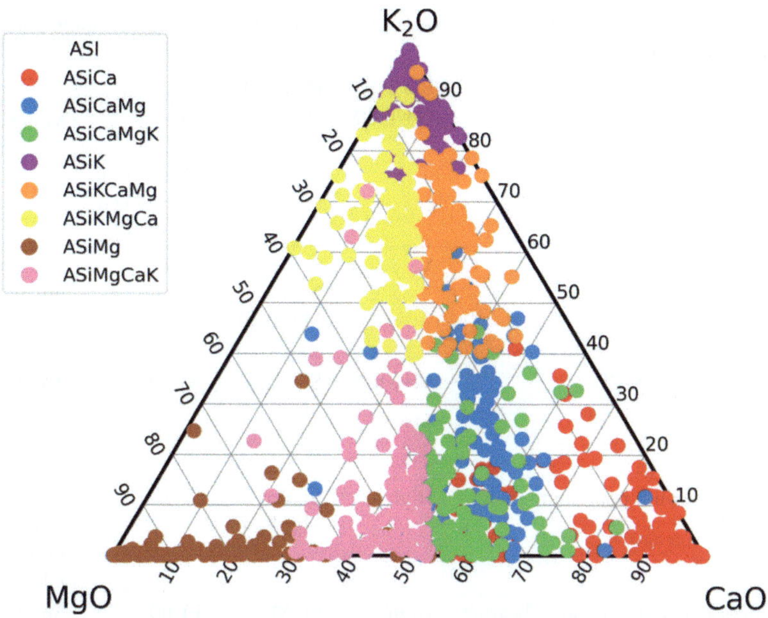

Fig. 8 Triangular diagram of CaO–MgO–K_2O for silicate agromineral (ASi) classes. Data extracted from the database referenced in Gard et al. (2019)

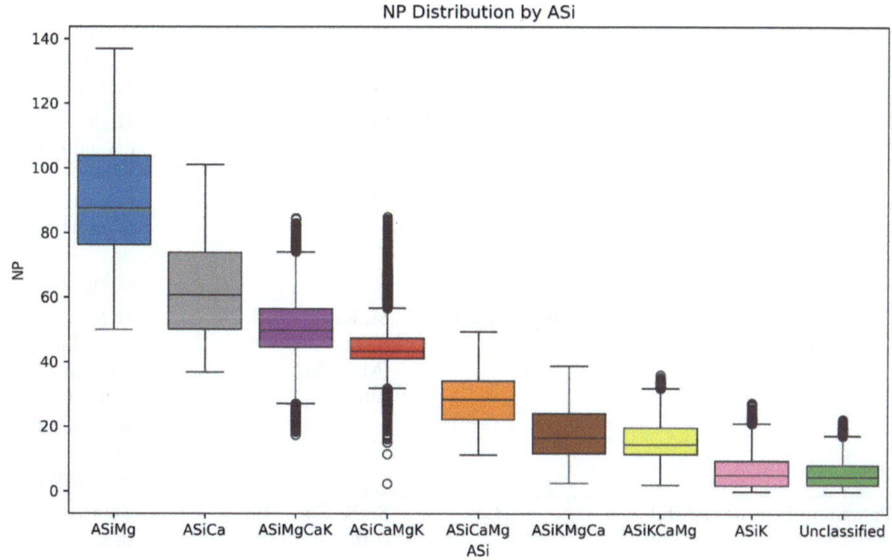

Fig. 9 Box plot displaying the neutralization power (NP), expressed as lime equivalent (%equivalent $CaCO_3$, = %CaO*1.79 + %MgO*2.48), for each silicate agromineral class (ASi). Data extracted from the database referenced in Gard et al. (2019)

Table 2 Main silicate agrominerals, relative reactivity, and soil effects. Elements in parentheses occur in smaller proportions

Class	Relative reactivity	Direct effects		
		Nutrient sources	pH correction (% NP) lime equivalence potential*	Mineral formation
Magnesium-bearing (ASiMg)	Very high	Si, Mg, Ni, Fe, Mn, (Co)	70–110	Short-range order minerals (Si, Fe)
Calcium-bearing (ASiCa)	Very high	Si, Ca, Zn	50–70	Short-range order minerals (Si, Al)
Calcium-magnesium-potassium-bearing (ASiCaMgK)	High	Si, Ca, Mg, K, Zn	50–70	Short-range order minerals (Si, Al, Fe)
Magnesium-calcium-potassium-bearing (ASiMgCaK)	High	Si, Mg, Ca, K, Zn	50–60	Short-range order minerals (Si, Al, Fe)
Calcium-magnesium-bearing (ASiCaMg)	High	Si, Ca, Mg, Mn, Cu, (Zn)	25–35	Short-range order minerals (Si, Al, Fe)
Potassium-magnesium-calcium-bearing (ASiKMgCa)	Median	Si, K, (Mg), (Ca), Zn, Mn	15–25	High crystallinity phase (vermiculite from biotite or phlogopite)
Potassium-calcium-magnesium-bearing (ASiKCaMg)	Median	Si, K, (Ca), (Mg), Zn, Mn	15–20	High crystallinity mineral (vermiculite from biotite or phlogopite)
Potassium-bearing (ASiK)	Low	Si, K, Zn, (Mn)	15	Short-range order minerals (Si, Al, Fe)

% NP—% of neutralization power. *NP calculated as potential lime equivalence, derived from total CaO and MgO by formula, $\%eqCaCO_3 = \%CaO * 1.79 + \%MgO * 2.48$

Fig. 10 Neutralization potential (NP) concerning K_2O levels of silicate agromineral (ASi) classes. Data extracted from the database referenced in Gard et al. (2019)

close to field capacity over a period of several days, show significant changes in pH modifications, increased availability of Ca and Mg, and a decrease in exchangeable Al (Conceição et al. 2022). In this case, hydrolysis reactions begin solely with the soil's acidity, without the need for an active biological process, such as the action of the plant rhizosphere. Several studies also demonstrate that biological activity significantly accelerates this dissolution process (Conceição et al. 2022). Basalts contain minerals with relative dissolution rates related to K-feldspar (RDR) exceeding 10^2.

Several studies have shown that minerals with RDR values between 10^0 and 10^1 can be weathered within the timeframe of agricultural management. However, they require biologically intensive management to make these processes efficient. Biotite and phlogopite are rapidly weathered through the action of the plant rhizosphere (Hinsinger et al. 1991, 1992; Jafari et al. 2021) but do not show significant effects with only incubation in acidic soils (Krahl et al. 2022).

One strategy to enhance the efficiency of potassium silicate fertilizers rich in K-feldspar is to reduce the grain size of the product to increase the specific surface area of the minerals and enable a higher dissolution rate under agricultural conditions, such as, for example, 100% smaller than 0.075 mm for phonolite (Soratto et al. 2021).

Thus, the improvement of silicate agrominerals' classification needs to include their mineralogical composition and their qualification regarding weathering potential and possible agronomic behaviors.

Table 3 Main rock minerals with relative dissolution rates relative to that of K-feldspar (modified from Manning and Theodoro 2020)

Mineral family	Mineral	Formula	Releases	Typical content wt %	Relative dissolution rate
Potassium silicates					
Feldspathoid	Nepheline	$(Na, K)AlSiO_4$	Na, K	11.8, 6.7	$2{,}14 * 10^7$
Mica	Glauconite	$(K, Na)(Fe^{3+}, Al, Mg)_2(Si, Al)_4O_{10}(OH)_2$	K, Mg	5.5, 2.3	$1.82 * 10^5$
Feldspathoid	Leucite	$KAlSi_2O_6$	K	17.9	$1.15 * 10^4$
Mica	Phlogopite	$KMg_3(AlSi_3O_{10}(OH)_2$	K, Mg	9.3, 17.4	$5.81 * 10^0$
Mica	Biotite	$K(Mg, Fe)_3(AlSi_3O_{10})(OH)_2$	K, Mg	9.0, 14.0	$1.66 * 10^0$
K-feldspar	Orthoclase	$KAlSi_3O_8$	K	14.1	$1 * 10^0$
Mica	Muscovite	$KAl_2(AlSi_3O_{10})(OH)_2$	K	9.8	$1.6 * 10^{-2}$
Other silicates					
Plagioclase	Anorthite	$CaAl_2Si_2O_8$	Ca	13.7	$3.63 * 10^6$
Olivine	Fayalite	$(Fe^{2+})_2SiO_4$	Fe	54.8	$1.82 * 10^5$
Wollastonite	Wollastonite	$CaSiO_3$	Ca	34.5	$4.90 * 10^4$
Serpentine	Lizardite	$Mg_3Si_2O_5(OH)_4$	Mg	26.3	$2.29 * 10^4$
Plagioclase	Bytownite	$(Ca, Na)(Si, Al)_4O_8$	Ca, Na	11.7, 1.7	$1.62 * 10^4$
Pyroxene	Diopside	$CaMgAlSiO_3$	Ca, Mg	18.5, 11.2	$5.01 * 10^3$
Tourmaline	Dravite	$NaMg_3Al_6B_3Si_6O_{30}(OH)$	Mg, B	7.8, 3.5	$3.63 * 10^3$
Pyroxene	Augite	$(Ca, Na)(Mg, Fe, Al, Ti)(Si, Al)_2O_6$	Ca, Mg	15.3, 9.3	$1.74 * 10^3$
Olivine	Forsterite	Mg_2SiO_4	Mg	34.5	$1.62 * 10^3$
Amphibole	Hornblende	$Ca_2Mg_4Al(Al_8Si_4O_{22})(OH)_2$	Ca, Mg	8.6, 7.8	$1.15 * 10^3$
Plagioclase	Labradorite	$(Ca, Na)(Si, Al)_4O_8$	Ca, Na	8.9, 3.4	$1.55 * 10^2$
Serpentine	Tremolite	$Ca_2Mg_5Si_8O_{22}(OH)_2$	Mg, Ca	15.0, 9.9	$4.57 * 10^1$
Plagioclase	Andesine	$(Na, Ca)(Si, Al)_4O_8$	Ca, Na	6.0, 5.1	$1.52 * 10^1$
Pyroxene	Enstatite	$Mg_2Si_2O_6$	Mg	24.2	$1.01 * 10^1$
Plagioclase	Oligoclase	$(Na, Ca)(Si, Al)_4O_8$	Na, Ca	6.9, 3.0	$2.45 * 10^0$
Plagioclase	Albite	$NaAlSi_3O_8$	Na	8.3	$7.94 * 10^{-1}$
Chlorite	Clinochlore	$(Mg, Fe^{2+})_5Al(Si_3Al)O_{10}(OH)_8$	Mg	15.3	$8.91\ 10^{-2}$

(continued)

Table 3 (continued)

Mineral family	Mineral	Formula	Releases	Typical content wt %	Relative dissolution rate
Amphibole	Anthophyllite	$Mg_7Si_8O_{22}(OH)_2$	Mg	21.8	$1.32 * 10^{-2}$
Quartz	Quartz	SiO_2	Si	46.7	$5 * 10^{-3}$
Carbonates					
Carbonate	Calcite	$CaCO_3$	Ca	40.0	$5.75 * 10^9$
Carbonate	Dolomite	$CaMg(CO_3)_2$	Ca, Mg	21.7, 13.2	$7.41 * 10^6$
Carbonate	Magnesite	$MgCO_3$	Mg	28.8	$4.79 * 10^3$
Phosphates					
Phosphate	Hydroxyapatite	$Ca_5(PO_4)_3(OH)$	P, Ca	18.5, 40.0	$5.89 * 10^5$
Phosphate	Fluoroapatite	$Ca_5(PO_4)_3F$	P, Ca	18.4, 39.7	$2.14 * 10^6$

5 Concluding Remarks

The silicate agromineral (ASi) classification presented here aims to simplify the complex compositional variability of silicate sources into manageable classes for both research and practical applications. This proposal should be evaluated in relation to agronomic application strategies based on ASi classes, to assess the three main effects: nutrient supply, contribution to soil acidity correction, and formation of new functional minerals.

The ASi classification was based on the proportions of the total sum of bases from the silicate source ($SBr = CaO + MgO + K_2O$). Two main groups were defined: one with SBr exceeding 20%, and another with SBr between 9 and 20%. The sources with SBr above 20% were classified into four categories: ASiMg, ASiCa, ASiMgCaK, and ASiCaMgK. Those within the 9% to 20% range were also divided into four classes: ASiK, ASiCaMg, ASiKMgCa, and ASiKCaMg.

Data extracted from the database referenced in Gard et al. (2019) indicate that all types of ASi are present on all continents, demonstrating that these sources are used regionally. The ASiCaMg class is the most abundant, accounting for two-thirds of the data in the database used. The data suggest a relationship between the neutralization potential of ASi and their reactivity, with ASiMg representing one extreme and ASiK the other, showing an inverse correlation between neutralization potential and K_2O content.

On the other hand, the weathering behavior of silicate minerals in agricultural soils depends on how the elements are organized in different minerals. There are minerals with high weathering potential, rich in magnesium, such as olivine, pyroxene, and lizardite. However, there are also magnesium silicates with low reactivity, including clinochlore and anthophyllite. Similarly, some potassium silicates, such as kalsilite and nepheline, are easily weathered in agricultural soils, while others like muscovite are highly resistant. Therefore, it is necessary to have a deep understanding of the mineralogical composition to determine its behavior in soil fertility management. The interpretation of chemical and mineralogical composition becomes the foundation for advancing the characterization of silicate agromineral classes.

Finally, the ASi classification has the potential to be helpful in the development of remineralizer products and natural fertilizers, facilitating the understanding of vast geodiversity for use as agrominerals.

Acknowledgements The authors gratefully acknowledge the financial support provided by Embrapa and by FINEP—Fundo Setorial Mineral through the project 'Rede de PD&I sobre remineralizadores e agrominerais (REMAg)' (Project No. 1541/2022), which made possible the development of this book chapter.

AuthorContribution Author's Contribution Éder de Souza Martins conceived the initial concepts that led to the classification of silicate agrominerals, interpreted the results extracted from the rock database, and conducted a critical review of the text. Elzino Rodrigues Júnior collected and processed the geochemical database data and generated the Figures. Giuliano Marchi, Gisele Freitas Vilela, Philipp Swoboda, and Carlos Augusto Posser Silveira discussed the various versions of the silicate agromineral classification until reaching the version presented in this chapter and reviewing the final text. Eduardo de Souza Martins and Pablo Hardoim contributed with the biological perspective on the transformation process of silicate minerals in soil environments, aiding in the conception of the silicate agromineral classification proposed here. Suzi Huff Theodoro performed a critical review and contributed to the significant improvement of the final text.

Declaration of Competing Interest The authors declare that they have no known competing financial interests or personal relationships that could have appeared to influence the work reported in this paper.

References

Beerling DJ, Kantzas EP, Lomas MR, Wade P, Eufrasio RM, Renforth P, Sarkar B, Andrews MG, James RH, Pearce CR, Mercure J-F, Pollitt H, Holden PB, Edwards NR, Khanna M, Koh L, Quegan S, Pidgeon NF, Janssens IA, Hansen J, Banwart SA (2020) Potential for large-scale CO_2 removal via enhanced rock weathering with croplands. Nature 583(7815):242–248. https://doi.org/10.1038/s41586-020-2448-9

Brazil (2016) Instrução Normativa No 5, de 10 de março de 2016. Ministério da Agricultura, Pecuária e Abastecimento (MAPA). Gabinete da Ministra. http://www.agricultura.gov.br/assuntos/insumos-agropecuarios/insumos-agricolas/fertilizantes/legislacao/in-5-de-10-3-16-remineralizadores-e-substratos-para-plantas.pdf

Brasil (2018) Zoneamento Agrogeológico do Brasil, 1:1,000,000. Embrapa, Geological Survey of Brazil. Ministry of Mining. https://www.gov.br/mme/pt-br/assuntos/secretarias/geologia-mineracao-e-transformacao-mineral/publicacoes-1/eventos/zoneamento-agrogeologico-do-brasil.pdf

Chiwona AG, Cortés JA, Gaulton RG, Manning DAC (2020) Petrology and geochemistry of selected nepheline syenites from Malawi and their potential as alternative potash sources. J Afr Earth Sc 164:103769

Conceição LT, Silva GN, Holsback HMS, Figueiredo Oliveira C, Marcante NC, de Souza Martins É, de Souza Santos FL, Santos EF (2022) Potential of basalt dust to improve soil fertility and crop nutrition. J Agric Food Res 10:100443. https://doi.org/10.1016/j.jafr.2022.100443
Corrêa JDS, Araújo DADO, Rodrigues ACPS, Brito OG, Gusmão GR, Rodrigues TTMS (2025) Rock powder application combined with bacterial inoculation enhances the early growth of coffee plants. Ciência e Agrotecnologia 49:e024124
Denny DMT, Cerri CEP, Cherubin MR, Burnquist HL (2023) Carbon farming: nature-based solutions in Brazil. Green Low-Carbon Econ 1(3):130–137
Ferreira BC, Gualberto CAC, Fourtaine G, Pereira HS, Korndörfer GH, Spolidorio E, Clerot LC (2020) Kamafugite from the Alto Paranaíba province as a multi-nutrient fertilizer. J Agri Sci 12:213
Gard M, Hasterok D, Halpin JA (2019) Global whole-rock geochemical database compilation. Earth Syst Sci Data 11(4):1553–1566. https://zenodo.org/records/3359791
Gupta P, Bagchi A (2024) Python language basics. In: Essentials of Python for artificial intelligence and machine learning. Springer Nature Switzerland, Cham, pp 87–126
Hinsinger P, Jaillard B, Dufey JE (1992) Rapid weathering of a trioctahedral mica by the roots of Ryegrass. Soil Sci Soc Am J 56(3):977–982. https://doi.org/10.2136/sssaj1992.03615995005600030049x
Hinsinger P, Dufey JE, Jaillard B (1991) Biological weathering of micas in the rhizosphere as related to potassium absorption by plant roots. In: Developments in agricultural and managed forest ecology, vol 24. Elsevier, pp 98–105
Ikotun AM, Ezugwu AE, Abualigah L, Abuhaija B, Heming J (2023) K-means clustering algorithms: a comprehensive review, variants analysis, and advances in the era of big data. Inf Sci 622:178–210. https://doi.org/10.1016/j.ins.2022.11.139
Jafari F, Khademi H, Shahrokh V, Angel FAZ, Acosta JA, Khormali F (2021) Biological weathering of phlogopite during enriched vermicomposting. Pedosphere 31(3):440–451
Krahl LL, Marchi G, Paz SPA, Angélica RS, Sousa-Silva JC, Valadares LF, Martins ÉS (2022) Increase in cation exchange capacity by the action of maize rhizosphere on Mg or Fe biotite-rich rocks. Pesqui Agropecuária Trop 52:e72376. https://doi.org/10.1590/1983-40632022v5272376
Manning DAC, Theodoro SH (2020) Enabling food security through use of local rocks and minerals. Extr Ind Soc 7(2):480–487. https://doi.org/10.1016/j.exis.2018.11.002
Martins ÉS, Theodoro SH, Bernardez F, Siqueira DS, Caixeta AP, Curtis JCD, Luchese AV (2024) Produção Brasileira de Remineralizadores e Fertilizantes Naturais: 2019 a 2022. Revista Novo Solo. 3.10-19. Available: https://abrefen.org.br/2023/06/29/producao-brasileirade-remineralizadorese-fertilizantes-naturais-2019-a-2022/
Newton P, Civita N, Frankel-Goldwater L, Bartel K, Johns C (2020) What is regenerative agriculture? A review of scholar and practitioner definitions based on processes and outcomes. Front Sustain Food Syst 4:577723
Parra JRP (2023) Biological control in Brazil: state of art and perspectives. Sci Agric 80:e20230080
Persano A (2024) One health: the future of ecosystem health and where to start. C Univ J Glob Health 14(1)
Ramkumar D, Marty A, Ramkumar J, Rosencranz H, Duvvuri R, Goldman M, Meyer E, Steinmetz J, Weckle A, Bloedorn K, Rosier C (2024) Food for thought: making the case for food produced via regenerative agriculture in the battle against non-communicable chronic diseases (NCDs). One Health 18:100734
Silva AC, Cara DVC, Silva EMS, Leal GS, Machado AM, Silva LM (2019) Apatite flotation using saponified baker's yeast cells (Saccharomyces cerevisiae) as a bioreagent. J Market Res 8(1):752–758
Soares MDA, Charlo HCDO, Carvalho M, Paiva PE, Coelho VPDM (2023) Biostimulants increase the yield of greenhouse-grown tomato plants in summer under a tropical climate. Revista Caatinga 36(1):96–105

Soratto RP, Crusciol CAC, Campos M de, et al (2021) Efficiency and residual effect of alternative potassium sources in grain crops. Pesqui Agropecu Bras 56:. https://doi.org/10.1590/s1678-3921.pab2021.v56.02686

Swoboda P, Döring TF, Hamer M (2022) Remineralizing soils? The agricultural usage of silicate rock powders: A review. Sci Total Environ 807:150976. https://doi.org/10.1016/j.scitotenv.2021.150976

Theodoro SH, Manning DAC, Carvalho AMX, Ferrão FR, Almeida GR (2022) Soil remineralizer: a new route to sustainability for Brazil, a giant exporting agro mineral commodities. In: Routledge handbook of the extractive industries and sustainable development. Taylor and Francis, pp 261–277. https://doi.org/10.4324/9781003001317-16

Toledo MCM, Lenharo SL, Ferrari VC, Fontan F, De Parseval P, Leroy G (2004) The compositional evolution of apatite in the weathering profile of the Catalão I alkaline-carbonatitic complex, Goiás. Brazil Can Mineral 42(4):1139–1158

Van Straaten P (2022) Distribution of agromineral resources in space and time–a global geological perspective. Pesq Agrop BrasPesq Agrop Brasileira 57:e01453

Impact of Silicate Agrominerals on Charge Generation in Cerrado Soil

Giuliano Marchi, João Paulo Guimarães Soares, Éder de Souza Martins, Luise Lottici Krahl, Arminda Moreira de Carvalho, Luiz Fernando dos Santos, Priscila Brelaz da Silva, and Elis Marina de Freitas

Abstract This chapter examines the chemical and mineralogical changes, as well as carbon accumulation in soils, with a focus on charge generation using silicate agrominerals in soil from the Brazilian Cerrado biome. The presented case study includes data from a long-term field experiment with silicate agrominerals: biotite schist, syenite, basalt, phonolite, kamafugite, and vermiculite. Dosages were calculated to supply potassium for forage crops and to increase the theoretical soil charge by 0.3–1 $cmol_c\ kg^{-1}$. These studies were conducted on soils cultivated with *Urochloa brizantha* cv. BRS Paiaguás. Although accelerated mineral weathering and the formation of organo-mineral complexes have been observed, a gap remains in understanding the field effects of silicate agrominerals on charge generation and carbon accumulation in tropical soils. Measuring the carbon balance in tropical soils treated with silicate agrominerals is crucial for estimating the potential for long-term carbon retention in agricultural areas. This mechanism is vital for mitigating greenhouse gases, as it also promotes nutrient release, charge generation, organic matter stabilization, stimulates soil organism activity, and reduces costs for agricultural producers. Additionally, the charges generated in the soil increase cation exchange capacity,

G. Marchi (✉) · J. P. G. Soares · É. de Souza Martins · L. L. Krahl · A. M. de Carvalho · E. M. de Freitas
Agrogeological Laboratory, Empresa Brasileira de Pesquisa Agropecuária—Embrapa Cerrados, BR 020, km 18, Planaltina, Distrito Federal 73310970, Federal District, Brasil
e-mail: giuliano.marchi@embrapa.br

J. P. G. Soares
e-mail: jp.soares@embrapa.br

É. de Souza Martins
e-mail: eder.martins@embrapa.br

A. M. de Carvalho
e-mail: arminda.carvalho@embrapa.br

L. F. dos Santos · P. B. da Silva
Universidade de Brasília (UnB) Campos Universitário Darcy Ribeiro ICC Centro, Brasília, Distrito Federal, Brasil
e-mail: luiz.santos1@educa.go.gov.br

É. de Souza Martins and S. Huff Theodoro (eds.), *Soil Remineralizers and Silicate Fertilizers*, Frontier Studies in Soil Science,
https://doi.org/10.1007/978-3-032-14656-4_4

enhancing the retention of cations that might otherwise be lost through leaching during agricultural cultivation.

Keywords Regional agromineral · Soil pH · Mg-silicates

1 Introduction

Brazil has 18.94% of its territory covered by pasture areas, corresponding to approximately 160 million hectares. Of this total, 52% exhibit some level of degradation (approximately 89 million hectares), particularly in the Amazon and Cerrado regions (FGV 2022). To recover and restore all degraded pasture areas, approximately US$67.8 billion would be required.

Even though technologies for recovering degraded pastures have the potential to generate revenues exceeding their costs, such as increased livestock productivity or higher crop yields, Brazilian producers face average costs of US$253.29 ha^{-1} for pasture recovery/reformation technologies. Furthermore, the Brazilian market lacks essential inputs, such as high-quality seeds, fertilizers, and machinery, which hampers producers' ability to implement pasture recovery technologies effectively (FGV 2022).

Applying products derived from silicate agrominerals (ASi), such as soil remineralizer (REM) and silicate fertilizer (FSi), to the soil can serve as a sustainable, long-term local strategy to combat soil degradation (Theodoro et al. 2021) and benefit degraded pasture areas. This approach can also enhance soil fertility by gradually releasing essential plant nutrients, such as potassium, calcium, and magnesium (Nunes et al. 2014; Ramos et al. 2017; Manning et al. 2017; Ferreira et al. 2020). The adoption area of ASi in Brazil increased from 366,000 hectares in 2020 to 2.4 million hectares in 2021, and it could reach 7 million hectares depending on technical factors and farmers' willingness to adopt these inputs (Martins et al. 2023). Since 2019, sectoral surveys have indicated that ASi production has reached 8.9 million tons (Martins et al. 2024). The economic benefit has exceeded US$176 million over the years surveyed. The rapid expansion of adoption in Brazil is primarily driven by reduced production costs and a greater availability of registered products across various regions. Prospective maps illustrating the distribution of ASi, categorized into mineral groups based on their mineralogical composition, indicate the presence of these mineral sources throughout Brazil (Magalhães et al. 2024). Notably, the potential application of ASi in the Central-West region, particularly in pasture areas, spans approximately 47.5 million hectares.

Geological materials, such as silicate rocks, capture and sequester carbon in the soil as they undergo reactions related to accelerated weathering (Hartmann et al. 2013; Lefebvre et al. 2019; Kelland et al. 2020; Ramos et al. 2022). Brazil is the global leader in this process (Zhang et al. 2021). Additionally, these minerals are potential stabilizers of organic matter, contributing to carbon sequestration in the terrestrial compartment (Buss et al. 2024). Similarly, the use of agronomically relevant minerals has shown potential to promote the formation and stabilization of soil aggregates through interactions with organic matter and existing soil minerals (Mao et al. 2020).

The significance of ASi extends beyond merely providing nutrients (Beerling et al. 2018; Theodoro et al. 2020). Evidence shows that these rocks can exhibit relatively fast dissolution rates when influenced by plants, making them capable of releasing nutrients to crops (Anda et al. 2015; Basak et al. 2018; Burghelea et al. 2015; Krahl et al. 2022; Li et al. 2015; Manning et al. 2017; Wang et al. 2000). The weathering of silicate minerals by the rhizosphere and associated organisms, termed biological weathering or bioweathering, not only releases nutrients but also has the potential to form 2:1 clay minerals and short-range order minerals (Chevallier et al. 2019). It occurs within the annual crop cycle timeframe of 1–10 years (Hinsinger et al. 1993; Mohammed et al. 2014; Naderizadeh et al. 2010; Norouzi and Khademi 2010). Plant-induced weathering rates depend on the rock's mineralogy, soil characteristics, and crop species (Burghelea et al. 2015; Harley and Gilkes 2000; Mohammed et al. 2014; Wang et al. 2018). These factors determine the type and proportion of newly formed minerals. A feature of the study of silicate agromineral dissolution is that some minerals can transform into other minerals (chlorite, vermiculite, smectite, and short-range order minerals) or dissolve entirely. Given that the formation of new mineral stages can increase the soil's cation exchange capacity (CEC), it is proposed that the successive application of ASi and the rhizosphere weathering process by cultivated plants will generate emergent properties. These properties can improve soil physical and chemical attributes over time, especially in highly weathered, nutrient-depleted soils such as some Ferralsols. In the medium to long term, this practice could lead to the development of resilient, sustainable, and highly productive soils.

This chapter presents a study on the effects of chemical, physical, and mineralogical changes, as well as carbon content, using silicate agrominerals and fertilizers under the conditions of the Brazilian Cerrado. These effects will be evaluated in soils under *Urochloa brizantha* cv. BRS Paiaguás pasture through one long-term field experiment using biotite schist, syenite, basalt, phonolite, kamafugite, and vermiculite, with dosages designed to increase soil theoretical permanent charge up to 1 $cmol_c\ kg^{-1}$.

2 Silicate Agrominerals and Their Relationship with Soil Charges

The Brazilian Cerrado is one of the largest biomes in the country, covering 204 million hectares, and holds great importance in both national and global agricultural contexts, while simultaneously serving as a significant reservoir of biodiversity (Sano et al. 2008). It is characterized by dry winters and rainy summers, and a tropical humid (Köppen Aw) climate. The biome presents two well-defined seasons, with the dry season extending from April to September, and a mean annual precipitation of approximately 1500 mm, ranging from 750 to 2000 mm, concentrated between

October and March (rainy season). The vegetation of the Cerrado biome includes forest, savanna, and grassland formations.

The soils in the Cerrado region are predominantly deep and highly weathered, with Ferralsol being the most representative type. These soils possess excellent physical properties but are marked by high acidity, elevated levels of exchangeable aluminum, and low natural fertility. In these soils, it is still necessary to implement strategies to increase the CEC and improve fertilizer efficiency.

One alternative to enhance the fertility of these soils is to utilize local natural resources, such as ASi like biotite schist, syenite, and basalt. These materials contain abundant bases and primary minerals (Martins et al. 2008).

Certain silicate minerals found in rocks provide essential nutrients for plant growth. These nutrients are primarily released through the action of the plant root system. The main nutrients released by the rocks include Ca, Mg, K, and micronutrients such as B, Cu, Mn, and Zn (Marchi et al. 2020). Additionally, they reduce soil acidity by dissolving carbonates and silicates (Jones 1998; Krahl et al. 2020, 2022; Rodrigues et al. 2024).

Numerous studies have shown that agrominerals have a positive effect on crop yields (Burghelea et al. 2015; Luchese et al. 2021; Mohammed et al. 2014; Soratto et al. 2021; Swoboda et al. 2022; Theodoro et al. 2021). Unlike synthetic fertilizers, which are highly soluble, ASi require natural weathering processes for their dissolution. Therefore, nutrients are released gradually during plant growth. Soil microorganisms and root activity significantly enhance the weathering of silicate minerals (Jones 1998; Krahl et al. 2022; Naderizadeh et al. 2010; Yu 2018). In the soil environment, the plant rhizosphere and associated microbes promote mineral dissolution by generating acidity, absorbing and releasing elements, and producing low molecular weight organic acids (Hinsinger et al. 1993; Jones 1998; Kong et al. 2014; Uroz et al. 2015; White et al. 2017; Yu 2018).

Therefore, there are essentially three direct effects of ASi to consider: nutrient supply, soil acidity correction, and the formation of new mineral phases. For the optimal use of these materials in agriculture, it is essential to understand their chemical composition as well as their constituent minerals. The chemical composition of rocks enables inferences about the potential capacity of each rock to provide essential nutrients. The identification of mineral constituents enables inferences regarding their relative resistance to dissolution by weathering processes once they are incorporated into the soil (Ribeiro et al. 2010), also, as to the type of dissolution: congruent (complete dissolution) or incongruent (partial dissolution) (Krahl et al. 2022). Other beneficial effects include stabilizing organic matter, promoting carbon sequestration through soil alkalinization, and stable organomineral complexes. Additionally, soil reactions and mineral transformations in the soil contribute to these benefits. Residual effects and impacts on soil and plant growth are also influenced by the mineralogical composition of each material (Singh et al. 2018).

Particles produced by the incongruent dissolution of minerals can form new mineralogical phases, such as hydrobiotite and other clay minerals, with a larger surface area and permanent charges, which can directly enhance the soil's CEC (Krahl et al.

2022). The formation of 2:1 clay minerals from biotite schist and syenite was demonstrated (Krahl et al. 2022). Other studies have shown that the weathering of these primary minerals also forms short-range order minerals, resulting in a high specific surface area and charge density (Wu et al. 2021).

The studies also emphasize the interaction between soil permanent and variable charges in the dynamics of CEC. The proportion of permanent charges in Ferralsols varies depending on their mineralogy (Cunha et al. 2014). In some of these soils, variable charges are more prevalent, but the addition of ASi can increase permanent charges, thereby contributing to the greater chemical stability of the soil.

The Cs^+ adsorption methodology, which involves saturating the soil with CsCl solutions followed by ion exchanges to measure charges, demonstrates the practical application of this theory (Anderson and Sposito 1991). Despite its effectiveness, this technique is rarely used, with few studies reported in Brazilian soils. For example, Weber et al. (2005) determined that the permanent charge in the surface layer of a Yellow Acric Latosol was 5.8 $mmol_c$ kg^{-1}; for a Clayey Red Acric Latosol, it was 11.7 $mmol_c$ kg^{-1}; and for a Medium-Textured Red Acric Latosol, it was 10.4 $mmol_c$ kg^{-1}. In contrast, Busato et al. (2012) found 5.0 $mmol_c$ kg^{-1} of permanent charge in a typical Eutrophic Yellow Latosol. Additionally, studies like that of Cunha et al. (2014) revealed 65.6 $mmol_c$ kg^{-1} of permanent charges in Red Argisols and 10.9 $mmol_c$ kg^{-1} in a Yellow Latosol, highlighting the variability in soil charge capacity, with a tendency for a higher proportion of variable charges in intensely weathered tropical soils (Fontes et al. 2001). Silicate agrominerals, such as biotite syenite and others, can increase the CEC, reducing cation nutrient losses by leaching.

Studies conducted on Ferralsols from the Cerrado treated with biotite syenite have shown a significant increase in permanent charges after the first maize cultivation cycle (Santos et al. 2021). This effect was attributed to bioweathering, which releases ions from the mineral structure and forms new minerals with higher cation exchange capacity (Krahl et al. 2022).

Finally, ASi represent a promising alternative for improving fertility and sustainability in tropical soils, especially in Ferralsols of the Cerrado. Additional studies should investigate the long-term interactions between silicate agrominerals, organic matter, soil organisms, and agricultural management, thereby consolidating the role of these materials in revitalizing tropical soils (Guo et al. 2023; Medeiros et al. 2023; Singh et al. 2017; Theodoro et al. 2021).

3 Case Study: Charge Generation in Soils Treated with Silicate Agrominerals

3.1 Experimental Area, Chemical and Mineralogical Characterization of Silicate Agrominerals

An experiment was established in 2016 with the implementation of *Urochloa brizantha* cv. The BRS Paiaguás pasture was studied over an eight-year period. The experimental area, located at Embrapa Cerrados experimental field in Planaltina—DF (CPAC), at an altitude of 1171 m, coordinates 15°36′28″S and 47°44′54″W, is classified as a Xanthic Ferralsol (IUSS Working Group WRB 2022)/ dystrophic Yellow Latosol with a clayey texture—LAd (Reatto et al., 2004; Santos et al. 2018). Its chemical characteristics are as follows: pH = 5.63; Al = 0.04 $cmol_c$ dm^{-3}; Ca + Mg = 5.25 $cmol_c$ dm^{-3}; P = 0.23 mg dm^{-3}; and K = 36 mg dm^{-3}. The local climate, according to the Köppen-Geiger classification, is tropical savanna (Aw) (Alvares et al. 2013; Beck et al. 2018).

Based on the guidelines of Normative Instruction 05 (IN 05) from the Ministry of Agriculture, Livestock, and Supply (MAPA), which defined silicate agrominerals as new mineral inputs for managing soil fertility in agriculture Brazil (2016) and set minimum criteria for their chemical, mineralogical, and granulometric composition for registration and supervision, various ground rock materials were selected: biotite schist (Canastra Group, unit B, from Abadiânia—GO); phonolite (Poços de Caldas—MG), basalt (Araguari—MG), biotite syenite (Ceraíma—BA), kamafugite (Santo Antônio da Barra—GO), and vermiculite (São Luiz de Montes Belos—GO), according to their chemical (Table 1) and mineralogical composition (Table 2).

3.2 Design and Adopted Practices

The experimental design was a randomized complete block design (RCBD), with 14 treatments and 3 replicates, totaling 42 experimental units. However, 5 of the 14 treatments were installed a year later (Table 3). The plots had an area of 9 m^2 (3 × 3 m) per experimental unit (Fig. 1).

Before applying the treatments, soil acidity was corrected by broadcasting 500 kg ha^{-1} of dolomitic limestone and 1 t ha^1 of gypsum, incorporated to a depth of 20 cm. During planting, 120 kg ha^{-1} of P_2O_5 was applied, using magnesium thermophosphate as the source, and 100 kg ha^{-1} of FTE BR12 (containing 3.9% sulfur, 1.8% boron, 2.0% manganese, and 9.0% zinc) was applied as a micronutrient source. Subsequently, the respective treatments were incorporated into the soil with a rotary hoe cultivator to a depth of 0–10 cm.

The treatments included a single application of silicate agromineral sources, except for the treatment using vermiculite, which consisted of incorporating 300 g

Table 1 The chemical composition of the silicate agrominerals was determined using X-ray fluorescence spectrometry (XRF) and inductively coupled plasma optical emission spectroscopy (ICP-OES)

	SiO_2	Al_2O_3	Fe_2O_3	CaO	MgO	TiO_2	P_2O_5	Na_2O	K_2O	MnO	LOI[2]
	%										
Biotite schist	62.6	16.8	7.8	1.44	3.2	0.86	0.2	1.93	3.19	0.12	2.17
Biotite syenite	55.9	16	8.16	2	1.7	1.49	0.24	1.19	11.3	0.11	0.49
Phonolite	54.4	20.3	3.8	1.7	0.4	0.6	0.1	6.6	8.3	0.2	3.8
Kamafugite	37.6	1.8	10.6	15.1	16.9	7.8	0.93	2.5	2.65	0.17	3.6
Basalt	49.5	11.6	15.6	9.8	5.6	3.2	0.4	2.2	1.2	0.2	1.1
Vermiculite	48.2	6.4	8.3	8.4	20.4	0.9	0.1	0.5	0.3	0.2	6.4

[1] Analyses conducted by SGS Geosol laboratory (https://www.sgsgeosol.com.br) and the CRTI laboratory (https://crti.ufg.br); 2LOI—loss on ignition

Table 2 Mineralogical composition of silicate agrominerals using X-ray diffraction (DRX) and Rietveld refinement

	LCM[1]	Cc[2]	Px[3]	Pl[4]	FK[5]	Bt[6]	Fl[7]	Mu[8]	Ct[9]	Ap[10]	Am1	St[12]	Vm[13]	Fp[14]	Qz[15]	Im[16]	Mt[17]	Hm[18]	Ep[19]	Ze[20]	Ol[21]	Me[22]	Tb[23]	Total
Biotite schist	0	0	1.7	22.1	0	11.2	0	17.9	11.3	0	0	0	0	0	35.7	0	0	0	0	0	0	0	0	99.9
Biotite syenite	0	0	13.5	9.3	58.1	15.4	0	0	0	0	3.6	0	0	0	0	0	0	0	0	0	0	0	0	99.9
Phonolite	0	0	13	5.1	48.7	2.5	0	0	0	0	0	0	0	28	0	2.5	0	0	0	0	0	0	0	99.8
Kamafugite	13.9	1.77	1.77	0	1.72	0	18.1	0	2.03	2.7	0	0	1.15	8.57	3.0	0	2.9	1.07	2.8	2.16	22.1	2.2	3.39	91.3*
Basalt	11.1	0	35	35	5	0	0	0	0	0.9	0	2.5	0	0	2.1	6.8	1.1	0.6	0	0	0	0	0	100.1
Vermiculite	0	7	0	0	0	0	0	0	0	0.2	0	0	86	0	6.3	0.7	0		0	0	0	0	0	100.2

[1] Mineralogical analysis were performed by CRTI (https://crti.ufg.br) and Laboratory at Federal University of Pará. The materials analyzed included low-crystallinity minerals (LCM1), calcite (Cc_2) $CaCO_3$, pyroxene (Px_3) (Ca, Na)(Mg, Fe, Al, Ti)(Si, $Al)_2O_6$, plagioclase (Pl_4) (Na, Ca)(Si, $Al)_4O_8$, potassium feldspar (FK_5) $KAlSi_3O_8$, biotite (Bt_6) K(Mg, $Fe^{2+})_3[AlSi_3O_{10}(OH, F)_2]$, phlogopite (Fl7) $KMgSi_3O_{10}(OH, F)_2$, muscovite (Mu8) $KAl_2(Si_3Al)O_{10}(OH, F)_2$, chlorite (Ct9) (Mg, $Fe^{2+})_5Al(Si_3Al)O_{10}(OH)_8$, apatite (Ap10) $Ca_5(PO_4)_3(OH, F, Cl)$, amphibole (Am_{11}) A0-1B2C5T8O22 (OH, F)2 (where: A = Na +, K +; B = Ca^{2+}, Na^+, Mg^{2+}, Fe^{2+}; Mn^{2+}, Li^+; C = Mg^{2+}, Fe^{2+}, Al^{3+}, Fe^{3+}; Mn^{2+}, Zn^{2+}, Cr^{3+}, Li^+, or Ti^{4+}; T = Si, Al), smectite (St_{12}) (Na, $Ca)_{0.3}(Al, Mg)_2Si_4O_{10}(OH)_2 \cdot n(H_2O)$, vermiculite ($Vm_{13}$) (Mg, Fe^{2+}, $Al)_3(Al, Si)_4O_{10}(OH)_2 \cdot 4(H_2O)$, feldspathoid ($Fp_{14}$) (Na, K)$AlSiO_4$, quartz ($Qz_{15}$) SiO_2, ilmenite (Im_{16}) $Fe^{2+}TiO_3$, magnetite (Mt_{17}) $Fe^{2+}Fe^{3+}{}_2O_4$, hematite (Hm_{18}) Fe_2O_3, epidote (Ep_{19}) (Ca, Ce, La, Y, Th, Fe^{2+}, Mn^{2+}, $Mn^{3+})_2$(Al, Fe^{3+}, Mn^{3+}, Mn^{2+}, Ti, $Cr)_3O(SiO_4)(Si_2O_7)(OH)$, zeolite chabazite ($Ze_{20}$) (Ca, K_2, $Na_2)_2[Al_2Si_4O_{12}]_2 \cdot 12H_2O$, olivine ($Ol_{21}$) (Mg, $Fe)_2SiO_4$, melilite (Me_{22}) (Ca, $Na)_2$(Mg, Al)(Si, $Al)_2O_7$, and tobermorite (Tb_{23}) $Ca_5Si_6O_{16}(OH)_2 \cdot 4H_2O$. * 8.7% as trace minerals1

Table 3 Sources and doses of silicate agrominerals used in the field experiment

ID	Treatment description	Total % K_2O	kg per plot of silicate agromineral	Equivalent to kg ha^{-1} of silicate agromineral	kg per plot of total K_2O	Equivalent to kg ha^{-1} of total K_2O
Control	No K	0	0	0	0	0
KCl 1	150 g of KCl per plot	60	0.15	167	0.09	100
KCl 2	Yearly—300 g KCl per plot	60	0.3	333	0.18	200
KCl annual#	150 g of KCl per plot for 3 yrs	60	0.15	167	0.09	100
BAS 1	Basalt	1.2	7.5	8333	0.09	100
BAS 2	Basalt (0.010–0.013 $cmol_c\ kg^{-1}$ of soil)*	1.2	36	40,000	0.432	480
BSC 1	Biotite schist	3.19	2.78	3,089	0.089	99
BSC 2	Biotite schist (1.73 $cmol_c\ kg^{-1}$ of soil)*	3.19	136	151,111	4.34	4820
FHON	Phonolite	8	1.41	1567	0.11	125
Vermiculite + KCl	Vermiculite + KCl (1.72 $cmol_c\ kg^{-1}$)*	0.18 (60 in KCl)	12.6 (+ 0.3 as KCl)	14,0 (+ 333 kg of KCl)	0.023 (+ 0.18 as KCl)	25 (+ 200 as KCl)
BSY 1#	Biotite syenite	3.5	2.78	3089	0.10	108
BSY 2#	Biotite syenite (2.37 $cmol_c\ kg^{-1}$)*	3.5	136	151,111	4.76	5289
KAM 1#	Kamafugite	1.2	7.5	8333	0.09	100
KAM 2#	Kamafugite (rich in 2:1 clay minerals)	1.2	36	112,853	0.432	1354

* Theoretical charge after total dissolution (congruent or incongruent) of selected minerals, calculated using Eq. 1; # Treatments implemented after one year

of KCl applied once a year for the first three years. The treatments aimed to provide an agronomic dose of potassium, considering the total amount of K_2O in each material equal to 100 kg of K_2O, disregarding the fact that not all minerals present in the selected agrominerals release K; and a higher dose, aiming to increase the theoretical soil permanent charge ranging from 0.12 to 4.89 $mmol_c\ kg^{-1}$ in the 0–10 cm layer (Table 3).

Fig. 1 Experimental area with cultivation of *Urochloa brizantha* cv. BRS Paiaguás: **a** application of treatments, **b** newly established pasture in 2017, and **c** consolidated pasture in 2018

The calculation methodology for the theoretical charges of silicate agrominerals was determined as follows (Eq. 1): the amount of agromineral applied per plot (Table 3) was multiplied by the percentage of biotite in the agrominerals biotite schist and biotite syenite (11.2% and 15.4%, respectively; Krahl et al. 2022). The resulting value represents the amount of biotite mineral applied per plot. Considering that all biotite minerals will be converted to trioctahedral vermiculite with a charge of 150 $cmol_c\ kg^{-1}$ (Barnhisel and Bertsch 1989), in a proportion of 3:2 for biotite:vermiculite (0.78, Acker and Bricker 1992), the charge added per plot is found by multiplying the weight (kg) of biotite added per plot by the vermiculite charge. Dividing $cmol_c$ added per plot by 1.053 m^3 of soil [3×3 m $\times$ 0.1 m, per plot, bulk density = 1.14 (s = 0.08) g cm^{-3}], we find approximately 1.73 and 2.37 $cmol_c\ kg^{-1}$ of soil for biotite schist and biotite syenite, respectively.

$$\sigma Pt = (ApMpStCoP)/(\text{kg}soilperplot) \quad (1)$$

where:

σPt = Theoretical permanent charge applied in the soil ($cmol_c\ kg^{-1}$ of soil);

Ap = Amount of silicate agromineral applied by plot (kg per plot);

Mp = Percentage of primary mineral (biotite, vermiculite, or smectite) in the silicate agromineral, or its purity on the silicate agromineral (%);

St = Stoichiometric coefficient for the transformation from biotite to vermiculite, when applicable;

CoP = Charge of the secondary mineral or the mineral providing the charge, i.e. vermiculite, smectite ($cmol_c\ kg^{-1}$).

The theoretical permanent charge in the soil treated with vermiculite was calculated similarly, resulting in 1.72 $cmol_c\ kg^{-1}$ of soil. The same approach was used for basalt, considering the amount applied per plot, 2.5% of smectite in the basalt (Table 3), and a smectite charge between 11.56 and 15.22 $cmol_c\ kg^{-1}$ (Cheng and Heidari 2018). This resulted in a range from 0.010 to 0.013 $cmol_c\ kg^{-1}$ of soil. Permanent charges for kamafugite were not calculated, as charges would be generated from short-range order minerals resulting from primary mineral dissolution in the soil, thus generating variable charges.

The planting of *Urochloa brizantha* cv. BRS Paiaguás was carried out by broadcasting on January/25/2016, with the recommendation of 3 kg ha^{-1} of SPG, keeping the pasture without grazing throughout the experiment. Nitrogen was top-dressed in the form of ammonium sulfate (20% N) 30 days after sowing (DAS) at a dose of 50 kg ha^{-1}.

During the first two agricultural years, biomass cuts of the shoot portions were conducted in July and December. The biomass was removed from the plots to facilitate better regrowth. In subsequent years, cutting occurred annually during the rainy season, with the aerial biomass left on the soil surface in their respective experimental units. Cuts were made at a height of 10 cm above ground level within a randomly placed square of 1.0 m^2 at three different locations within each plot. Following this, the entire plot was mowed uniformly at 10 cm above ground, and the resulting biomass was removed from the area (Salman et al. 2006). The percentage of fresh weight of the grass plants was then calculated. Samples were oven-dried at 65 ºC until a constant weight was achieved, after which they were weighed, milled, and analyzed chemically.

Composite soil samples were collected at a depth of 0–10 cm in each plot and analyzed chemically (Donagema et al. 2011). Permanent and variable negative charges were determined using the Cs method (Anderson and Sposito 1991). Cation exchange capacity was determined by magnesium sorption (BS EN ISO 11260 2011). Reference samples identified as KGa-1b and STx-1 from the Source Clays Repository of The Clay Mineral Society (CMS) were used as standard references for the measured charges in treated soils.

4 Study Results

4.1 Forage Productivity

The following results refer to the first two years (2017 and 2018) of evaluations in the CPAC experimental area (Fig. 2). The evaluation of forage productivity was estimated only for the treatments implemented at the beginning of the experiment (Table 3).

Dry matter weight results for the KCl treatments were statistically higher than those of the other treatments. Notably, the treatments BSC 2, BAS 2, and the treatment

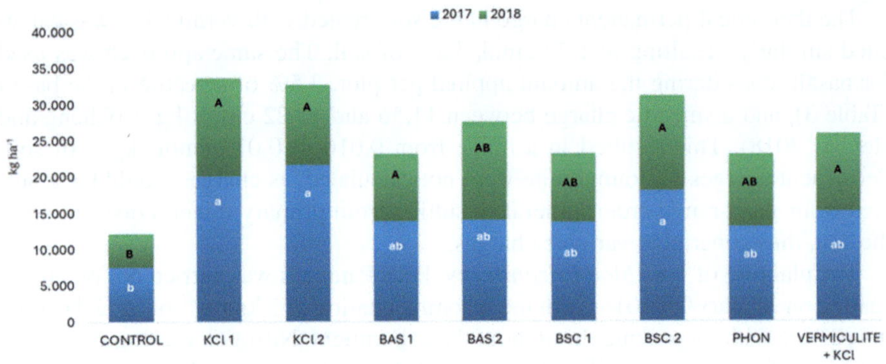

Fig. 2 Dry matter weight (DMW) of *Urochloa brizantha* cv. BRS Paiaguás in 2017 and 2018. Distinct letters between bars indicate significant differences according to the Scott-Knott test at 5% probability

with vermiculite exhibited statistically superior performance compared to the other ASi treatments, with average production values statistically comparable to those of the two KCl doses (Fig. 2).

The higher production of the grass as compared to the control suggests that minerals from silicate agrominerals added were dissolved and released nutrients. Above all, K was the only nutrient lacking in control. The experiments were evaluated only on the cited dates. However, every year, the grass was cut and removed from the area. The growth of grass over the years has impacted soil characteristics, including nutrient availability, organic matter accumulation, and the increase in the soil's permanent and variable charge, as evaluated in samples collected in 2023.

4.1.1 Results of Soil Chemical Attributes After 8 years of Cultivation with Urochloa Brizantha Cv. BRS Paiaguás and Silicate Agrominerals

The varied nutrient release mechanisms for each treatment led to alterations in the chemical and physicochemical properties of the soil. The results showed that BSY had the most favorable outcomes among the evaluated agrominerals. BSY 2 increased the levels of available K, P, and Fe in the soil (Tables 4 and 5).

All silicate minerals classified as K sources, except for muscovite, exhibit potential to release K (Martins et al. 2023). K-feldspar is more stable and less prone to weathering than biotite or nepheline. While BSY predominantly consisted of K-feldspar (51.8%), including orthoclase and microcline, the principal source of potassium release into the soil was identified as the mineral biotite, comprising 15.4% of the sample.

Table 4 Analysis of macro and micronutrients in the CPAC area in soil treated with silicate agrominerals and cultivation of *Urochloa brizantha* cv. BRS Paiaguás after 8 years

	K	P	Ca	Mg	Zn	Mn	Cu	Fe
	mg kg^{-1}		$cmol_c$ kg^{-1}		mg kg^{-1}			
Control	36.7 ± 1.17	6.21 ± 0.7	2.18 ± 0.2	1.31 ± 0.4	11.01 ± 3.9	13.3 ± 2.7	6.00 ± 0.2	47.93 ± 12.2
KCl 1	60 ± 27	5.79 ± 0.2	1.46 ± 0.2	1.06 ± 0.3	5.39 ± 1.5	9.09 ± 0.3	5.50 ± 0.5	57.5 ± 9
KCl 2	54 ± 19.3	5.91 ± 0.1	1.86 ± 0.2	1.3 ± 0.4	10.11 ± 4.8	11.67 ± 0.4	5.89 ± 0.6	52.6 ± 3.8
KCl annual	61.33 ± 25	6.16 ± 0.2	2.48 ± 0.1	1.69 ± 0.2	15.68 ± 7.9	16.71 ± 0.5	6.14 ± 0.9	48.5 ± 11.4
BAS 1	44.66 ± 11	5.77 ± 0.2	2.18 ± 0.5	1.26 ± 0.2	8.49 ± 4.51	11.15 ± 2.6	5.79 ± 0.4	55.65 ± 11.3
BAS 2	54 ± 7.21	5.77 ± 0.1	1.7 ± 0.3	1.03 ± 0.4	7.06 ± 2.96	10.51 ± 3.0	5.73 ± 0.1	48.75 ± 3.0
BSC 1	47.33 ± 9.02	5.69 ± 0.2	1.99 ± 0.9	1.23 ± 0.4	10.22 ± 8.1	13.59 ± 8.3	5.68 ± 0.5	52.68 ± 7.54
BSC 2	47.33 ± 6.43	5.94 ± 0.2	2.1 ± 0.2	1.38 ± 0.2	9.48 ± 2.62	13.12 ± 2.9	5.88 ± 0.2	49.25 ± 7.97
FON	48.7 ± 9.24	5.73 ± 0.1	1.49 ± 0.2	1.07 ± 0.2	6.52 ± 1.81	9.19 ± 1.01	5.71 ± 0.1	53.60 ± 15.6
VERM + KCl	62.7 ± 23.1	5.82 ± 0.1	1.88 ± 0.2	1.2 ± 0.2	11.10 ± 5.4	11.97 ± 3.5	5.87 ± 0.3	44.6 ± 8.1
BSY 1	51.33 ± 9.87	6.43 ± 0.8	2.66 ± 0.5	1.88 ± 0.2	16.55 ± 10.2	19.12 ± 7.0	6.15 ± 0.6	58.8 ± 5.84
BSY 2	208.66 ± 97.9	17.29 ± 4.5	2.67 ± 0.3	1.33 ± 0.2	16.55 ± 7.4	21.28 ± 2.2	6.43 ± 0.7	120.7 ± 12.4
KAM 1	80.66 ± 54.1	7.64 ± 2.9	2.41 ± 0.1	1.9 ± 0.1	16.95 ± 7.2	18.63 ± 2.2	6.09 ± 0.1	73.7 ± 31.7
KAM 2	104.67 ± 65.2	8.40 ± 3.6	2.52 ± 0.1	1.75 ± 0.3	18.60 ± 0.8	19.62 ± 1.1	6.37 ± 0.1	72.9 ± 36.2

Table 5 Analysis of organic matter and acidity components in the CPAC Area in soil treated with silicate agrominerals and cultivated with *Urochloa brizantha* cv. BRS Paiaguás after 8 years

	OM	pH H_2O	pH $CaCl_2$	H + Al	CECt	CECe	SB	BS
	%			$cmol_c\ kg^{-1}$				%
Control	2.98 ± 0.4	6.44 ± 0.5	5.60 ± 0.6	3.82 ± 1.3	7.41 ± 0.5	3.59 ± 1.0	3.59 ± 1.0	48.9 ± 14.6
KCl 1	3.02 ± 0.1	5.77 ± 0.1	4.92 ± 0.1	5.16 ± 0.3	7.84 ± 0.3	2.68 ± 0.3	2.67 ± 0.3	34.14 ± 3.4
KCl 2	3.15 ± 0.5	5.91 ± 0.3	5.19 ± 0.3	4.81 ± 0.7	8.16 ± 0.7	3.35 ± 0.5	3.35 ± 0.5	41.16 ± 6.1
KCl annual	3.17 ± 0.2	6.33 ± 0.3	5.67 ± 0.1	3.67 ± 0.5	7.97 ± 0.7	4.30 ± 0.2	4.30 ± 0.2	54.07 ± 2.4
BAS 1	3.35 ± 0.2	5.97 ± 0.3	4.98 ± 0.1	5.44 ± 0.4	8.55 ± 0.7	3.11 ± 0.3	3.10 ± 0.3	36.3 ± 0.9
BAS 2	3.15 ± 0.2	6.08 ± 0.2	5.08 ± 0.2	4.85 ± 0.4	7.70 ± 0.3	2.84 ± 0.7	2.84 ± 0.7	36.9 ± 7.60
BSC 1	3.02 ± 0.3	6.16 ± 0.6	5.30 ± 0.5	4.62 ± 0.9	7.96 ± 0.3	3.35 ± 1.3	3.34 ± 1.3	41.6 ± 14.4
BSC 2	3.17 ± 0.7	6.19 ± 0.2	5.37 ± 0.3	4.35 ± 0.9	7.95 ± 0.8	3.60 ± 0.4	3.59 ± 0.4	45.5 ± 6.92
FON	3.12 ± 0.3	5.99 ± 0.4	4.94 ± 0.2	5.13 ± 0.8	7.80 ± 0.6	2.68 ± 0.2	2.68 ± 0.2	34.5 ± 4.87
VERM + KCl	3.02 ± 0.2	6.24 ± 0.1	5.27 ± 0.2	4.48 ± 0.5	7.76 ± 0.3	3.28 ± 0.4	3.27 ± 0.4	42.2 ± 5.6
BSY 1	3.41 ± 0.5	6.32 ± 0.1	5.62 ± 0.2	3.98 ± 0.7	8.64 ± 0.7	4.67 ± 0.6	4.66 ± 0.6	54 ± 6.02
BSY 2	2.86 ± 0.2	6.47 ± 0.5	5.73 ± 0.3	3.38 ± 0.8	7.67 ± 0.1	4.29 ± 0.7	4.29 ± 0.7	55.9 ± 9.50
KAM 1	3.06 ± 0.3	6.45 ± 0.4	5.71 ± 0.1	3.64 ± 0.3	8.17 ± 0.3	4.52 ± 0.2	4.52 ± 0.2	55.4 ± 2.6
KAM 2	3.25 ± 0.1	6.46 ± 0.1	5.73 ± 0.1	3.47 ± 0.1	8.02 ± 0.5	4.55 ± 0.2	4.54 ± 0.1	56.8 ± 4.2

* OM: organic matter; SB: sum of bases; BS: base saturation of soil, CECt: total cation exchange capacity; CEC: effective cation exchange capacity

The biotite in BSC and BSY agrominerals show varying MgO and FeO content (Krahl et al. 2022). The differences in Fe and Mg content, along with element substitution in the silicate sheets (e.g., Si^{4+} by Al^{3+} and Fe^{3+} in tetrahedron site or Al^{3+} by Mg^{2+} in octahedron sites), result in different weathering rates (Murakami et al. 2004).

The higher content of Fe_2O_3 in BSY, especially in the smaller size fractions, is crucial as a higher structural concentration of Fe^{2+} accelerates the alteration and release of K from biotite (Gilkes et al. 1973). During the dissolution process, Fe^{2+} exits the mineral structure and is oxidized to Fe^{3+}, which accelerates weathering and releases K, as observed in soils treated with BSY.

When silicate rock powders are incorporated into soils, a portion of their Mn^{2+} and Fe^{2+} content undergoes oxidation. This process can occur either after dissolution in the soil solution or within the mineral crystals themselves, facilitated by solid-state diffusion of electrons to electron acceptors in the soil solution at the mineral surface (Gilkes and McKenzie 1988). The oxidation of Mn^{2+} and Fe^{2+} potentially serves as an alternative electron source for microorganisms, apart from reduced organic materials, thereby potentially reducing the carbon expended by plants exuding organic acids to acquire Fe, Mn, and P. Oxidized Fe(II) that is released from minerals forms short-range order minerals in soil (Krahl et al. 2020) such such as ferrihydrite (Colombo et al. 2014; Yu et al. 2017), which are more sensitive to redox reactions.

A significant aspect of the short-range order minerals in weathered soils, such as Ferralsols, is their relationship with P. A substantial fraction of soil P is known to be bound to redox-sensitive Fe minerals. Fe-redox reactions likely play a crucial role in regulating soil phosphorus cycling (Lin et al. 2018). The adsorption of P in short-range order Fe^{3+}-mineral surfaces hinders the re-adsorption of P onto crystalline Fe^{3+} and Al^{3+} surfaces (Fageria et al. 1990). Ferrydrite is perhaps the most effective Fe mineral in terms of P adsorption due to its small particle size, high surface area, and gel-like form (Rhoton and Bigham 2005).

Grasses produce small molecules known as phytosiderophores, which have a high affinity for Fe^{3+}. These compounds are excreted into the rhizosphere, where they chelate Fe^{3+}, facilitating its solubilization. Additionally, localized acidification in the rhizosphere converts crystalline Fe and Al oxides into water-soluble forms of iron and aluminum. During wet-dry cycles, re-alkalization leads to the hydrolysis of these soluble forms, resulting in the formation of short-range order Fe-Al hydroxides (Guo et al. 2012).

The kamafugite (KAM 1 and KAM 2) treatment showed high levels of K and Fe. The elevated levels of K in soil suggest the release of these elements mainly from phlogopite [$KMg_3Si_3AlO_{10}(OH)_2$], which makes up 18.1% of the sample; nepheline, a feldspathoid comprising 6.43% of the sample, and short-range order minerals. The substantial Fe_2O_3 content originates from diopside and low-crystallinity minerals. The reduced proportion of Al_2O_3 in the minerals present in KAM increases reactivity and promotes the formation of short-range order minerals associated with variable charges (Sparks et al. 2024).

BSY and KAM enhance nutrient availability through bioweathering by the soil microbiota in the rhizosphere. These modifications enhance the short-range order of mineral fractions and create new phases with high surface areas, thereby stabilizing organic matter. This long-term effect-generating variable and permanent charges promote the accumulation of organic matter in soils. Given the extensive use of ASi in Brazilian agriculture, tracking carbon retention in the soil and identifying where organic matter is bound is crucial for reducing greenhouse gas emissions. In Ferralsols treated with ASi, the most significant fraction is the organic carbon associated with minerals, as these minerals may stabilize organic carbon in the soil (Kleber et al. 2015; Sarkar et al. 2018; Singh et al. 2017; Wang et al. 2018). This stabilization aids in retaining carbon in the soil after it has been sequestered from the atmosphere.

The primary challenge is to enhance our understanding of carbon sequestration processes. Instead of measuring total soil organic carbon stocks, the focus should be on the dynamics and interactions between carbon and soil minerals. Understanding how carbon is linked and stored will inform more effective soil management practices using agrominerals. The adsorption of soil minerals and occlusion within soil aggregates might stabilize SOM against decomposition (Angst et al. 2018; Sarkar et al. 2018; Singh et al. 2017). The type and amount of clay minerals strongly influence SOM stabilization in soils, primarily due to their structural and physicochemical properties. Li et al. (2025) found that mineral-associated organic carbon (MAOC) is enriched in ecosystems dominated by allochthonous inputs and abundant active minerals, leading to an increased proportion of persistent organic molecules. Soils with significant allochthonous inputs (> 50%) exhibit a substantial contribution of MAOC (> 70%) to total SOC, with notable preservation of lipids (36%) across coastal wetlands. In some carbon ecosystems, the carbon pool is stabilized by binding to mineral surfaces or being enclosed within microaggregates, typically resulting in the formation of MAOC (Li et al. 2025).

In tropical environments, applications of silicate rocks in agricultural soils can add raw minerals to the system. Besides improving soil quality, this is a promising alternative to stabilize the SOM by the bioweathering process and newly formed minerals (Edwards et al. 2017; Beerling et al. 2018). These neoformed minerals (vermiculites, smectites, and short-range order minerals) with high CEC can lead to adsorption or occlusion of SOM and can contribute to SOC stabilization.

4.1.2 Soil Carbon After 8 Years of Cultivation with Urochloa Brizantha Cv. BRS Paiaguás and Silicate Agrominerals

To evaluate the organic carbon content of the soil, analyses were conducted to quantify total levels and to assess the fractionation of soil organic matter (SOM) using both physical and chemical methods. These analyses were performed in samples from the CPAC experiment under treatments designed to enhance the soil's permanent and variable charge in the 0–10 cm soil layer: BAS 2, BSC 2, KAM 2, and in KCl2.

The assessment of soil Total Organic Carbon (TOC) was performed by analyzing each sampled layer individually. Initially, the TOC content in the soil was measured using a CHNS elemental analyzer. For carbon quantification via the dry combustion method, subsamples were oxidized at 900 °C utilizing the Vario Macro Cub CHNS organic elemental analyzer provided by Elementar. This method is widely recognized as a standard in scientific research due to its superior precision among analytical techniques. Soil organic matter (SOM) evaluation was conducted employing the Walkley and Black (1934) wet oxidation method, adapted by Embrapa (2017). This approach aimed at quantifying the easily oxidizable or decomposable carbon in the soil through a redox reaction.

The fractionation of soil organic matter (SOM) involved both physical and chemical methods. Physical fractionation was performed using Cambardella and Elliot's (1992) method, with modifications by Nascimento et al. (2010). Particulate organic carbon (POC) and mineral-associated organic carbon (MAOC) were measured at depths of 0–10, 10–20, and 20–30 cm.

Chemical fractionation separated humin (H), humic acids (HA), and fulvic acids (FA) using the solubility method described by Swift (1996) and adapted by Benites (2003). Samples from the same depths were analyzed in triplicate. Organic carbon levels in humic substances were determined using Yeomans and Bremner's (1988) protocol, with adaptations by Benites et al. (2017). The carbon was oxidized via wet digestion in a potassium dichromate ($K_2Cr_2O_7$) solution in an acidic medium. These fractions are essential for generating cation exchange capacity (CEC) in highly weathered soils such as Ferralsols. Their lower point of zero charge (PZC) and high concentration of carboxyl and hydroxyl groups enable them to retain and gradually release cations that are vital for plant nutrition.

The study of treatments with biotite schist, kamafugite, and basalt sources at a depth of 30 cm indicates that silicate agrominerals were able to maintain average carbon content (CT) levels similar to the conventional source (KCl) (Table 6). Regarding the easily oxidizable carbon (TOC), which is more susceptible to decomposition and acts as a promoter of other carbon fractions in the soil, kamafugite exhibited a higher average trend (20 g kg^{-1}) compared to both the control and KCl (17.9 g kg^{-1}) in the 0–10 cm layer, which is considered the most active for soil organic matter (SOM).

The physical fractionation of soil organic matter (SOM) indicates that biotite schist tends to yield higher levels of particulate organic carbon (POC) (15 g kg^{-1}) compared to other treatments. This fraction is crucial for soil health, as it represents SOM in its initial stages. Essentially, this signifies a reserve of organic material that, as it transitions into new fractions, can supply energy and nutrients to soil biota and plants.

The findings indicate that kamafugite exhibited higher concentrations than the controls across all three fractions, with values of 2.8 g kg^{-1} for C-HA, 4.6 g kg^{-1} for C-FA, and 11.6 g kg^{-1} for C-HUM. Notably, basalt also demonstrated a trend to increase the levels of humic acid carbon and fulvic acid carbon to 2.8 g kg^{-1} and 4.6 g kg^{-1}, respectively.

Table 6 Average levels of total carbon (CT), total organic carbon (TOC), particulate organic carbon fractions (POC), mineral-associated organic carbon (MAOC), humic acid carbon (C-HA), fulvic acid carbon (C-FA), and humin carbon (C-HUM) in the experimental area of CPAC after 8 years of cultivation

	Prof	CT	POC	MAOC	TOC	C-HA	C-FA	C-HUM
	cm	g kg^{-1}						
	0–10	24.6	14.6	10.0	18.2	2.5	4.3	10.9
Control	10–20	21.0	10.7	10.3	14.8	2.4	3.7	9.1
	20–30	17.23	6.31	10.92	11.90	1.79	2.73	6.68
	0–10	23.4	14.5	8.9	17.9	2.7	4.2	10.1
KCl 2	10–20	19.4	10.4	9.0	15.2	2.5	3.6	9.0
	20–30	15.16	5.44	9.71	11.36	1.78	2.70	6.19
BAS 2	0–10	22.5	13.2	9.3	18.0	2.8	4.6	9.8
	10–20	19.3	9.1	10.2	15.4	2.7	3.7	8.4
	20–30	16.35	6.24	10.11	11.38	1.91	2.75	5.83
BSC 2	0–10	23.1	15.0	8.1	18.2	2.3	4.1	10.5
	10–20	20.4	10.8	9.6	14.2	2.3	3.6	9.2
	20–30	16.87	6.37	10.50	12.19	1.81	2.86	6.44
KAM 2	10	22.5	14.3	8.2	20.0	2.8	4.6	11.6
	20	19.4	10.8	8.6	14.7	2.7	3.6	9.0
	30	16.60	7.09	9.51	11.99	1.66	2.75	6.76

Silicate agrominerals generally exhibited comparable or superior performance relative to the control and conventional sources (KCl). This suggests that these materials indirectly contribute to carbon stabilization in the soil. As a result, the sequestration of carbon in the terrestrial compartment, specifically the soil, in various forms, enhances the soil's carbon residence time and mitigates the impacts of climate change.

These findings suggest that the application of silicate agrominerals, such as kamafugite and biotite schist, can improve soil carbon fractions, especially in the topsoil. These treatments can be effective in augmenting the levels of various forms of organic carbon, which contribute to improving the physicochemical quality of the soil.

4.1.3 Permanent and Variable Charges in Soils After 8 Years of Cultivation with Urochloa Brizantha Cv. BRS Paiaguás and Silicate Agrominerals

The mean values of variable (σH), permanent (σp), and total (σT = σH + σp) electric charges, along with the Cation Exchange Capacity (CEC) in CPAC soil treated with agrominerals under the cultivation of *Urochloa brizantha* cv. BRS Paiaguás, enable an assessment of how various agrominerals and their respective doses influence the cation exchange capacity and nutrient retention dynamics in the soil (Table 7). The comparative analysis between treatments underscores the importance of electric charges in maintaining the fertility of tropical soils and promoting agricultural sustainability.

Selected ASi, after undergoing bioweathering processes, presented secondary minerals with high permanent charge and specific surface area (Santos et al. 2021; Krahl et al. 2022). It is expected that these newly formed secondary minerals will influence the chemical properties of the soil. Anticipated changes in soil, particularly for Ferralsols, include increased stability of organic matter and an increase in the mineral-associated organic carbon fraction in soils treated with silicate agrominerals. Additionally, there may be an increase in permanent charges due to the presence of secondary minerals, as well as variable charges resulting from increased organic matter and short-range order minerals composed of polymerized aluminosilicates, Fe, and Mn.

Table 7 Results of CEC, variable (σH), permanent (σp), and total surface charges (σT)[1], in the soils of the CPAC experiment after 8 years of cultivation of *Urochloa brizantha* cv. BRS Paiaguás with silicate agrominerals[2]

	CEC[3]	σH	σp	σT
——$cmol_c\ kg^{-1}$——				
Control	7.47 ± 0.2	1.07 ± 0.15	0.53 ± 0.07	1.60
KCl 1	5.02 ± 0.1	1.34 ± 0.51	0.44 ± 0.07	1.78
KCl 2	5.54 ± 0.3	1.40 ± 0.50	0.44 ± 0.05	1.84
KCl annual	5.75 ± 0.4	1.36 ± 0.38	0.44 ± 0.05	1.80
BAS 1	7.58 ± 0.6	0.75 ± 0.13	0.56 ± 0.06	1.31
BAS 2	8.55 ± 0.9	0.65 ± 0.17	0.63 ± 0.03	1.28
BSC 1	6.79 ± 0.9	0.79 ± 0.44	0.84 ± 0.07	1.63
BSC 2	6.94 ± 0.3	1.14 ± 0.33	0.85 ± 0.09	1.99
PHON	6.91 ± 0.2	0.50 ± 0.18	0.76 ± 0.10	1.26
VERM + KCl	5.67 ± 0.5	1.30 ± 0.41	0.35 ± 0.04	1.65
BSY 1	6.93 ± 0.3	1.38 ± 0.06	0.85 ± 0.01	2.23
BSY 2	7.01 ± 0.3	1.52 ± 0.83	0.90 ± 0.07	2.42
KAM 1	6.24 ± 0.8	1.73 ± 0.34	0.53 ± 0.09	2.26
KAM 2	7.03 ± 0.4	2.06 ± 0.06	0.46 ± 0.04	2.52
STX-1b[4]	> 50[6]	12.93 ± 0.11	72.43 ± 0.161	85.36
KGA-1b[5]	5.55	1.39 ± 0.0	1.03 ± 0.05	2.44

[1]Permanent and variable charges (Anderson and Sposito 1991)
[2]Average ± standard deviation (n = 9) and for the certified standards (n = 3)
[3]BS EN ISO 11260 (2011)
[4]certified standard Texas Montmorillonite STx-1 (Gonzales County, Texas, USA)
[5]certified standard kaolinite KGA-1b.[6] Value above the detection limit of the method

The ASi had a significant impact on various soil attributes, including pH, organic matter, potential acidity, and cation exchange capacity (Tables 4, 5, 6 and 7). These parameters are crucial for understanding nutrient retention and availability in the soil, particularly in relation to the dynamics of permanent and variable electrical charges. Variable charge is influenced by changes in pH and cation concentration in the soil solution, whereas permanent charge is inherently linked to the mineralogical composition (Alleoni et al. 2009; Manning and Theodoro 2020).

The transformation of vermiculite from biotite weathering, which is found in BSC and BSY, occurs due to the progressive dissolution of biotite under aerobic conditions. This process results in the gradual release of ions such as K^+, Fe^{2+}, Mg^{2+}, and Si^{4+} from the crystalline structure. During this transformation, cation exchange occurs, wherein ions such as H^+ or other hydrated ions replace K^+ in the interlayer, thereby increasing the space between mineral layers (Holgersson et al. 2024). Initially, octahedral ions such as Fe and Mg are released, followed by the substitution of tetrahedral cations like Si and Al, culminating in the formation of vermiculite—a mineral with high surface area, porosity, and CEC (Barshad 1948; Churchman and Lowe 2012). This secondary mineral, a hydrated aluminum silicate, retains part of the laminar structure of biotite, gradually transforming into hydrobiotite and subsequently into vermiculite, which eventually evolves into kaolinite (Coleman et al. 1963; Fordham 1990; Holgersson et al. 2024). The process is directly influenced by pH and soil solution composition, which impacts the generation of permanent and variable charges, thereby influencing the nutrient retention capacity of soils.

Some elements released during this process are essential for physiological functions, such as photosynthesis and protein synthesis, in addition to enhancing plant resistance to environmental stresses (Sposito 2008; Santos et al. 2021). Furthermore, the results indicated that treatments such as KCl 1, 2, annual, and vermiculite + KCl resulted in a significant reduction of up to 32.8% in the CEC in the experiment compared to the control (Table 7). In contrast, ASi demonstrated a lesser decrease, except for BAS 2, which exhibited a 14.5% increase in CEC relative to the control. These findings support the hypothesis that these materials enhance soil fertility by increasing both permanent and variable charges (Krahl et al. 2022).

Direct measurements of variable and permanent charges indicated that the use of KCl increased the variable charge by up to 30.8%. In comparison, the measurement of permanent charge decreased by 17.0% compared to the control (Table 7). Treatments BAS 1, 2, BSC 1, and PHON reduced variable charge by up to 53.3%, while increasing permanent charges by up to 58.5% compared to the control. The most significant increases in variable charge were observed with the application of VERM + KCl, BSY, and KAM, showing an increase from 21.5% to 92.5% compared to the control. KAM releases significant low-range order minerals, leading to variable charge generation. Noteworthy increases in permanent charges were observed with the use of BSC and BSY, resulting in an increase from 58.5 to 69.8% compared to the control.

To assess the potential contribution of these silicate agrominerals in generating permanent charge in soils, it is essential to estimate the amount of charge provided per ton of applied material and its distribution per hectare within the topsoil layer

(0–0.1 m). This calculation requires consideration of: (i) the biotite fraction in the ASi (Table 2); (ii) the weathering reaction stoichiometry, in which three moles of biotite can be converted into two moles of vermiculite; and (iii) the permanent charge of vermiculite, estimated at 150 $cmol_c\ kg^{-1}$ (Acker and Bricker 1992; Barnhisel and Bertsch 1989). Based on these premises, a quantitative assessment of permanent charge generation in a field experiment revealed values of 2414 $cmol_c\ t^{-1}$ for BSC and 0.0027 $cmol_c\ t^{-1}$ BSC ha^{-1}, while for BSY, it exhibited slightly higher values, reaching 2791 $cmol_c\ t^{-1}$ BSY and 0.0032 $cmol_c\ t^{-1}$ BSY ha^{-1}. The total potential theoretical permanent charge, assuming all biotite transforms into vermiculite, is 13025 $cmol_c\ t^{-1}$ of BSC and 0.015 $cmol_c\ t^{-1}$ of BSC ha^{-1}. For BSY, it is 17910 $cmol_c\ t^{-1}$ of BSY and 0.020 $cmol_c\ t^{-1}$ of BSY ha^{-1}. Permanent charges were formed at estimated rates of 18.5% and 15.6%, respectively, for BSC and BSY. This process highlights the relevance of ASi mineralogy in modifying the soil's electrochemical properties, particularly in relation to CEC. Given biotite's role in K availability for agricultural systems, it is recommended that fertilization strategies consider exclusively the fraction of K derived from this mineral in the BSC and BSY. This is because other mineral constituents in these materials do not release K as efficiently within timeframes compatible with agricultural management.

The results indicate that sure ASi, particularly BSC, BSY, for permanent charges, and KAM for variable charges, may be effective in enhancing the fertility of Ferralsols. The use of these materials provides essential nutrients over time, improving properties such as CEC, pH, and nutrient retention capacity. These factors are crucial for promoting sustainable and efficient agriculture, potentially reducing reliance on chemical inputs and minimizing the environmental impacts associated with fertilizer use (Freire et al. 2023; Marchi et al. 2024). Additionally, incorporating agrominerals into soil management practices may positively influence agricultural activities in tropical regions by promoting soil conservation and resilience to climate change (Medeiros et al. 2023), contributing to regenerative agriculture.

5 Conclusions

- Silicate agrominerals like biotite schist, syenite, basalt, phonolite, kamafugite, and vermiculite contribute to improving soil attributes in tropical savannah-like regions such as the Brazilian Cerrado.
- The application of silicate agrominerals promotes the formation of new mineral phases, such as vermiculite, and short-range order mineral forms, which have high surface area and porosity.
- These silicate agrominerals increase the soil's CEC by generating both permanent and variable charges, which help in nutrient retention and reduce leaching.
- Silicate agrominerals aid in the stability of organic matter and carbon storage, helping to stabilize organic carbon in the soil and potentially contribute to mitigating climate change.

- Experimental results indicate that treatments with higher doses of silicate agrominerals, such as biotite syenite and kamafugite, show significant improvements in soil fertility and productivity compared to control treatments.
- Further research is needed to fully understand the long-term interactions between silicate agrominerals, soil organic matter, and agricultural management practices for optimal use in various edaphoclimatic conditions.

Giuliano Marchi led the writing of the chapter and was primarily responsible for its conception and data interpretation. João Paulo Guimarães Soares and Éder de Souza Martins contributed to the experimental design, data interpretation, and writing. Luise Lottici Krahl was responsible for methodology, data interpretation, and contributed to the writing. Arminda Moreira de Carvalho contributed to the writing and critically reviewed the final version of the chapter. Luiz Fernando dos Santos and Priscila Brelaz da Silva conducted the laboratory analyses, contributed to data interpretation, and participated in writing. Elis Marina de Freitas performed the statistical analyses and contributed to the writing.

Acknowledgements The authors gratefully acknowledge the financial support provided by Embrapa and by FINEP—Fundo Setorial Mineral through the project 'Rede de PD&I sobre remineralizadores e agrominerais (REMAg)' (Project No. 1541/2022), which made possible the development of this book chapter.

AuthorContribution Author's Contributions Giuliano Marchi led the writing of the chapter and was primarily responsible for its conception and data interpretation. João Paulo Guimarães Soares and Éder de Souza Martins contributed to the experimental design, data interpretation, and writing. Luise Lottici Krahl was responsible for methodology, data interpretation, and contributed to the writing. Arminda Moreira de Carvalho contributed to the writing and critically reviewed the final version of the chapter. Luiz Fernando dos Santos and Priscila Brelaz da Silva conducted the laboratory analyses, contributed to data interpretation, and participated in writing. Elis Marina de Freitas performed the statistical analyses and contributed to the writing.

Declaration of Competing Interest The authors declare that they have no known competing financial interests or personal relationships that could have appeared to influence the work reported in this paper.

References

Acker JG, Bricker OP (1992) The influence of pH on biotite dissolution and alteration kinetics at low temperature. Geochim Cosmochim Acta 56(8):3073–3092

Alleoni LRF, Mello JWV, Rocha WSD (2009) Eletroquímica, adsorção e tro-ca iônica no solo. In: Alleoni LRF, Melo VF (eds) Química e Mineralogia do Solo—parte II. Viçosa-MG: Sociedade Brasileira de Ciência do Solo, pp 69–130

Alvares CA, Stape JL, Sentelhas PC, Gonçalves JLM, Sparovek G (2013) Köppen's climate classification map for Brazil. Meteorol Z 22(6)

Anda M, Shamshuddin J, Fauziah CI (2015) Improving chemical properties of a highly weathered soil using finely ground basalt rocks. CATENA 124:147–161

Anderson SJ, Sposito G (1991) Cesium-adsorption method for measuring accessible structural surface charge. Soil Sci Soc Am J 55(6):1569–1576

Angst G, Messinger J, Greiner M, Häusler W, Hertel D, Kirfel K, Kögel-Knabnera I, Leuschner C, Rethemeyer J, Mueller CW (2018) Soil organic carbon stocks in topsoil and subsoil controlled by parent material, carbon input in the rhizosphere, and microbial-derived compounds. Soil Biol Biochem 122:19–30. https://doi.org/10.1016/j.soilbio.2018.03.026

Barshad I (1948) Vermiculite and its relation to biotite as revealed by base exchange reactions, X-ray analyses, differential thermal curves, and water content. Am Mineral: J Earth Planet Mater 33(11–12):655–678

Basak BB, Sarkar B, Sanderson P, Naidu R (2018) Waste mineral powder supplies plant available potassium: evaluation of chemical and biological interventions. J Geochem Explor 186:114–120. https://doi.org/10.1016/j.gexplo.2017.11.023

BS EN ISO 11260 (2011) Soil Quality: determination of Effective Cation Exchange Ca-pacity and Base Saturation Level using Barium Chloride Solution. B. Standards

Barnhisel RI, Bertsch PM (1989) Chlorites and hydroxy-interlayered vermiculite and smectite. Miner Soil Environ 1:729–788

Beerling DJ, Leake JR, Long SP, Scholes JD, Ton J, Nelson PN, Bird M, Kantzas E, Taylor LL, Sarkar B, Kelland M, DeLucia E, Kantola I, Müller C, Rau G, Hansen J (2018) Farming with crops and rocks to address global climate, food and soil security. Nature Plants. 4:138–147. https://doi.org/10.1038/s41477-018-0108-y

Beck HE, Zimmermann NE, McVicar TR, Vergopolan N, Berg A, Wood EF (2018) Present and future. Köppen-Geiger climate classification maps at 1-km resolution. Sci Data 5

Benites VM, Madari B, Machado PLOA (2003) Comunicado técnico nº 16: Extração e fracionamento quantitativo de substâncias húmicas do solo – um procedimento simplificado de baixo custo. Embrapa, Brasília. https://www.embrapa.br/busca-de-publicacoes/-/publicacao/337076/extracao-e-fracionamento-quantitativo-de-substancias-humicas-do-solo-um-procedimento-simplificado-de-baixo-custo

Brazil. Ministério da Agricultura, Pecuária e Abastecimento. (2016). Instrução Normativa nº 05, de 10 de março de 2016: Estabelece critérios para definição, classificação, registro e uso de remineralizadores e outros insumos minerais para uso agrícola. Diário Oficial da União.

Burghelea C, Zaharescu DG, Dontsova K, Maier R, Huxman T, Chorover J (2015) Mineral nutrient mobilization by plants from rock: influence of rock type and ar-buscular mycorrhiza. Biogeochemistry 124:187–203

Busato JG, Leão TP, Baldotto MA, Canellas LP (2012) Organic matter quality and dynamics in tropical soils amended with sugar industry residue. Revista Brasilei-Ra De Ciência Do Solo 36:1179–1188. https://doi.org/10.1590/S0100-06832012000400012

Buss W, Hasemer H, Ferguson S, Borevitz J (2024) Stabilisation of soil organic matter with rock dust partially counteracted by plants. Glob Change Biol 30(1):e17052. https://doi.org/10.1111/gcb.17052

Cambardella CA, Elliott ET (1992) Particulate soil organic-matter changes across a grassland cultivation sequence. Soil Sci Soc Am J 56:777–783

Carlos SM, Assad ED, Estevam CG, Lima CZ, Pavão EM, Pinto TP (2022) Custos da recuperação de pastagens degradadas nos estados e biomas brasileiros. Observatório de Conhecimento e Inovação em Bioeconomia, Fundação Getulio Vargas – FGV-EESP. https://eesp.fgv.br/centros/observatorios/bioeconomia

Cheng K, Heidari Z (2018) A new method for quantifying cation exchange capacity in clay minerals. Appl Clay Sci 161(1):444–455

Chevallier T, Fujisaki K, Roupsard O, Guidat F, Kinoshita R, de Melo Viginio Filho E, Lehner P, Albrecht AA (2019) Short-range-order minerals as powerful factors explaining deep soil organic carbon stock distribution: the case of a coffee agroforestry plantation on Andosols in Costa Rica. Soil 5(2):315–332

Churchman GJ, Lowe DJ (2012) Alteration, formation, and occurrence of minerals in soils. In: Huang PM, Li Y, Sumner ME (eds) Handbook of soil sciences, vol 1. CRC Press, Boca Raton, pp 20.1–20.72. https://doi.org/10.1201/b11267

Coleman N, Leroux F, Cady J (1963) Biotite-hydrobiotite-vermiculite in soils. Nature 198:409–410

Colombo C, Palumbo G, He J-Z, Pinton R, Cesco S (2014) Review on iron availability in soil: interaction of Fe minerals, plants, and microbes. J Soils Sed 14:538–548. https://doi.org/10.1007/s11368-013-0814-z

Cunha JC, Ruiz HA, Freire MBG, Alvarez VH, Fernandez RB (2014) Quantification of permanent and variable charges in reference soils of the state of Pernambuco, Brazil. Rev Bras Ciênc Solo 38:1162–1169. https://doi.org/10.1590/S0100-06832014000400012

Donagema GK, Campos DVBD, Calderano SB, Teixeira WG, Viana JHM (2011) Manual de métodos de análise de solo. 2.ed. rev. Rio de Janeiro: Embrapa Solos, p 230 (Embrapa Solos. Documentos, 132)

Edwards DP, Lim F, James RH, Pearce CR, Scholes J, Freckleton RP, Beerling DJ (2017) Climate change mitigation: potential benefits and pitfalls of enhanced rock weathering in tropical agriculture. Biol Lett 13:1–7. https://doi.org/10.1098/rsbl.2016.0715

Embrapa (2017) Manual de métodos de análise de solo. Rio de Janeiro, RJ: Embrapa Solos

Fageria NK, Baligar VC, Wright RJ (1990) Nutrição de ferro das plantas: química e fisiologia da sua deficiência e toxicidade. Pesquisa Agropecuária Brasileira, 553–570

Ferreira BC, Gualberto CAC, Faroutine G, Pereira HS, Korndörfer GH, Eduardo Spolidorio E, Clerot, L.C (2020) Kamafugite from the Alto Paranaíba province as a multi-nutrient fertilizer. J Agric Sci 12(10):213

Fontes MPF, Camargo OA, Sposito G (2001) Eletroquímica das partículas co-loidais e sua relação com a mineralogia de solos altamente intemperizados. Scientia Agricola 58:627–646. https://doi.org/10.1590/S0103-90162001000300029

Freire YBD, Fonseca IS, Santos RA, Valicheski RR, de Oliveira SSC, Stürmer SLK (2023) Uso do pó de rocha nefelina-sienito como fonte alternativa de potássio no cultivo da soja. Revista Em Agronegócio e Meio Ambiente 16(4):1–19. https://doi.org/10.17765/2176-9168.2023v16n4e11565

Fordham AW (1990) Weathering of biotite into dioctahedral clay minerals. Clay Miner 25(1):51–63. https://doi.org/10.1180/claymin.1990.025.1.06

Fundação Getúlio Vargas- FGV (2021) Potencial de mitigação de gases de efeito estufa das ações de descarbonização da pecuária até 2030. Escola de economia de São Paulo, São Paulo-SP, 2012. Portaria no 52, de 23 de março de 2021, Brasília, DF. https://eesp.fgv.br/sites/eesp.fgv.br/files/ocbio_potencial_de_mitigacao_de_gee_pecuaria_2112.pdf

Gilkes RJ, McKenzie RM (1988) Geochemistry of manganese in soil. In: Graham RD, Hannan RJ, Uren NC (eds) Manganese in soils and plants. Kluwer Academic Publishers, Dordrecht, pp 23–34. https://doi.org/10.1007/978-94-009-2817-6

Gilkes RJ, Young RC, Quirk JP (1973) Artificial weathering of oxidized biotite: I. Potassium removal by sodium chloride and sodium tetraphenylboron solutions. Soil Sci Soc Am Proc 37:25–28. https://doi.org/10.2136/sssaj1973.03615995003700010013x

Guo F, Sun H, Yang J, Zhang L, Mu Y, Wang Y, Wu F (2023) Improving food security and farmland carbon sequestration in China through enhanced rock weathering: field evidence and potential assessment in different humid regions. Sci Total Environ 903:166118. https://doi.org/10.1016/j.scitotenv.2023.166118

Guo X, Drury CF, Yang X, Reynolds WD, Zhang R (2012) Impacts of wet–dry cycles and a range of constant water contents on carbon mineralization in soils under three cropping treatments. Soil Sci Soc of Am J, 76(2):485–493

Harley AD, Gilkes RJ (2000) Factors influencing the release of plant nutrient elements from silicate rock powders: a geochemical overview. Nutr Cycl Agroecosyst 56:11–36

Hartmann J, West AJ, Renforth P, Köhler P, De La Rocha CL, Wolf-Gladrow DA, Dürr HH, Scheffran J (2013) Enhanced chemical weathering as a geoengineering strategy to reduce atmospheric carbon dioxide, supply nutrients, and mitigate ocean acidification. Rev Geophys 51:113–149. https://doi.org/10.1002/rog.20004

Hinsinger P, Barros ONF, Benedetti MF, Noack Y, Callot G (2001) Plant-induced weathering of a basaltic rock: experimental evidence. Geochim Cosmochim Acta 65(1):137–152

Hinsinger P, Elsass F, Jaillard B, Robert M (1993) Root induced irreversible transformation of a trioctahedral mica in the rhizosphere of rape. J Soil Sci 44(3):535–545

Holgersson S, Drake H, Karlsson A, Krall L (2024) Biotite dissolution kinetics at pH 4 and 6.5 under anaerobic conditions and the release of dissolved Fe (II). Chem Geol 122204. https://doi.org/10.1016/j.chemgeo.2024.122204

IUSS Working Group WRB (2022) World reference base for soil resources. In: International soil classification system for naming soils and creating legends for soil maps, 4th edn. International Union of Soil Sciences (IUSS), Vienna, Austria.

Jones DL (1998) Organic acids in the rhizosphere – a critical review. Plant Soil 205:25–44. https://doi.org/10.1023/A:1004356007312

Kelland ME, Wade PW, Lewis AL, Taylor LL, Sarkar B, Andrews MG, Pearce CR (2020) Increased yield and CO_2 sequestration potential with the C_4 cereal Sorghum bicolor cultivated in basaltic rock dust-amended agricultural soil. Glob Chang Biol 26:3658–3676

Kleber M, Eusterhues K, Keiluweit M, Mikutta C, Mikutta R, Nico PS (2015) Mineral-Organic associations: formation, properties, and relevance in soil environments. Adv Agron 130:1–140. https://doi.org/10.1016/bs.agron.2014.10.005

Kong M, Huang L, Li L, Zhang Z, Zheng S, Wang MK (2014) Effects of oxalic and citric acids on three clay minerals after incubation. Appl Clay Sci 99:207–214

Krahl LL (2020) Mineral formation and element release from aluminosilicate rocks promoted by maize rhizosphere. Universidade de Brasília, Tese de Doutorado

Krahl LL, Marchi G, Paz SPA, Angélica RS, Sousa-Silva JC, Valadares LF, Martins ES (2022) Increase in cation exchange capacity by the action of maize rhizosphere on Mg or Fe biotite-rich rocks. Pesqui Agropecuária Trop 52. https://doi.org/10.1590/1983-40632022v5272376

Krahl LL, Paz SPA, Angélica RS, Valadares LF, Souza-Silva JC, Marchi G, Martins EDS (2020) Successive offtake of elements by maize grown in pure basalt powder. Afr J Agric Res V15(2):229–239

Lefebvre D, Goglio P, Williams A, Manning DAC, Carlos de Azevedo A, Bergmann M, Meersmans J, Smith P (2019) Assessing the potential of soil carbona-tion and enhanced weathering through Life Cycle Assessment: A case study for Sao Paulo State. Brazil. J. Clean Prod. 233:468–481

Lin CW, Tian G, Pai CW, Chiu CY (2018) Characterization of phosphorus in subtropical coastal sand dune forest soils. Forests, 9(11):710

Li T, Wang H, Zhou Z, Chen X, Zhou J (2015) A nano-scale study of the mechanisms of non-exchangeable potassium release from micas. Appl Clay Sci, 118:131–137.https://doi.org/10.1016/j.clay.2015.09.013

Li Y, Fu C, Ye C, Song Z, Kuzyakov Y, Vancov T, Guo L, Luo Z, Van Zwieten L, Wang Y, Luo Y, Wang W, Zeng L, Han G, Wang H, Luo Y (2025) In-creased mineral-associated organic carbon and persistent molecules in allochthonous Blue Carbon Ecosystems. Glob Change Biol. https://doi.org/10.1111/gcb.70019

Luchese AV, Pivetta LA, Batista MA, Steiner F, Giaretta APS, Curtis JCD (2021) Agronomic feasibility of using basalt powder as soil nutrient remineralizer. Afr J Agric Res 17(3):487–497

Magalhães LA, Martins ÉDS, Bergmann M, Blaskowski AE, Sander A, Fioravanti TN (2024) Mapas prospectivos de ocorrência potencial de agrominerais silicáticos no Brasil

Manning DAC, Baptista J, Limon MS, Brandt K (2017) Testing the ability of plants to access potassium from framework silicate minerals. Sci Total Environ 574:476–481

Manning DA, Theodoro SH (2020) Enabling food security through use of local rocks and minerals. Extr Ind Soc 7(2):480–487. https://doi.org/10.1016/j.exis.2018.11.002

Mao X, Van Zwieten L, Zhang M, Qiu Z, Yao Y, Wang H (2020) Soil parent material controls organic matter stocks and retention patterns in subtropical China. J Soils Sediments 20:2426–2438

Martins ES, Martins ES, Hardoim PR (2023) Princípios geoquímicos, mineralógicos e biológicos do manejo de remineralizadores de solos. Inf Agropecuário 44(321):26–39

Martins ÉDS, Oliveira CGD, Resende ÁVD, Matos MSFD (2008) Agrominerais-Rochas silicáticas como fontes minerais alternativas de potássio para a agricultura. In: Agrominerais para o Brasil. Brasília: CETEM/MCT, 2014. p. 109–154

Martins ES, Theodoro SH, Bernardez FFG, Luchese AV, Bergmann M, Siqueira DS, Teixeira A, Azevedo A, Curtis JCD (2024) Produção Brasileira de Remineralizadores e Fertilizantes Naturais: 2019 a 2023. Revista Novo Solo, 11/19

Marchi G, Guelfi-Silva DR, Malaquias JV, Guilherme LRG, Spehar CR, Martins ES (2020) Solubility and availability of micronutrients extracted from silicate agrominerals. Pesquisa Agropecuária Brasileira 55:e00807. Thematic Issue—Regional Agrominer-Als

Marchi G, Silva JCS, Soares JPG, de Abrantes EG, de Souza Martins É (2024) Growth of *Urochloa* grass in an Oxisol treated with powdered silicate materials. Peer Rev 6(2):248–272

Medeiros FDP, Carvalho AMD, Gindri Ramos C, Dotto GL, Cardoso IM, Theodoro SH (2023) Rock powder enhances soil nutrition and coffee quality in agroforestry systems. Sustainability 16(1):354. https://doi.org/10.3390/su16010354

Mohammed SMO, Brandt K, Gray ND, White ML, Manning DAC (2014) Comparison of silicate minerals as sources of potassium for plant nutrition in sandy soil. Eur J Soil Sci 65(5):653–662

Murakami T, Ito J, Utsunomiya S, Kasama T, Kozai N, Ohnuki T (2004) Anoxic dissolution processes of biotite: implications for Fe behavior during Archean weathering. Earth Planet Sci Lett 224:117–129. https://doi.org/10.1016/j.epsl.2004.04.040

Naderizadeh Z, Khademi H, Arocena JM (2010) Organic matter induced mineralogical changes in clay-sized phlogopite and muscovite in alfalfa rhizosphere. Geoderma 159(3–4):296–303. https://doi.org/10.1016/j.geoderma.2010.08.003

Nascimento PC, Bayer C, de Silva Netto L, Vian AC, Vieiro F, Macedo VR, Marcolin É (2010) Sistemas de Manejo e a matéria Orgânica de solo de Várzea com cultivo de Arroz. Rev Bras Ciênc Solo 33(6):1821–1827. https://doi.org/10.1590/s0100-06832009000600030

Nascimento M, Lapido-Loureiro FE (2004) Fertilizantes e sustentabilidade: o potássio na agricultura brasileira. CETEM/MCT, Rio de Janeiro, p 59 (Série estudos e documentos, 61)

Norouzi S, Khademi H (2010) Ability of alfalfa (Medicago sativa L.) to take up potassium from different micaceous minerals and consequent vermiculitization. Plant Soil 328:83–93

Nunes J, Kautzmann R, Oliveira C (2014) Evaluation of the natural fertilizing potential of basalt dust wastes from the mining district of Nova Prata (Brazil). J Clean Prod 84:649–656

Ramos CG, Querol X, Oliveira MLS, Silva LF.O, Schneider IAH (2017) Evaluation of the potential of volcanic rock waste from southern Brazil as a natural soil fertilizer. J Clean Prod, 142:2700–2706

Ramos CG, Hower JC, Blanco E, Oliveira MLS, Theodoro SH (2022) Possibilities of using silicate rock powder: an overview. Geosci Front 13(1):101185. https://doi.org/10.1016/j.gsf.2021.101185

Reatto A, Martins ES, Farias MFR, Silva AV, Carvalho Júnior OA (2004) Mapa pedológico digital—SIG atualizado do Distrito Federal escala 1:100.000 e uma síntese do texto explicativo. Brasília, DF: Embrapa Cerrados

Ribeiro LS, Santos AR, Souza LFS, Souza JS (2010) Rochas silicáticas portadoras de potássio como fontes do nutriente para as plantas solo. Rev Bras Ciênc Solo 34(3):891–897

Rodrigues M, Silva Junges LF, Mozorovicz C, Ziemmer GS, Neto CK, An-drade EA, Melo Teixeira L (2024) Paraná basin basalt powder: a multinutrient soil amendment for enhancing soil chemistry and microbiology. J S Amn Earth Sci 104957. https://doi.org/10.1016/j.jsames.2024.104957

Rhoton FE, Bigham JM (2005) Phosphate adsorption by ferrihydrite-amended soils. J Envion Qual, 34(3):890–896.

Salman AKD, Soares JPG, Canesin RC (2006) Métodos de amostragem para avaliação quantitativa de pastagens. Circular Técnica, Porto Velho, p 84

Sano SM, Almeida SP, Ribeiro JF (eds) 2008 Cerrado: ecologia e flora. Embrapa Informação Tecnológica, DF, Brasília, vol 2, p 1.279

Santos HG, Jacomine PKT, Anjos LHC, Oliveira VA, Lumbreras JF, Coelho MR, Almeida JA, Araújo Filho JC, Oliveira JB, Cunha TJF (2018) Sistema Brasileiro de Classificação de Solos. 5. ed., rev. e ampl. Embrapa, Brasília, p 356

Santos LFD, Sodré FF, Martins ÉDS, Figueiredo CCD, Busato JG (2021) Effects of biotite syenite on the nutrient levels and electrical charges in a Brazilian Savanna Ferralsol. Pesqui Agropecuária Trop 51. https://doi.org/10.1590/1983-40632021v5166691.

Sarkar B, Singh M, Mandal S, Churchman GJ, Bolan NS (2018) Clay minerals—organic matter interactions in relation to carbon stabilization in soils. In: Garcia C, Nannipieri P, Hernandez T (eds) The future of soil carbon. Academic Press, Elservier, London, pp 71–86. https://doi.org/10.1016/B978-0-12-811687-6.00003-1

Singh M, Sarkar B, Biswas B, Bolan NS, Churchman GJ (2017) Relationship between soil clay mineralogy and carbon protection capacity as influenced by temperature and moisture. Soil Biol Biochem 109:95–106. https://doi.org/10.1016/j.soilbio.2017.02.003

Singh VK, Singh AL, Singh R, Kumar A (2018) Iron oxidizing bacteria: insights on diversity, mechanism of iron oxidation and role in management of metal pollution. Environmental Sustainability, 1(3):221–231

Soratto RP, Crusciol AAC, Campos M, Costa CHM, Gilabel AP, Castro GSA, Neto JF (2021) Silicate rocks as an alternative potassium fertilizer for upland rice and common bean crops. Pesq Agropec Bras 56:e01411

Sparks DL, Singh S, Siebecker MG (2024) An introduction to environmental soil chemistry. In: Environmental soil chemistry, 3rd edn., pp 1–38

Sposito, G. (2008). The chemistry of soils, 2nd edn. Oxford University Press, Oxford, p 277

Swoboda P, Döring TF, Hamera M (2022) Remineralizing soils? The agricultural usage of silicate rock powders: a review. Sci Total Environ 807:150976

Theodoro SH, Medeiros FP, Ianniruberto M, Jacobson TKB (2021) Soil remineralization and recovery of degraded areas: an experience in the tropical region. J S Amn Earth Sci 107

Uroz S, Kelly LC, Turpault M-P, Lepleux C, Frey-Klett P (2015) The mineralo-sphere concept: mineralogical control of the distribution and function of mineral-associated bacterial communities. Trends Microbiol 23:751–762

Walkley A, Black IA (1934) An examination of the Degtjareff method for determining soil organic matter, and a proposed modification of the chromic acid titration method. Soil Sci 37:29–38

Wang X, Yoo K, Wackett AA, Gutknecht J, Amundson R, Heimsath A (2018) Soil organic carbon and mineral interactions on climatically different hillslopes. Geoderma 322:71–80. https://doi.org/10.1016/j.geoderma.2018.02.021

Weber OLS, Chitolina JC, Camargo OA, Alleoni LRF (2005) Cargas elétri-cas estruturais e variáveis de Solos tropicais altamente intemperizados. Revista Brasilei-Ra De Ciência Do Solo 29:867–873. https://doi.org/10.1590/S0100-06832005000600004

White RA, Rivas-Ubach A, Borkum MI, Koberl M, Bilbao A, Colby SM, Hoyt DW, Bingol K, Kim Y-M, Wendler JP, Hixson KK, Jansson C (2017) The state of rhizospheric science in the era of multi-omics: a practical guide to omics technologies. Rhizosphere 3:212–221. https://doi.org/10.1016/j.rhisph.2017.05.003

Wu S, Liu Y, Southam G, Robertson LM, Wykes J, Yi Q, Hall M, Li Z, Sun Q, Saha N, Chan T, Lu Y, Huang L (2021) Rhizosphere drives biotite-like mineral weathering and secondary Fe–Si mineral formation in Fe ore tailings. Earth Space Chem. 5:618–631. https://doi.org/10.1021/acsearthspacechem.0c00331

Yeomans JC, Bremmer JM (1988) A rapid and precise method for routine determination of organic carbon in soil. Commun Soil Sci Plant Anal 19:1467–1476

Yu G (2018) Root exudates and microbial communities drive mineral dissolution and the formation of nano-size minerals in soils: implications for soil carbon storage. In: Giri B, Prasad RAV (eds) Root Biology, vol 52. Springer International Publishing, pp 143–166. https://doi.org/10.1007/978-3-319-75910-4_5

Zhang S, Bai X, Zhao C, Tan Q, Luo G, Wang J, Li Q, Wu L, Chen F, Li C, Deng Y, Yang Y, Xi H (2021) Global CO_2 consumption by silicate rock chemical weathering: its past and future. Earth's Futur 9(5)

Potential of Soil Remineralization to Carbon Dioxide Removal in Agricultural Soils

Antonio Carlos de Azevedo

Abstract This chapter explores the potential of Enhanced Rock Weathering (ERW) as a mechanism for Carbon Dioxide Removal (CDR) in agricultural soils. It highlights three primary pathways for CDR: (A) the leaching of bicarbonate ions into oceans, (B) the formation of Mineral-Associated Organic Matter (MAOM), and (C) the precipitation of carbonate minerals. While bicarbonate leaching and MAOM synthesis show promise, carbonate precipitation is less feasible in tropical agricultural soils due to unfavorable conditions. In bicarbonate leaching, CO_2 reacts with water and silicate minerals to form bicarbonate ions, which are transported to the oceans and stored as carbonates for millennia. Increasing soil pH and using remineralizers rich in basic cations (e.g., basalt) can enhance this process. In the MAOM formation, organic matter binds to fresh mineral surfaces and amorphous phases neoformed due to the weathering of rock powder, promoting MAOM synthesis and stabilizing carbon, thereby extending its residence time in soils. In carbonate precipitation, which is limited in tropical soils due to low pH, this pathway becomes less viable for CDR in agricultural settings. The challenges of measuring CDR rates in soils remain complex due to variability in soil conditions, biological activity, and geochemical interactions. Current efforts focus on developing robust measurement, reporting, and verification (MRV) protocols to certify carbon credits from bicarbonate leaching. Therefore, soil remineralization offers significant potential for climate-smart agriculture by enhancing CDR mechanisms, particularly bicarbonate leaching and MAOM synthesis. However, further interdisciplinary research is needed to address challenges in measurement and implementation, especially in tropical environments.

Keywords Carbon capture · Carbonates · Inorganic soil carbon

A. C. de Azevedo (✉)
Luiz de Queiroz College of Agriculture, University of São Paulo, Piracicaba, Brazil
e-mail: aazevedo@usp.br

É. de Souza Martins and S. Huff Theodoro (eds.), *Soil Remineralizers and Silicate Fertilizers*, Frontier Studies in Soil Science,
https://doi.org/10.1007/978-3-032-14656-4_5

1 Overview of Carbon in Soils

Soil carbon content is almost 3 times greater than the carbon content in the atmosphere (2500 Pg versus 880 Pg of C) (Lal et al. 2021). As a result, there is an extreme interest in studying the mechanisms of carbon flow and stock in this reservoir to facilitate the development of technologies that increase its carbon content. Due to its potential to achieve global scale, agricultural management and other climate-smart agriculture technologies are of particular interest. Climate smart agriculture takes in account the climate changes outcomes and its impacts in agriculture by considering three foundational concepts: the contribution to sustainable and equitable increase in agricultural productivity and incomes; the increase in adaptation and resilience of food systems from the farm to climate change; and the reduction, or removal, of greenhouse gas emissions, where and when possible (De Pinto et al. 2020).

Atmospheric carbon can be stored in soil in two pools: the organic (Soil Organic Carbon—SOC) and the inorganic (Soil Inorganic Carbon—SIC) (Manning et al. 2024). Soil organic carbon (SOC) is the carbon fraction of the Soil Organic Matter (SOM). The SOM is comprised of biological cells, isolated or in tissues, at different stages of decomposition, as well as their previous metabolic products, that remain in the soil in varying states of decomposition. In addition to Carbon (C), SOM contains elements such as nitrogen (N) and sulfur (S). The SOM is quite complex and has been divided into several fractions, mostly operationally defined by the method used to obtain them. Recently, Lavallee et al. (2020) suggested to split SOM into Particulate Organic Matter (POM) and Mineral-Associated Organic Matter (MAOM) for simplicity and standardization, allowing comparison across different studies.

The POM is the lightweight, "isolated" fraction of SOM, while MAOM refers to the C-dominant molecules or small organic particles that are bonded and protected by the mineral soil fraction, and therefore, are denser (Lavallee et al. 2020). Although operationally defined as a function of size or density for practical purposes, it is likely that there is an overlap between POM and MAOM fractions (Yu et al. 2022).

The Soil Inorganic Carbon (SIC) is the inorganic C pool in soils. Globally, it is composed mainly of carbonates, with a residence time in the order of 10^3 to 10^6 years, while for SOC it corresponds to 10 to 10^4 years (Zamanian et al 2021). The SIC pool connects carbon from the fast biotic cycle to the slow geological cycle. Precipitation of pedogenic carbonates occurs under very particular conditions, which are dominant in semi-arid and arid soils. However, there exist precipitation-promoting microorganisms that can widen the range of precipitation conditions significantly (Ferris et al. 2009; Martinez et al. 2013).

In contrast to the pedogenic carbonates, carbonic acid is the main acid acting in the global terrestrial weathering; therefore, bicarbonate and carbonate anions are present in the soil solution of all soils.

2 Mechanisms of Carbon Dioxide Removal (CDR) in Soils Using Remineralizers

2.1 *The Earth Thermostat*

Weathering of silicate rocks is the primary mechanism of atmospheric CO_2 buffering on Earth, at the geological timescale (Penman et al. 2020). This buffering occurs because CO_2 is simultaneously a greenhouse gas and the source of carbonic acid to Earth's surface waters. Carbonic acid is the main acid in the weathering of the rocks exposed to Earth's surface.

Simply put, the increase in CO_2 intensifies the greenhouse effect, which in turn increases the temperature and accelerates rock weathering. This, in turn, increases the consumption of CO_2, leading to a decrease in the greenhouse effect and consequently decelerating the weathering process. This negative feedback mechanism is known as the Earth's thermostat (Voosen 2019; Brantley et al. 2023; Hilton 2023).

Earth's thermostat is, in fact, more complicated than that. The feedback mechanism is also related to a set of intertwined factors in addition to temperature, such as rain distribution, plant transpiration, and erosion intensity, that are difficult to evaluate independently (Hilton 2023). For example, the increase in CO_2 induces the plant stomata to shut, reducing transpiration. If local precipitation remains the same, soil moisture and leaching should increase. In this case, the flow of bicarbonate into the groundwater would increase, contributing to the negative feedback mechanism of the Earth's thermostat (Eq. 1).

$$3H_2O + 2CO_2 + CaSiO_3 \leftrightarrow Ca^{+2} + 2HCO^{3-} + H_4SiO_4 \quad (1)$$

The geological record shows that, for example, volcanic eruptions that significantly increased atmospheric CO_2 levels were subsequently followed by a rise in the sediment load transported to the oceans by rivers draining the continents. This change was driven by increased erosion and rock weathering processes (Brantley et al. 2023).

The products of the weathering reaction (Eq. 1) can be leached from soils by the water surplus at least some time of the year, except in areas where evapotranspiration is greater than precipitation, yielding no water surplus to be leached. Due to the small kinetic constant of most weathering reactions (not to mention incongruent dissolution), the Earth's thermostat works on a geological timescale. The rapid global increase in CO_2, caused by the mining and burning of fossil carbon deposits (coal and oil) and changes in land use (including the expansion of agriculture), took only two centuries to increase the concentration of CO_2 from 0.03 to 0.04% in the Earth's atmosphere (Calvin et al. 2023).

Three main mechanisms of the Earth's thermostat are of interest for Carbon Dioxide Removal (CDR) using remineralizers. They are:

(A) The leaching of bicarbonates to the oceans.

(B) The formation of Mineral Associated Organic Matter (MAOM).
(C) The precipitation of carbonate phases in soil (pedogenic carbonates).

As will be discussed, A and B can be increased by speeding up the weathering engine of the Earth's thermostat. The increase in the C path is more challenging to achieve in agricultural soils, as they are more prone to man-made soils (Technosols).

To boost the Earth's thermostat, the rate of mechanisms A to C can be increased by:

- Increasing the rate of weathering reactions by increasing the Specific Surface Area (SSA) of minerals.
- Exposing fresh mineral surfaces to the higher partial pressure of CO_2 in the soil pores (the soil atmosphere), which can be from 5 to 50 times greater than that of the atmosphere (Flechard et al. 2007).
- Increasing bioweathering by exposing fresh mineral surfaces to the action of the soil microbiome, particularly active in the soil surficial horizons.

Enhanced Rock Weathering (ERW) aims to achieve these goals. Igneous basic rocks are preferred because of their large cation content, quick weathering (as compared to most of the other silicate rocks), the abundance of basic igneous provinces, and their location close to important agricultural areas. In addition, these rocks are commonly comminuted to be used as construction material (construction aggregates), which inevitably produces about 15% of rock dust, in some cases up to 30% (Mitchell et al. 2008). Frequently, the rock dust piles up in the quarry's backyard, potentially causing environmental and logistical issues.

Most of the quarry fines heaps, resulting from the comminution of igneous basic rocks, fulfill the requirements to be registered in the Brazilian Ministry of Agriculture as remineralizers. In such cases, carbon dioxide removal can be considered as an additional function of them. It is worth highlighting, however, that issuing carbon credits to remineralizers is not covered by Brazilian regulations. It is a separate process, involving measurements, reporting, and verification protocols (MRV protocols) that are still in development (Bellassen et al. 2015).

(A) **Leaching of Bicarbonates**

This is the only CDR mechanism considered so far in ERW as a source of certified carbon credits.

As previously discussed, in this pathway, CO_2 from the gas phase (atmosphere and/or soil-atmosphere) dissolves in water, producing carbonic acid (H_2CO_3). This acid reacts with silicate minerals in rocks and produces bicarbonate anion in the soil solution. If the soil water balance allows for water surplus events throughout the year, the excess water leaches towards the groundwater reservoir, taking the bicarbonate ion with it. From there it goes to rivers and finally to the oceans, where it precipitates as carbonate and is stored for thousands to millions of years (Berner et al. 1983). During the transfer of bicarbonate by surface waters to the oceans, a fraction of it can be lost back into the atmosphere (leaking or backflux), but this is modelled as less than 10% in most scenarios (Zhang et al. 2021).

The pH is a strong driver of the speciation in the $H_2CO_3/HCO_3^-/CO_3^{-2}$ system. Using the speciation equations proposed by Butler (1991), the amount of bicarbonate in solution can increase a hundred times from pH 4 to 6. Therefore, increasing soil pH is a good strategy to increase carbonate leaching. In tropical agricultural practice, this is mainly done using lime, which is milled carbonate rock applied to soils. This introduces, however, carbonate and bicarbonate ions from the dissolution of lime. In addition, lime is also fossil carbon being exhumed and exposed to the atmosphere. Stoichiometrically, half of the carbon in lime is released back into the atmosphere during the lime reaction in soil. However, the production of agro-commodities such as soybeans and corn in vast tropical agricultural areas, such as the Brazilian Cerrado, would not be possible with present-day technology without its use (Fageria and Nascente 2014).

The leaching of carbonate/bicarbonate must satisfy the balance of charge in solution (e.g., Gaston et al. 1992). Therefore, the activity of soluble, mobile mono- and divalent cations, such as Na^+, K^+, Ca^{2+}, and Mg^{2+}, is also important (Hudson 1995). Remineralizers of igneous basic rocks, such as basalt, diabase, and gabbro, provide plenty of these by the dissolution of pyroxenes, amphiboles, and plagioclase present in their composition (Gill 2011).

(B) **Formation of Mineral-Associated Organic Matter (MAOM)**

MAOM (Mineral Associated Organic Matter) is the association between organic colloids or even just organic molecules and mineral phases. The study of such materials is not trivial, as Kleber and co-authors put (Kleber et al. 2015). The importance of MAOM from the CDR perspective is that it may protect organic substances from microbial oxidation, allowing them to return as CO_2 to the atmosphere. This mechanism potentially increases the amount of carbon and its residence time in soil. On average, the amount of mineral-associated carbon in MAOM is small, about 1.3%, but it comprises 90% of the SOC in soils and increases carbon turnover time in soil by four times (Kleber et al. 2015).

Highly weathered soils may lack the availability of free surfaces to associate to newly added organic carbon sources to the soil. This relates to the controversial concept of soil carbon saturation, in which soil properties would limit the amount of carbon a particular soil can store (Stewart et al. 2007). Slessarev et al. (2021) modeled soil carbon and weathering on a continental scale and pointed out that soil under high weatheriln tropical agricultural prectice, increase in soil pH is mainly done using lime, which is milled carbonate rock applied to soils. thering products, particularly amorphous mineral phases. Remineralizers can boost MAOM synthesis by providing fresh mineral surfaces of both the rock's primary minerals and those from secondary phases resulting from remineralizer weathering. Amorphous secondary phases are particularly of interest because of their high surface reactivity and the more intimate bonds provided by the co-precipitation of organic and mineral phases (Masiello et al. 2004; Chadwick et al. 2003; Rasmussen et al. 2018a, b). In this context, basic igneous rocks are rich in Fe, Al, and Si minerals that weather quickly, favoring the precipitation of significant amounts of amorphous phases. Basalt powder increased the soil amorphous phases as much as 0.13 g kg^{-1} (soil) of amorphous FeAOD oxides

(FeAOD stands for Ammonium Oxalate in the Dark Extractable Fe) and 0.16 g kg^{-1} (soil) of amorphous AlAOD oxides, in a column experiment (Silva et al. 2024). For comparison, soils derived from basalt in tropical environments have from 0.1 to 0.6 g kg^{-1} (soil) FeAOD amorphous iron oxides (Jones et al. 2000).

Private companies have recently started to apply remineralizers in large tropical agricultural areas in South America, Africa, and Eastern Asia (the hot spots for ERW suggested by (Strefler et al. 2018) with the primary goal of producing carbon credits by the bicarbonate leaching path, but in the future, they may also be able to measure if new MAOM was synthesized, and how much.

The formation of MAOM is quite promising as a mechanism for capturing and stabilizing carbon in the soil, although it is not restricted to the SIC pool, which is the present-day focus of ERW. A significant challenge in accounting for the additivity of carbon captured through MAOM synthesis, resulting from remineralizer weathering, is the establishment of a robust and scalable measurement methodology.

(C) **Carbonate Precipitation**

Carbonate minerals are found in soils in several mineral phases and in widely varying amounts. In tropical humid soils, they usually do not form, except in soils developed in calcareous rocks, where they are found as residual carbonate phases (Doner and Lynn 1992).

The conditions for carbonate precipitation in soils under a tropical climate are generally not favorable. Taking only pH, pCO_2, and pCa as variables, even considering that the pCO_2 in the soil can reach as high as 10 times the pCO_2 in the atmosphere, it still requires a pH close to 7 for precipitation to occur. These conditions are not common in large areas in the tropics, although they may occur at depth, near the saprolite zone, as Santos et al. (2018) described in Northeast Brazil.

In general, the occurrence of pedogenic carbonates at low latitudes is associated with arid and semi-arid climate or paleoclimate, or carbonate rocks as parent materials. These conditions are currently not favorable for implementing the ERW based on remineralizer application.

3 Challenges in Measuring Cdr by Remineralizers

Approximately 40% of the Earth's land area is composed of silicate rocks, resulting in an average flux of 1.67 t km^{-2} $year^{-1}$ of CO_2. Brazil has the most significant silicate carbon sink on the planet (although not the greatest flux), in the evaluation of Zhang et al. (2021), constituted by the Amazon and the Parana rivers basins.

Measuring carbon flux in draining streams is not the best choice on a farm scale because remineralizers are not applied uniformly across drainage basins. Therefore, leaching flows do not necessarily flow into the same stream, and likewise, the streams do not necessarily receive only water from remineralized areas. Even when such an arrangement is purposely done, the measurements of carbonate flow are very complex to interpret (Larkin et al. 2022). Therefore, approaches using geochemical modelling based on variables such as soil solution composition (e.g., solid phase dissolution, Reershemius et al. 2023), and acidity neutralization (e.g., Dietzen and Rosing 2023) have been tested.

Geochemical modelling is challenged by the immense number of variables to be considered, and by the difficulty of measuring and monitoring them. In addition, important variables such as biological activity, quantification of surfaces available for weathering, and organomineral interactions change from one context to another. Experimentally, field measurements are also difficult to obtain due to factors such as the small volume of extractable soil solution in the dry season, the significant dilution of the rock grains in the soil mass (1 ton of rock mixed in 1 hectare to the depth of 0–20 cm results in only 0.05% m/m), and the preservation of the collected solutions, avoiding precipitation or the transfer of carbon from the bicarbonate to other forms, such as CO_2, between the time of collection in the field and the quantification in the lab.

The potential (upper limit condition) of carbon capture by the ERW can be estimated stoichiometrically based on Steinor's equation (Renforth 2019), and it serves as an important reference for field studies. However, it informs the maximum amount of carbonate-bicarbonate a certain remineralizer could capture, but the actual amount depends on a plethora of variables. The upper limit condition assumes that all cations present in a remineralizer will be accompanied by carbonate-bicarbonate anions and leached. In a real scenario, these cations will be partitioned in the leaching solution, plant absorption, soil adsorption, and possibly in the precipitation of new solid phases. Additionally, the availability of carbonate-bicarbonate anions for leaching will also depend on pH and pCO_2, as previously discussed. The bicarbonate anions will also compete with inorganic (such as SO_4^{-2} and NO^{3-}) and organic (such as citrate and oxalate) anions for the cations to leach, obeying the principle of the electrostatic neutrality of the solution.

Using this procedure, Renforth (2019) estimated that basalts would have the potential to capture 0.3 ± 0.1 t CO_2 t rock^{-1}.

There is still much to be investigated in the use of ERW for carbon capture, and one of the main obstacles is the time needed to obtain robust experimental results in the field (Edwards et al. 2017).

4 Concluding Remarks

In a tropical environment, remineralizers used in agricultural fields can capture atmospheric carbon primarily through bicarbonate leaching and the synthesis of MAOM (Mineral-Associated Organic Matter).

Bicarbonate leaching occurs in most agricultural soils, and the Enhanced Rock Weathering concept involves feeding this process with weatherable minerals to increase the amount of bicarbonate leached. Since bicarbonate leaching represents a "rate" rather than a "state" (borrowing the terminology from Slessarev et al. 2021), the main challenge in using it as a source of carbon credit lies in accurately measuring this rate.

The MAOM path is related to the association of Soil Organic Matter (SOM) with the neogenesis of rapidly forming, poorly crystalline mineral phases, which are weathering products of remineralizers in soils. The association of carbon substances with minerals extends the carbon residence time in soils by stabilizing it and protecting it from microbial oxidation, which would otherwise send it back to the atmosphere.

Carbonate precipitation requires specific conditions that are usually not met in agricultural tropical soils. Frequently, small amounts of carbonate phases present in remineralizers are dissolved in soils due to soil acidity and microbial attack, indicating unfavorable conditions for carbonate precipitation.

Climate-smart soil management techniques should leverage the Carbon Dioxide Removal (CDR) mechanisms provided by remineralizers. Currently, efforts and investments are being made to establish reliable protocols for quantifying and certifying CDR via bicarbonate leaching, enabling its trade in the carbon market.

There are still plenty of research topics to be addressed regarding the use of remineralizers for CDR. For example, how would the CDR mechanism be affected in reduced soils such as those cultivated with rice in Eastern Asia? How is the bicarbonate leaching affected by the Anion Exchange Capacity (AEC) of tropical soils, particularly in Acric soils where AEC can be greater than CEC in subsurface horizons? How to identify and quantify remineralizer-derived MAOM from the native soil MAOM?

Remineralizer research is inherently an interdisciplinary field of study. Soil scientists, geologists, and several other branches of geoscience should interact to address primary weathering issues. Still, it can be related to areas such as genomics, plant sciences, ecology, and societal studies. Remineralizers can be sourced as a mining co-product at a low cost, making them accessible to family farmers, who are the leading producers of non-industrial food and are often most exposed to the drawbacks of climate change.

Declaration of Competing Interest

The authors declare that they have no known competing financial interests or personal relationships that could have appeared to influence the work reported in this paper.

List of Acronyms

AEC	Anion Exchange Capacity
CDR	Cabon Dioxide Removal
CEC	Cation Exchange Capacity
ERW	Enhanced Rock Weathering
MAOM	Mineral Associated Organic Matter
POM	Particulate Organic Matter

SIC Soil Inorganic Carbon
SOC Soil Organic Carbon
SOM Soil Organic Matter

References

Bellassen V, Stephan N, Afriat M, Alberola E, Barker A, Chang JP, Chiquet C, Cochran I, Deheza M, Dimopoulos C, Foucherot C, Jacquier G, Morel R, Robinson R, Shishlov I (2015) Monitoring, reporting and verifying emissions in the climate economy. Nat Clim Chang 5(4):319–328. https://doi.org/10.1038/nclimate2544

Berner RA, Lasaga AC, Garrels RM (1983) The carbonate-silicate geochemical cycle and its effect on atmospheric carbon dioxide over the past 100 million years. Am J Sci 283(7):641–683. https://doi.org/10.2475/ajs.283.7.641

Brantley SL, Shaughnessy A, Lebedeva MI, Balashov VN (2023) How temperature-dependent silicate weathering acts as Earth's geological thermostat. Science 379(6630):382–389. https://doi.org/10.1126/SCIENCE.ADD2922

Butler JN (1991) Carbon dioxide equilibria and their applications. Lewis Publishers, p 259

Calvin K, Dasgupta D, Krinner G, Mukherji A, Thorne PW, Trisos C, Romero J, Aldunce P, Barrett K, Blanco G, Cheung WWL, Connors S, Denton F, Diongue-Niang A, Dodman D, Garschagen M, Geden O, Hayward B, Jones C, Ha M (2023) IPCC, 2023: Climate change 2023: Synthesis Report. In: Lee H, Romero J (eds) Contribution of working groups I, II and III to the sixth assessment report of the intergovernmental panel on climate change. IPCC, Geneva, Switzerland. https://doi.org/10.59327/IPCC/AR6-9789291691647

Chadwick OA, Gavenda RT, Kelly EF, Ziegler K, Olson CG, Crawford Elliott W, Hendricks DM (2003) The impact of climate on the biogeochemical functioning of volcanic soils. Chem Geol 202(3–4):195–223. https://doi.org/10.1016/j.chemgeo.2002.09.001

De Pinto A, Cenacchi N, Kwon H-Y, Koo J, Dunston S (2020) Climate smart agriculture and global food-crop production. PLoS ONE 15(4):e0231764. https://doi.org/10.1371/journal.pone.0231764

Dietzen C, Rosing MT (2023) Quantification of CO_2 uptake by enhanced weathering of silicate minerals applied to acidic soils. Int J Greenh Gas Control 125:103872. https://doi.org/10.1016/J.IJGGC.2023.103872

Doner HE, Lynn WC (1992) Carbonate, halide, sulfate, and sulfide minerals. In Dixon JB, Weed SB (eds) Mineral in soil environments, 2nd edn. Soil Science Society of America, pp 279–330

Edwards DP, Lim F, James RH, Pearce CR, Scholes J, Freckleton RP, Beerling DJ (2017) Climate change mitigation: potential benefits and pitfalls of enhanced rock weathering in tropical agriculture. Biol Lett 13(4). https://doi.org/10.1098/RSBL.2016.0715

Fageria NK, Nascente AS (2014) Management of soil acidity of South American soils for sustainable crop production. In Advances in Agronomy, vol 128. Academic Press Inc., pp 221–275. https://doi.org/10.1016/B978-0-12-802139-2.00006-8

Ferris FG, Wiese RG, Fyfe WS (2009) Precipitation of carbonate minerals by microorganisms: Implications for silicate weathering and the global carbon dioxide budget. Geomicrobiol J 12(1):1–13. https://doi.org/10.1080/01490459409377966

Flechard CR, Neftel A, Jocher M, Ammann C, Leifeld J, Fuhrer J (2007) Temporal changes in soil pore space CO_2 concentration and storage under permanent grassland. Agric Meteorol 142(1):66–84. https://doi.org/10.1016/J.AGRFORMET.2006.11.006

Gaston LA, Mansell RS, Selim HM (1992) Predicting removal of major soil cations and anions during acid infiltration: model evaluation. Soil Sci Soc Am J 56(3):944–950. https://doi.org/10.2136/sssaj1992.03615995005600030043x

Gill R (2011) Igneous rocks and processes: a practical guide. Wiley & Sons, p 438
Hilton RG (2023) Earth's persistent thermostat. Science 379(6630):329–330. https://doi.org/10.1126/SCIENCE.ADF3379/ASSET/D88A3604-B26A-4482-AC31-B83CF758D5EF/ASSETS/GRAPHIC/SCIENCE.ADF3379-F2.SVG
Hudson BD (1995) Reassessment of Polynov's ion mobility series. Soil Sci Soc Am J 59(4):1101. https://doi.org/10.2136/sssaj1995.03615995005900040022x
Jones RC, Babcock CJ, Knowlton WB (2000) Estimation of the total amorphous content of Hawai'i soils by the Rietveld method. Soil Sci Soc Am J 64(3):1100–1108. https://doi.org/10.2136/sssaj2000.6431100x
Kleber M, Eusterhues K, Keiluweit M, Mikutta C, Mikutta R, Nico PS (2015) Mineral-organic associations: formation, properties, and relevance in soil environments. Adv Agron 130:1–140. https://doi.org/10.1016/bs.agron.2014.10.005
Lal R, Monger C, Nave L, Smith P (2021) The role of soil in regulation of climate. Philos Trans R Soc B: Biol Sci 376(1834):20210084. https://doi.org/10.1098/RSTB.2021.0084
Larkin CS, Andrews MG, Pearce CR, Yeong KL, Beerling DJ, Bellamy J, Benedick S, Freckleton RP, Goring-Harford H, Sadekar S, James RH (2022) Quantification of CO_2 removal in a large-scale enhanced weathering field trial on an oil palm plantation in Sabah Malaysia. Front Clim 4:959229. https://doi.org/10.3389/FCLIM.2022.959229/BIBTEX
Lavallee JM, Soong JL, Cotrufo MF (2020) Conceptualizing soil organic matter into particulate and mineral-associated forms to address global change in the 21st century. Glob Change Biol 26(1):261–273. https://doi.org/10.1111/GCB.14859
Manning DAC, Azevedo AC, Zani CF, Barneze AS (2024) Soil carbon management and enhanced rock weathering: the separate fates of organic and inorganic carbon. Eur J Soil Sci 75(4):e13534. https://doi.org/10.1111/EJSS.13534
Martinez BC, DeJong JT, Ginn TR, Montoya BM, Barkouki TH, Hunt C, Tanyu B, Major D (2013) Experimental optimization of microbial-induced carbonate precipitation for soil improvement. J Geotech Geoenvironmental Eng 139(4):587–598. https://doi.org/10.1061/(asce)gt.1943-5606.0000787
Masiello CA, Chadwick OA, Southon J, Torn MS, Harden JW (2004) Weathering controls on mechanisms of carbon storage in grassland soils. Global Biogeochem Cycles 18(4):1–9. https://doi.org/10.1029/2004GB002219
Mitchell CJ, Mitchell P, Pascoe RD (2008) Quarry fines minimisation: can we really have 10 mm aggregate with no fines?. In: Walton G (ed) 14th extractive industry geology conference (pp. 37–44). https://static1.squarespace.com/static/54199a46e4b05afa19b4e68c/t/565746c0e4b0f33a7ac974a4/1448560320478/pp37-44.+Mitchell+CJ+et+al.+Quarry+fines+minimisation..pdf
Penman DE, Caves Rugenstein JK, Ibarra DE, Winnick MJ (2020) Silicate weathering as a feedback and forcing in Earth's climate and carbon cycle. Earth Sci Rev 209:103298. https://doi.org/10.1016/J.EARSCIREV.2020.103298
Rasmussen C, Throckmorton H, Liles G, Heckman K, Meding S, Horwath WR (2018a) Controls on soil organic carbon partitioning and stabilization in the California Sierra Nevada. Soil Syst 2(3):1–18. https://doi.org/10.3390/SOILSYSTEMS2030041
Rasmussen C, Heckman K, Wieder WR et al (2018b) Beyond clay: towards an improved set of variables for predicting soil organic matter content. Biogeochemistry 137:297–306. https://doi.org/10.1007/s10533-018-0424-3
Reershemius T, Kelland ME, Jordan JS, Davis IR, D'Ascanio R, Kalderon-Asael B, Asael D, Suhrhoff TJ, Epihov DZ, Beerling DJ, Reinhard CT, Planavsky NJ (2023) Initial validation of a soil-based mass-balance approach for empirical monitoring of enhanced rock weathering rates. Environ Sci Technol 57(48):19497–19507. https://doi.org/10.1021/ACS.EST.3C03609/SUPPL_FILE/ES3C03609_SI_006.XLSX
Renforth P (2019) The negative emission potential of alkaline materials. Nat Commun 10(1):1–8. https://doi.org/10.1038/s41467-019-09475-5

Dos Santos JCB, Le Pera E, Souza Júnior VS, Oliveira CS, Juilleret J, Corrêa MM, Azevedo AC (2018) Porosity and genesis of clay in gneiss saprolites: the relevance of saprolithology to whole regolith pedology. Geoderma 319:1–13. https://doi.org/10.1016/j.geoderma.2017.12.031

Silva RC, Guimarães CB, Azevedo AC, Alves MR, Demattê JAL, Rabêlo HSF (2024) Precipitation of amorphous iron and aluminum during the weathering of rock dust in soil columns. Sci Agric 81:e20230219. https://doi.org/10.1590/1678-992X-2023-0219

Slessarev EW, Chadwick OA, Sokol NW, Nuccio EE, Pett-Ridge J (2021) Rock weathering controls the potential for soil carbon storage at a continental scale. Biogeochemistry 2021:1–13. https://doi.org/10.1007/S10533-021-00859-8

Stewart CE, Paustian K, Conant RT, Plante AF, Six J (2007) Soil carbon saturation: concept, evidence and evaluation. Biogeochemistry 86(1):19–31. https://doi.org/10.1007/S10533-007-9140-0/FIGURES/5

Strefler J, Amann T, Bauer N, Kriegler E, Hartmann J (2018) Potential and costs of carbon dioxide removal by enhanced weathering of rocks. Environ Res Lett 13(3):034010. https://doi.org/10.1088/1748-9326/aaa9c4

Voosen P (2019) Tropical uplift may set Earth's thermostat. Science 363(6422):13. https://doi.org/10.1126/SCIENCE.363.6422.13

Yu W, Huang W, Weintraub-Leff SR, Hall SJ (2022) Where and why do particulate organic matter (POM) and mineral-associated organic matter (MAOM) differ among diverse soils? Soil Biol Biochem 172:108756. https://doi.org/10.1016/J.SOILBIO.2022.108756

Zamanian K, Zhou J, Kuzyakov Y (2021) Soil carbonates: the unaccounted, irrecoverable carbon source. Geoderma 384:114817. https://doi.org/10.1016/J.GEODERMA.2020.114817

Zhang S, Bai X, Zhao C, Tan Q, Luo G, Wang J, Li Q, Wu L, Chen F, Li C, Deng Y, Yang Y, Xi H (2021) Global CO_2 consumption by silicate rock chemical weathering: its past and future. Earth's Futur 9(5):e2020EF001938. https://doi.org/10.1029/2020EF001938

Agronomic Characterization of Calcium Magnesium Remineralizers in Tropical Conditions

Augusto Vaghetti Luchese, Robson Fernando Missio, Antonio Carlos Saraiva da Costa, Ivan Granemann de Souza Junior, and Alessandra Algeri

Abstract Soil remineralizers derived from Ca and Mg-rich aluminosilicate rocks (ASiCaMg) contain 9–20% alkaline cation oxides ($CaO + MgO + K_2O$), with K_2O between 1 and 3%. Basalt is the primary rock used in Brazil, alongside gabbro, amphibolite, andesite, and certain gneisses. Studies show these remineralizers enhance soil fertility by increasing macronutrients (Ca, Mg, P, K) and micronutrients, with effects depending on rock type and soil properties. Basalt powder consistently elevates soil Ca^{2+}, Mg^{2+}, K^{+}, and P levels, even under high crop nutrient uptake. It also raises soil pH, reduces exchangeable Al3^{+}, and increases base saturation, improving crop growth and productivity in soybeans, corn, oats, and pastures. Basalt application also contributes to CO_2 sequestration through silicate weathering, with estimated capture rates of 0.6–1.6 Mg CO_2 ha^{-1} yr^{-1}, depending on climate and application rates. High doses (e.g., 50 Mg ha^{-1}) can sequester up to 10.5 Mg CO_2 ha^{-1} over four years. Additionally, replacing limestone with basalt reduces CO_2 emissions by up to 78.5 kg CO_2 ha^{-1} per Mg of applied material, offering dual environmental benefits. While remineralizers enhance soil fertility and crop yields, their effectiveness varies depending on soil type, rock composition, and management practices. Some experimental studies suggest combining remineralizers with microbial inoculants for optimal efficiency. Overall, ASiCaMg remineralizers provide a sustainable

A. V. Luchese (✉) · R. F. Missio
Department of Agronomic Sciences, Federal University of Paraná, Palotina, Brazil
e-mail: aluchese@ufpr.br

R. F. Missio
e-mail: rfmissio@ufpr.br

A. C. S. da Costa · I. G. de Souza Junior
Agronomy Department, Maringá State University, Maringá, Brazil
e-mail: acscosta@uem.br

I. G. de Souza Junior
e-mail: igsjunior@uem.br

A. Algeri
Federal Institute of Paraná, Quedas Do Iguaçu, Brazil
e-mail: alessandra.algeri@ifpr.edu.br; sandra.algeri.utfpr@gmail.com

É. de Souza Martins and S. Huff Theodoro (eds.), *Soil Remineralizers and Silicate Fertilizers*, Frontier Studies in Soil Science,
https://doi.org/10.1007/978-3-032-14656-4_6

approach to enhancing soil health, boosting agricultural productivity, and mitigating climate change through enhanced weathering and reduced liming emissions. This chapter presents the results of several studies conducted by the authors under a single greenhouse experimental protocol that evaluated various ASiCaMg remineralizers to determine their primary effects.

Keywords Basalt powder · Nutrient release · CO_2 sequestration · Enhanced weathering · Sustainable agriculture

1 Introduction

Soil remineralizers derived from rocks composed of Ca and Mg aluminosilicates (ASiCaMg) are characterized by presenting a sum of alkaline cations, represented in the form of oxides ($CaO + MgO + K_2O$) between 9 and 20% with K_2O contents greater than 1 and less than 3% (Martins et al. 2024). In Brazil, basalt is the main rock of this group, in addition to, to a lesser extent, other rocks such as gabbros, amphibolites, andesites, and certain gneisses that belong to this group.

There is a substantial body of literature on the topic, highlighting the effectiveness of remineralizers as inputs in agricultural systems. These studies demonstrate their ability to significantly enhance the levels of macronutrients such as calcium, magnesium, phosphorus, and potassium (Conceição et al. 2022; Crusciol et al. 2022; Luchese et al. 2021, 2023a, b; Martins et al. 2015; Moretti et al. 2019; Zhao et al. 2019) and micronutrients (Conceição et al. 2022; Rodrigues et al. 2024). The magnitude of increase varies depending on the type of rock or material used in the formulation of the remineralizer.

According to Decree 8384/2014 of the Ministry of Agriculture, Livestock, and Supply (MAPA), several rocks meet the chemical characteristics established for use as remineralizers. These rocks have been the subject of studies, including basalt, diabase, phonolite, syenite, dunite, serpentinite, siltstone, biotite schist, calc schist, and gneiss (Martins et al. 2015; Silva et al. 2017; Swoboda et al. 2022).

Among the rocks belonging to the group of ASiCaMg remineralizers, basalt has the most extensive coverage area and is the most explored for the production of this input in Brazil, leaving the other rocks with potential for use, such as some gneisses, with a production of less than 150 thousand tons in 2023, considerably below the 500 thousand tons originating from basalt (Martins et al. 2024).

The use of basalt remineralizers has promoted consistent responses with the amount applied, showing increases in the levels of Ca^{2+}, Mg^{2+}, K^+ and P in soils (Lewis et al. 2021; Luchese et al. 2021, 2023a, b; Martins et al. 2015; Rodrigues et al. 2024). This phenomenon persists even in circumstances where there is an escalation in the export of these nutrients by crops (Korchagin et al. 2022; Luchese et al. 2023a, b).

The increased export of nutrients is caused by their greater availability in the soil. This increase can be quantified by analyzing the leaf tissue of the crops studied, with an increase in the levels of calcium, magnesium, phosphorus and potassium in the leaf tissue of plants grown in soils fertilized with basalt powder being common, depending on the doses applied (Kelland et al. 2020; Luchese et al. 2021, 2023a, b; Martins et al. 2015; Skov et al. 2024).

In addition to nutrient release, other soil properties are improved, the most important being an increase in pH (Bauwhede et al. 2024; Kelland et al. 2020; Luchese et al. 2021, 2023a, b; Rodrigues et al. 2024; Toscani and Campos 2017).

The application of basalt powder results in an increase in the concentration of basic cations in the soil, which subsequently leads to changes in pH. This, in turn, promotes an increase in base saturation in the soils, resulting in a subsequent drop in the saturation of exchangeable aluminum or even a reduction in the Al^{3+} content of the soils (Korchagin et al. 2022; Martins et al. 2015; Rodrigues et al. 2024; Toscani and Campos 2017).

All the improvements in chemical properties associated with soil fertility promoted by the application of basalt powder culminate in better nutrient availability for crops, which favors their growth and development. Several studies in the literature show that several crops respond positively to the application of basalt powder. Among other improvements, we can mention better initial development in crops such as soybean and corn, resulting in greater production of shoot dry mass (Luchese et al. 2021, 2023b), the increase in leaf area of the bean crop (Toscani and Campos 2017), or even the increase in productivity parameters observed in cereals such as oats, soybean, and corn (Beerling et al. 2024; Skov et al. 2024) and pasture biomass (Korchagin et al. 2022; Martins et al. 2015).

In this chapter, we will discuss the primary improvements achieved through the use of ASiCaMg remineralizers, highlighting the enhancements they provide to soil chemical properties, as well as their environmental impacts and effects on crop productivity. For this purpose, we will use bibliographic data and results from a series of studies on remineralizers.

The research dataset was obtained from multiple trials conducted under a protocol to evaluate the agronomic efficiency of ASiCaMg remineralizers. These protocols are characterized by experiments conducted with two crops, soybeans and corn, and soils of varying textures. The group of soils classified as sandy has a clay content $<$ 250 g kg^{-1}, while soils classified as clayey have a clay content $>$ 500 g kg^{-1} (Luchese et al. 2021, 2023a, b, 2024a, b).

2 Dataset

The research dataset derives from multiple trials conducted by the authors following a protocol validated in peer-reviewed agronomic publications (Luchese et al. 2021, 2023a, b, 2024a, b) to evaluate the agronomic efficiency of ASiCaMg remineralizers. Experiments were performed in a greenhouse using 5 L pots filled with soils of contrasting textures and tested with two crops: soybean and maize.

Soils were collected from the 0–20 cm layer and classified into two texture groups: sandy (clay < 250 g kg^{-1}) and clayey (clay > 500 g kg^{-1}). All remineralizers were applied at four rates (0, 33, 66, and 99 Mg ha^{-1}), with four replicates per treatment. Application rates were calculated assuming a soil volume of 2,000,000 L ha^{-1}. After treatment application, pots were maintained and moistened for 60–90 days to promote remineralizer reactions prior to sowing.

For each pot, five seeds were sown; soybean seeds were pre-inoculated with *Bradyrhizobium sp.* Seedlings were thinned to two plants per pot 20 days after sowing. Plants were hand-irrigated daily throughout the experiment. For maize, urea was applied once at a rate equivalent to 100 kg N ha^{-1} at 30–35 days after sowing. Harvest occurred 45–50 days after sowing: aerial biomass was cut, soils were sampled for chemical analysis, and plant tissues were oven-dried in a forced-air oven at 50 °C for approximately 48 h, weighed for dry matter (DM) determination, and ground for foliar analysis.

Soil analyses followed Pavan et al. (1992): pH in $CaCl_2$ (potentiometry); H + Al by the SMP method (potentiometry); available P (spectrophotometry) and K (flame photometry) in Mehlich-I extracts; and Ca and Mg by volumetric determination in 1 M KCl extracts.

3 Basalt Remineralizers as a Source of Nutrients

3.1 Calcium and Magnesium

Calcium, magnesium, and iron are the three nutrients present in the highest concentrations within this group of remineralizers. The predominance of these elements, linked to their tendency to dissociate from their oxides or oxyhydroxides in silicate minerals, follows the sequence: CaO>MgO>>Al(OH$_3$ >FeOOH>SiO_2, which facilitates their enhanced release (Oelkers 2001; Swoboda et al. 2022).

Consequently, Ca^{2+} and Mg^{2+} are frequently the nutrients that, due to their abundance and favored release, tend to provide an increase in soil levels by this group of remineralizers.

The evaluation of Ca^{2+} levels in a series of studies conducted in pot experiments reveals a positive relationship between the CaO levels applied by remineralizers and the increase in the concentration of these elements in the soil. However, this pattern is not consistently observed for soil Mg^{2+} levels, particularly in more clayey soils (Fig. 1).

Increases in Ca^{2+} levels occur systematically when basic igneous rocks are applied to the soil. Luchese et al. (2021, 2023a, b) reported increases in Ca^{+2} levels in nearly all of the pot experiments conducted, with increases ranging from 0.6 to 2.9 $cmol_c$ dm^{-3}, depending on the applied dose, the type of rock, and the soil class. A similar pattern was observed in field experiments following the application of basalt powder (Korchagin et al. 2022; Rodrigues et al. 2024).

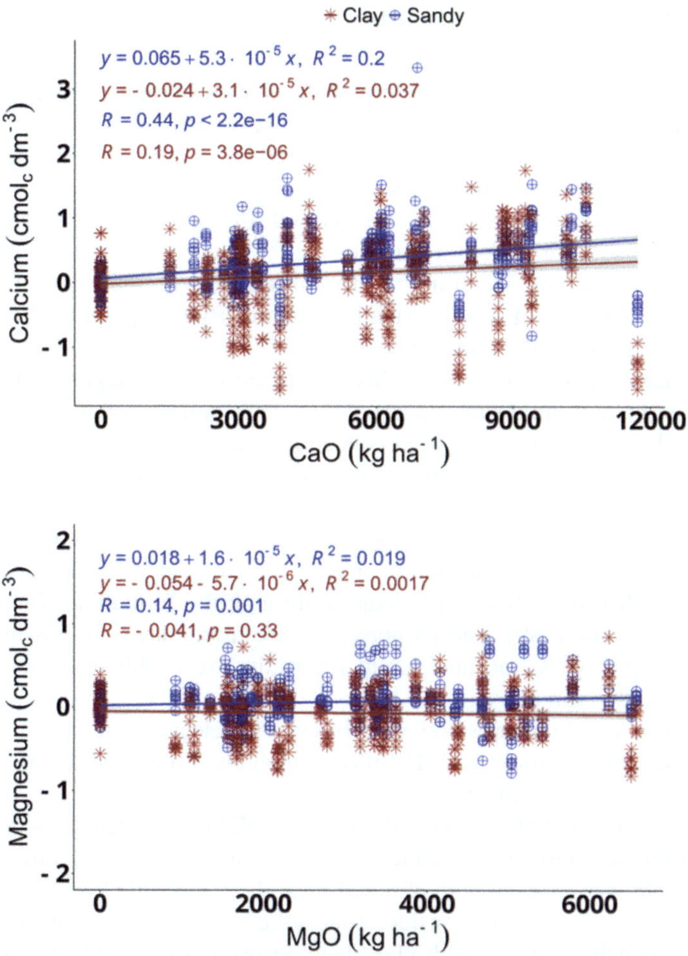

Fig. 1 Analysis of variation in Ca^{2+} **a** and Mg^{2+} contents **b** in soil after soybean and corn cultivation as a function of applied CaO and MgO contents, data from 72 pot trials. The shading of the regression line indicates the 95% confidence interval

The availability of Mg^{2+} to the soil by remineralizers has shown different results, as observed in the set of tests mentioned. While some studies have documented an increase in Mg^{2+} levels in soil following the application of remineralizers (Luchese et al. 2023a; Melo et al. 2012; Swoboda et al. 2022), others have not observed this trend. In fact, the findings suggest a potential decrease in Mg^{2+} levels in soil, depending on the dose of the remineralizer, the type of soil, the material applied, and the species cultivated (Luchese et al. 2021).

3.2 Phosphorus and Potassium

P is the macronutrient present at the lowest levels in these rocks, with P_2O_5 concentrations generally ranging from 0.2 to 0.5%. However, increases in the levels of this nutrient in the soil have been significant and rapid, especially when compared to the other macronutrients that make up rocks.

The combined analysis of tests in soils with different textures shows that the P levels in the soil are directly related to the P_2O_5 concentrations applied by the materials, distinguishing according to the soil texture with increases of 0.026 and 0.074 mg of P per dm^3 of soil for each kilogram of P_2O_5 applied by the materials, for clayey and sandy soils, respectively (Fig. 2).

These data were obtained from highly weathered soils in a region of Brazil characterized by a Cfa (humid subtropical) climate and semi-deciduous seasonal forest (Wrege et al. 2017), containing high concentrations of iron and aluminum oxyhydroxides with a strong capacity for phosphorus fixation. The observed variation in the available phosphorus levels in soils according to texture is indicative of the differing degrees of interaction of the element with the particle surface. In soils with higher clay content, there is a greater interaction of the element with the surface of the particles, which favors phosphorus fixation through the formation of inner sphere complexes (Sposito 1989). Furthermore, in these more clayey soils with a higher P buffering capacity (Novais and Smyth 1999), the extractable levels are generally lower compared to sandier soils.

The dissolution of minerals present in basalt also releases a significant amount of silicon in the form of silicic acid (H_4SiO_4), which plays a crucial role in enhancing plant growth and development.

The enhancement of phosphorus (P) levels in soil, facilitated by the remineralizer, can be attributed to three primary mechanisms documented in literature. Firstly, an increase in pH is associated with elevated P levels in the soil (Viviani et al. 2010). A second factor that warrants consideration is the introduction of silicon oxide into the system, which competes with phosphate adsorption sites (Alleoni et al. 2019; Schaeller et al. 2019). A third notable factor is the availability of P from the rock, often from the apatite mineral within the rocks, which contain P_2O_5 contents of less than 1%.

Numerous studies on remineralizers have disconnected pH from increases in soil P levels, thereby attributing this increase to the input of silicon (H_4SiO_4) and the addition of P by these materials (Luchese et al. 2021, 2023a, b).

However, although increases in P levels are contributed by the small concentrations of P_2O_5 in the remineralizer, there is still a relationship between P availability and the release of SiO_4^{2-}, which competes with the P adsorption sites of iron and aluminum oxides and hydroxides in soils, making this nutrient more available (Matichenkov and Bocharnikova 2001; Alleoni et al. 2019, Schaller et al. 2019).

Silicon (Si) has been identified as a beneficial element for a wide range of cultivated plants. Its presence plays a crucial role by enhancing the mechanisms that induce plant resistance to pests and diseases. In the soil environment, an increase in

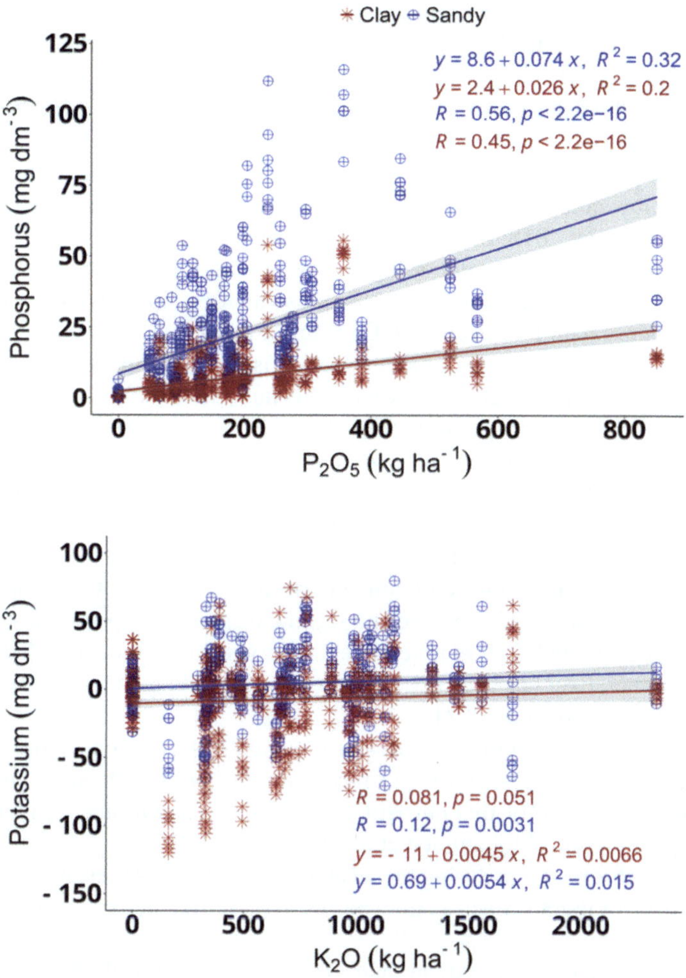

Fig. 2 Analysis of variation in P (mg dm^{-3}) **a** and K ($cmol_c$ dm^{-3}) contents **b** in soil after soybean and corn cultivation as a function of applied P_2O_5 and K_2O contents, data from 72 pot trials. The shading of the regression line indicates the 95% confidence interval

silicon (Si) concentration has been shown to influence the dynamics of phosphorus (P), with an enhancement in Si uptake frequently observed after the application of silicate rock remineralizers.

The contribution of SiO_4^{4-} to P availability has been demonstrated by Korchagin et al. (2022), who observed a decrease in P levels in native pasture soil, which had not received fertilization over the years, following the application of basalt powder. This outcome is attributed to the low release of P by basalt powder, combined with the increase in pasture mass, which promotes substantial extraction of the nutrient from the soil, and the minimal amount of P adsorbed to iron oxides. Consequently, the entry of SiO_4^{4-} into the system hinders any increase in P due to the limited availability of fixed P.

Notwithstanding the factors that contribute to the outcomes with P, there is a substantial body of literature that finds positive results for the increase in the levels of this nutrient in the soil promoted by the application of remineralizers, even when they present low concentrations (< 1%) in their composition (Luchese et al. 2021, 2023a; Martins et al. 2015; Silva et al. 2012; Viviani et al. 2010).

Potassium (K^+), even present at higher levels than P in the composition of silicate rocks of Ca and Mg, such as basalts, presents less expressive responses when compared mainly with P or even with Ca (Figs. 1 and 2).

However, a correlation, albeit modest yet statistically significant, exists between the amount of K_2O applied and the levels of this nutrient in the soil, even after uptake by plants and in short-term pot tests (Fig. 2).

Although it is not the nutrient that is preferentially released in the short term, some materials do provide increases in this nutrient in soils. The initial release of K^+ is likely to be preferentially associated with biotite, a mineral known for its increased solubility (Lewis et al. 2021; Mohammed et al. 2014).

Notably, the extant literature documents positive responses to elevated soil K^+ levels. Swoboda et al. (2022) observed no significant soil K^+ responses in only 32.5% of the studies evaluated with remineralizers. In some cases, the soil K^+ response to the same remineralizer has been observed to vary depending on the soil type and the crop planted (Luchese et al. 2021).

The release of potassium from basalts is typically slow because the nutrient is often contained in volcanic glass and forms strong bonds with SiO_4^- and AlO_4^- tetrahedral. The release of this nutrient varies due to several factors, including the rate of crop extraction (Mohammed et al. 2014).

It has been observed that during the weathering of minerals present in rock formations, a portion of the K^+ released may be absorbed or retained by the newly formed minerals.

Incubation time is also a critical factor in this process. Basak, Sarkar and Naidu (2021) applied basalt in an incubation test and found increases in exchangeable K^+ in the soil only after 12 months. This period is significantly longer than the 4 months from application to soil sample collection in short-term tests, which are generally conducted in pots. Lewis et al. (2021) demonstrated through the evaluation of basalts from diverse origins that the release rates of potassium depend on the particle size and mineralogical characteristics. Consequently, the findings from different studies have shown conflicting results, with some reporting an increase in K^+ levels in the soil (Martins et al. 2015; Korchagin et al. 2022; Rodrigues et al. 2024), and others documenting a decrease (Luchese et al. 2023a, b; Rodrigues et al. 2024). This variability is attributed to the distinct characteristics of each basalt sample, its interaction with the soil, and the influence of the crop.

4 Basalt as Soil Amendment—Ph and Cec

The increase in pH values in soils due to the application of silicate rock powders has been demonstrated in studies such as those by Anda et al. (2015), Martins et al. (2015), Luchese et al. (2023b), Rodrigues et al. (2024), and Skov et al. (2024). These studies indicate that the application of basalt powder leads to a reduction in the H^+ content of the soil and an improvement in its pH.

Beerling et al. (2018) confirm that there is an alkalizing effect on soils promoted by these rocks, which, according to Curi et al. (2005), is due to the consumption of H^+ during the weathering of silicate minerals.

It is worth noting that the effect of the remineralizer on this amendment does not occur in the same way as that of limestone, the primary acidity corrector utilized in agriculture. While limestone exerts its effect through the reaction of calcium and magnesium carbonates present in its composition, the effect of remineralizers is to increase pH by consuming H^+ in the weathering reaction of silicates, mainly plagioclase, which occurs as silicates with Ca^{2+} in these rocks, being characterized as more reactive, compared to Mg^{2+} and K^+ silicates (Anda et al. 2015; Kelland et al. 2020).

The effect on pH, although it presents a direct relationship with the CaO content of the remineralizers, confirms that Ca^{2+} silicates tend to be more reactive than the other minerals present in the remineralizers (Anda et al. 2015; Kelland et al. 2020). The relationship between the SiO_2 content in the remineralizers shows a greater correlation with the increase in pH when compared to the CaO content itself, demonstrating the response of other silicates even when present at lower concentrations. When evaluating the effect of pH increase in soils with different textures that received the application of basalt powder, a greater response is observed in sandier soils due to their lower buffering capacity (Fig. 3).

However, a decline in soil pH has been documented, particularly in acidic soils that have not received prior limestone application. Anda et al. (2015) also noted a decrease in soil pH, mainly during the initial months of the experiment, when basalt doses lower than 40 Mg ha^{-1} were applied.

The increase in CEC is crucial for improving agricultural soils, a fact verified in several scientific studies through the application of remineralizers (Daniell and Tonder 2023; Silva et al. 2017). According to part of the literature, this response is partly due to the formation of secondary 2:1-type minerals during weathering, which possess a high natural CEC (Mohammed et al. 2014). However, in tropical climates, the formation of these minerals does not occur frequently, making it more likely that this increase in charge occurs due to the formation of poorly crystalline forms of silicon that can develop negative charges on their surface (Anda et al. 2013).

As an example of rocks that contain primary minerals such as plagioclase, quartz, and alkali feldspar, these may form part of the weathered minerals that contribute to the formation of 2:1 clay minerals, thereby increasing the permanent charges of the soil (Medeiros et al. 2021).

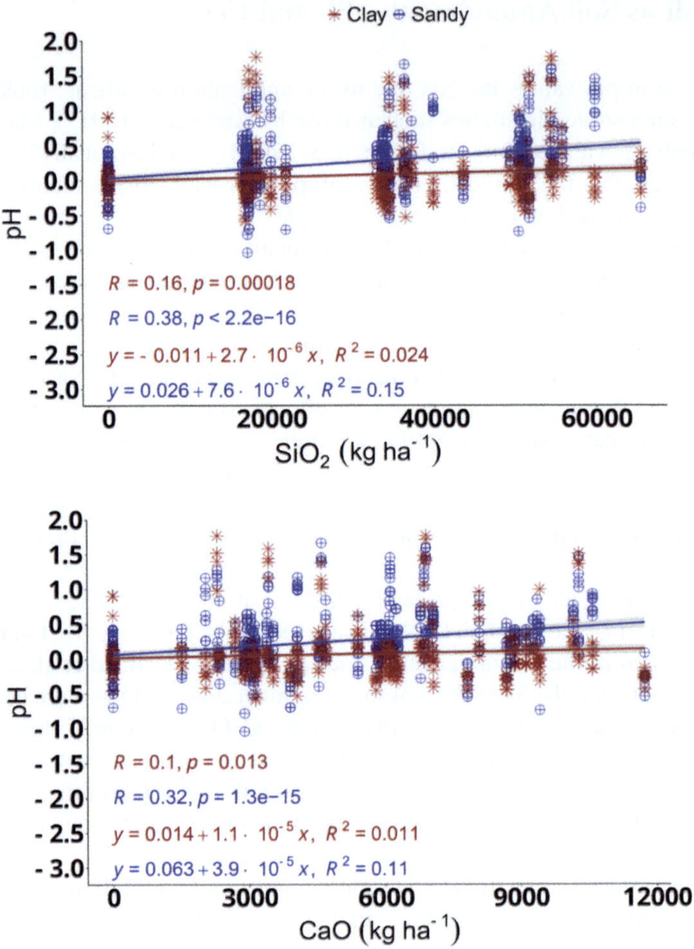

Fig. 3 Analysis of pH variation in soil after soybean and corn cultivation as a function of the SiO_2 **a** and CaO levels **b** applied, data from 72 pot trials. The shading of the regression line indicates the 95% confidence interval

The weathering process leads to the formation of secondary clay minerals of the 2:1 type, including vermiculite and smectite, through the transformation of the primary minerals that comprise the rocks. In these clay minerals, potassium can become firmly bound to their crystalline structure (Etesami et al. 2017).

5 Increase of Crop Production Potential Through the Application of Basalt Powder

Improved plant development is one of the benefits generated by using remineralizers in the soil, as shown by significant increases in several growth-related traits in the literature. This phenomenon has been observed to result in an augmentation of crop productivity, manifesting in increased plant height, stem diameter, leaf area, and even dry matter mass (Anda et al. 2013; Crusciol et al. 2019; Kelland et al. 2020; Mancuso et al. 2014; Medeiros et al. 2021; Mohammed et al. 2014; Moretti et al. 2019; Souza et al. 2017; Ramos et al. 2022; Swoboda et al. 2022).

However, the application of basalt powder does not always affect crop growth, as seen in soybeans and corn, where the increase in shoot dry mass due to basalt application may not occur uniformly across all trials. The lack of response to basalt powder application in some cases may be related to the effects (isolated or not) of the combination of the dose applied, the soil used, the type and intrinsic properties (particle size, mineralogy, chemical composition, etc.) of the material used as a remineralizer, and the crop used (Luchese et al. 2023a, b; Moretti et al. 2019).

In some crops, such as pastures, better growth has a direct impact on productivity. In this crop, the application of basalt powder can result in increases of 10–14 kg ha^{-1} of dry mass per Mg ha^{-1} of basalt powder applied (Korchagin et al. 2022).

Significant increases in corn and soybean productivity were also documented by Beerling et al. (2024), who applied high doses (50 Mg ha^{-1}) of basalt powder and observed increases of 12% and 16% for corn and soybean, respectively.

The utilization of basalt in oat crops has been demonstrated to yield enhanced productivity. A study by Skov et al. (2024) examined the application of basalt powder at a dose of 18.86 Mg ha^{-1}, resulting in increases of 20.5 and 9.3% compared to the control in no-till or plowed soil, respectively. The enhanced productivity observed in no-till systems is likely attributable to the presence of a diverse and abundant microbiota, which fosters the weathering of the applied silicate minerals.

However, some studies indicate that the enhancement of productivity or the attainment of desired plant traits is only achieved with the combined use of microorganisms that tend to favor the solubilization of remineralizers. This claim is substantiated by the outcomes of Ramos et al. (2023), who tested ground basaltic rock in conjunction with microorganisms in "Sweet Grape" and "Cereja" tomatoes and observed superior responses in certain variables when a biological treatment was employed in combination with basaltic rock powder.

6 Environmental Impact

The use of rocks abundant in Ca^{2+} and Mg^{2+} silicates, encompassing basic and ultrabasic rocks such as basalts and their analogues, holds considerable promise in mitigating CO_2 emissions and removing it from the atmosphere.

Current research focuses on the accelerated weathering of basalts, demonstrating the capacity of Ca^{2+} and Mg^{2+} silicates to sequester CO_2 during the weathering process of these minerals within the water flow.

Lewis et al. (2021) have delineated the processes that initiate with the reaction of Ca^{2+} and Mg^{2+} silicates with gaseous CO_2 in the presence of water, forming carbonic acid (Eq. 1). Depending on the soil pH, this can precipitate as carbonate (Eq. 2), which can be transported to the oceans, where it remains in a stable form for thousands of years.

$$CaMgSi_2O_6 + 4CO_2(g) + 6H_2O \rightarrow Mg^{2+} + Ca^{2+} + 2H_4SiO_4 + 4HCO_3^- \quad (1)$$

$$Ca^{2+} + 2HCO_3^- \rightarrow CaCO_3 + H_2O + CO_2(g) \quad (2)$$

As a conservative approach, Baek et al. (2023) estimated in a modeling study that the application of 10 Mg ha^{-1} of basalt results in CO_2 sequestration rates that vary according to the region, with areas in North America, South America and Europe/ Russia exhibiting rates of about 1 Mg CO_2 ha^{-1} year^{-1}, while rates in Australia are less than 0.6 Mg CO_2 ha^{-1} year^{-1}. Regions in Africa, Thailand, eastern China, and India have higher rates, ranging from 1.2 to 1.6 Mg CO_2 ha^{-1} year^{-1}. This approach suggests that basalt application may be an effective strategy for large-scale CO_2 removal, particularly in tropical and developing regions, where weathering efficiency is optimized.

However, doses as low as 10 Mg ha^{-1} year^{-1} are considered low and are not usually adopted. Several studies have employed doses as high as 50 Mg ha^{-1} in a single application, evaluating the effect over the years (Lewis et al. 2021; Vienne et al. 2022), or even the application of this same dose of 50 Mg ha^{-1} year^{-1}, being applied sequentially over several years (Beerling et al. 2024).

In studies with single-dose applications of 50 Mg ha^{-1}, the results were below 1 Mg CO_2 ha^{-1} year^{-1}, ranging from 1.3 to 8.5 Mg CO_2 ha^{-1}over 15 years (Lewis et al. 2021) or in a shorter time series of only 5 years, with 4.48 Mg CO_2 ha^{-1} captured (Vienne et al. 2022).

The reapplication of high doses of basalt, 50 Mg ha^{-1} year^{-1}, promotes a significant increase in CO_2 capture, reaching values of 10.5 ± 3.8 Mg CO_2 ha^{-1} in four years (Beerling et al. 2024).

These results indicate that high doses are crucial for achieving more effective responses in removing inorganic carbon from the atmosphere.

While basalt can sequester inorganic C, its potential to alkalize the environment can also be significant, especially in tropical and consequently acidic soils that necessitate liming to enhance productivity. This process releases CO_2 into the atmosphere due to the reaction of Ca^{2+} and Mg^{2+} carbonates present in limestone.

Estimates regarding the impact of applying rocks from the ASiCaMg class on the potential acidity of soils with varying textures (Fig. 4) indicate that this practice reduces CO_2 emissions from limestone by 1.23 and 7.92 kg ha^{-1} Mg^{-1}, for sandy and clayey soils, respectively, as observed over 105–110 days of testing. Consequently,

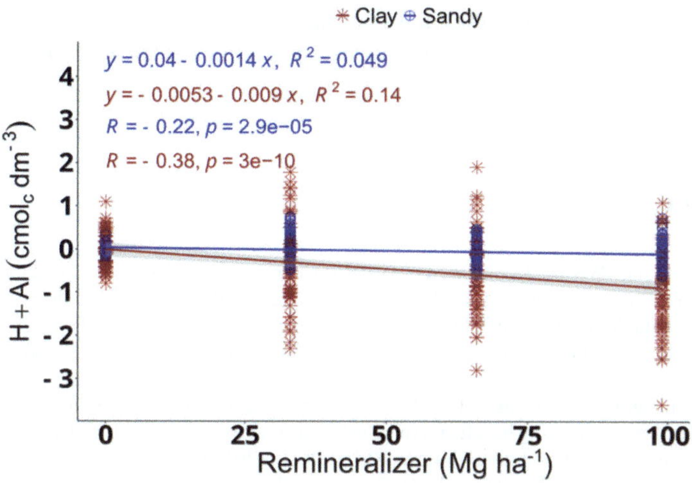

Fig. 4 Analysis of H + Al variation ($cmol_c\ dm^{-3}$) in soil after soybean and corn cultivation as a function of the applied doses of remineralizers, data from 42 pot trials. The shading of the regression line indicates the 95% confidence interval

the potential reduction in CO_2 emissions from agricultural activities can range from 4.09 to 26.29 kg CO_2 ha^{-1} for each Megagram of material applied.

If we consider the application in more clayey soils with a conservative dose of 10 Mg ha^{-1}, annual reductions close to 0.263 Mg CO_2 ha^{-1} $year^{-1}$ are expected. This value, although lower than the estimates for sequestration presented in several studies, is considerable, suggesting that the environmental impact of the application may be approximately 30% greater than expected from the use of rocks from the SiCaMg group.

7 Concluding Remarks

Remineralizers from the ASiCaMg class are effective in improving soil fertility by increasing the levels of calcium, magnesium, phosphorus, and potassium, while also amending soil acidity. These benefits vary with soil texture and rock composition, allowing for productivity gains.

In addition to agronomic benefits, remineralizers help sequester atmospheric CO_2 by accelerating weathering to form bicarbonate, which provide a stable storage mechanism for carbon. Replacing limestone with remineralizers also reduces CO_2 emissions associated with liming, expanding its environmental potential.

Although many studies have shown positive results, some highlight the need to combine remineralizers with specific microorganisms or practices to maximize their benefits, reinforcing the importance of research adapted to local conditions.

Declaration of Competing Interest

The authors declare that they have no known competing financial interests or personal relationships that could have appeared to influence the work reported in this paper.

References

Alleoni LRF, Mello JWV de, Rocha WSD da (2019) Eletroquímica, adsorção e troca iônica no solo In: Química e mineralogia do solo. Viçosa: Sociedade Brasileira de Ciência do Solo, 1381 p

Anda M, Shamshuddin J, Fauziah CI (2015) Improving chemical proprieties of a highly weathered soil usind finely ground basalt rocks. Catena 124:147–61. https://doi.org/10.1016/j.catena.2014.09.012

Anda M, Shamshuddin J, Fauziah CI (2013) Increasing negative charge and nutrient contents of a highly weathered soil using basalt and rice husk to promote cocoa growth under field conditions. Soil Tillage Res 132:1–1. https://doi.org/10.1016/j.still.2013.04.005

Baek SH, Kanzaki Y, Lora JM, Planavsky N, Reinhard CT, Zhang S (2023) Impact of climate on the global capacity for enhanced rock weathering on croplands. Earths Futur 11(8):e2023EF003698. https://doi.org/10.1029/2023EF003698

Basak BB, Sarkar B, Naidu R (2021) Environmentally safe release of plant available potassium and micronutrients from organically amended rock mineral powder. Environ Geochem Health 43:3273–3286. https://doi.org/10.1007/s10653-020-00677-1

Bauwhede RVD, Muys B, Vancampenhout K, Smolders E (2024) Accelerated weathering of silicate rock dust predicts the slow-release liming in soils depending on rock mineralogy, soil acidity, and test methodology. Geoderma 441:116734. https://doi.org/10.1016/j.geoderma.2023.116734

Beerling et al. (2018) Farming with crops and rocks to address global climate, food and soil security. Nat Plants 4(3):138–47. https://doi.org/10.1038/s41477-018-0108-y

Beerling et al. (2024) Enhanced weathering in the US Corn Belt delivers carbonremoval with agronomic benefits. PNAS. https://doi.org/10.1073/pnas.2319436121

Conceição et al (2022) Potential of basalt dust to improve soil fertility and crop nutrition. J Agric Food Res 10:100443. https://doi.org/10.1016/j.jafr.2022.100443

Crusciol CAC, Moretti LG, Bossolani JW, Moreira A, Micheri PH, Rossi R (2019) Can dunite promote physiological changes, magnesium nutrition and increased corn grain yield?. Commun Soil Sci Plant Anal 50(18):2343–53. https://doi.org/10.1080/00103624.2019.1659304

Crusciol CAC, Soratto RP, Gilabel AP, Costa CHM da, Campos M de, Castro GSA, Ferrari Neto J (2022) Broadcast application of ground silicate rocks as potassium sources for grain crops. Pesquisa Agropecuária Brasileira 57:e02443. https://doi.org/10.1590/S1678-3921.pab2022.v57.02443

Curi N, Kämpf N, Marques JJ (2005) Mineralogia e formas de potássio em solos do Brasil. In: Potássio na agricultura brasileira, Yamada T, Roberts TL. Associação Brasileira para a Pesquisa da Potassa e do Fosfato: Piracicaba, Brasil, pp 71–92

Daniell A, Tonder DM Van (2023) Opportunity for increasing the soil quality of non-arable and depleted soils in south africa: a review. J Soil Sci Plant Nutr 23(2):2476–87. https://doi.org/10.1007/s42729-023-01205-7

Etesami H, Emami S, Alikhani HÁ (2017) Potassium solubilizing bacteria (KSB):: mechanisms, promotion of plant growth, and future prospects a review. J. Soil Sci. Plant Nutr 17(4):897–911. https://doi.org/10.4067/S0718-95162017000400005

Kelland et al (2020) Increased yield and CO2 sequestration potential with the C4 cereal Sorghum bicolor cultivated in basaltic rock dust-amended agricultural soil. Glob Chang Biol 26(6):3658–76. https://doi.org/10.1111/gcb.15089

Korchagin J, Escoteguy PAV, Bostoluzzi EC (2022) Nutrient transfer in rangelands under rock powder amendment. J Plant Nutr Soil Sci 185(5):656–67. https://doi.org/10.1002/jpln.202200059

Lewis et al (2021) Effects of mineralogy, chemistry and physical properties of basalts on carbon capture potential and plant-nutrient element release via enhanced weathering. Appl Geochem 132:105023. https://doi.org/10.1016/j.apgeochem.2021.105023

Luchese AV, Leite IJGC, Giaretta APS, Alves ML, Pivetta LA, Missio RF (2023a) use of quarry waste basalt rock powder as a soil remineralizer to grow soybean and maize. Heliyon 9(3). https://doi.org/10.1016/j.heliyon.2023.e14050

Luchese AV, Leite IJG de C, Alves ML, Viecelle JP dos S, Pivetta LA, Missio RF (2023b) Can basalt rock powder be used as an alternative nutrient source for soybeans and corn?. J Soil Sci Plant Nutr 23(3):4044–54. https://doi.org/10.1007/s42729-023-01322-3

Luchese AV, Missio RF, Algeri A, Giaretta AP, Ludwig AL, Potrich G (2024a) Disponibilidade e absorção de nutrientes liberados por remineralizador em duas classes de solo. Revista Em Agronegócio E Meio Ambiente. https://doi.org/10.17765/2176-9168.2024v17n4e12118

Luchese AV, Missio RF, Algeri A, Giaretta AP, Ludwig AL, Potrich G (2024b) Remineralizador: o efeito de basalto em propriedades química de solos distintos. Ciências Agrárias e Meio Ambiente. https://doi.org/10.29327/5345677.2-4

Luchese AV, Pivetta LA, Batista MA, Steiner F, Giaretta APS, Curtis JCD (2021) Agronomic feasibility of using basalt powder as soil nutrient remineralizer. Afr J Agric Res 17(3):487–97. https://doi.org/10.5897/AJAR2020.15234

Mancuso MA, Soratto RP, Crusciol CAC, Castro GSA (2014) Effect of potassium sources and rates on arábica coffee yeld, nutrition, and macronutrient export. Rev. Bras. de Ciênc. Solo 38:1448–56. https://doi.org/10.1590/S0100-06832014000500010

Martins ÉS, Theodoro SH, Bernardez, F.F.G. Luchese AV, Bergmann M, Siqueira DS, Teixeira A, Azevedo A, Curtis JCD (2024) Produção brasileira de remineralizadores e fertilizantes naturais 2019–2023. Revista Novo Solo 6:14–29

Martins V, Silva DRG, Marchi G, Leite MCA, Martins E de S, Gonçalves ASF, Guilherme LRG (2015) Effect of alternative multinutrient sources on soil chemical proprieties. Rev Bras de Ciênc Solo 39:194–204. https://doi.org/10.1590/01000683rbcs20150587

Matichenkov VV, Bocharnikova EA (2001) The relationship between silicon and soil physical and chemical proprieties. In: Datnoff LE, Snyder GH, Kondörfer GH Silicon in agriculture. Elservier Science. https://doi.org/10.1016/S0928-3420(01)80017-3

Melo VF, Uchôa ACP, Dias F de O, Barbosa GF (2012) Doses de basalto moído nas propriedades químicas de um Latossolo Amarelo distrófico da savana de Roraima. Acta Amaz 42:471–6. https://doi.org/10.1590/S0044-59672012000400004

Medeiros DS, Sanchotene DM, Ramos CG, Oliveira LFS, Sampaio CH, Kautzmann RM (2021) Soybean crops cultivated with dacite rock by-product: a proof of a cleaner technology to soil remineralization. J Environ Chem Eng 9(6):106742. https://doi.org/10.1016/j.jece.2021.106742

Mohammed SMO, Brandt K, Gray ND, White ML, Manning DAC (2014) Comparison of silicate minerals as sources of potassium for plant nutrition in sandy soil. Eur J Soil Sci 65(5):653–62. https://doi.org/10.1111/ejss.12172

Moretti LG, Bossolani JW, Crusciol CAC, Moreira A, Micheri PH, Rossi R, Imaizumi C (2019) Dunite in agriculture: physiological changes, nutritional status and soybean yield. Commun Soil Sci Plant Anal 50(14):1775–84. https://doi.org/10.1080/00103624.2019.1635143

Novais RF, Smyth TJ (1999) Fósforo em solos e planta em condições tropicais. Viçosa, MG:UFV, 399 p

Oelkers E (2001) General kinetic description of multioxide silicate mineral and glass dissolution. Geochim Cosmochim Acta 65(21):3703–19. https://doi.org/10.1016/S0016-7037(01)00710-4

Pavan MA, Bloch MF, Zempulski HC, Miyazawa M, Zocoler DC (1992) Manual de análise química de solo e controle de qualidade. Circular Técnica 76:38

Ramos CG, Hower JC, Blanco E, Oliveira MLS, Theodoro SH (2022) Possibilities of using silicate rock powder: an overview. Geosci Front 13(1):101185. https://doi.org/10.1016/j.gsf.2021.101185

Ramos JFK, Luchese AV, Alonso JM, Missio RF, Mai KT, Ziech AR (2023) Efeito da adubação orgânica na produção e parâmetros de qualidade de cultivares de tomate. In: Ranst VE, Coninck de F (2002) Evaluation of ferrolysis in soil formation. European Journal of Soil Science. https://doi.org/10.1046/j.1365-2389.2002.00475.x

Rodrigues et al. (2024) Paraná basin basalt powder: a multinutrient soil amendment for enhancing soil chemistry and microbiology. J S Amn Earth Sci 141:104957. https://doi.org/10.1016/j.jsames.2024.104957

Schaeller J, et al (2019) Silicon increases the Phosphorus availability of artic soils. Sci Rep 9(1):449. https://doi.org/10.1038/s41598-018-37104-6

Silva, DRG, Merchi G, Spehar CR, Guilherm LR, Rein TA, Soares DA, Ávila FW (2012) Characterization and nutrient release from silicate rocks and influence on chemical changes in soil. Rev Bras de Ciênc Solo 47:e20160982. https://doi.org/10.1590/S0100-06832012000300025

Silva RC, Cury ME, Ieda JJC, Sermarini RA, Azevedo AC de (2017) Chemical atributes of a remineralized oxisol. Ciência Rural 47:e20160982. https://doi.org/10.1590/0103-8478cr20160982

Skov et al. (2024) Initial agronomic benefits of enhanced weathering using basalt: a study of spring oat in a temperate climate. PLOS One 19(3):e0295031. https://doi.org/10.1371/journal.pone.0295031

Souza F, Oliveira CG, Martins ES, Alves JM (2017) Efeitos condicionador e nutricional de um remineralizador de solos obtido de resíduos de mineração. Agri-Environ Sci 3(1):1–4. https://doi.org/10.36725/agries.v3i1.204

Sposito G (1989) The chemistry of soils. Oxford Univ Press, New York, 277 p

Swoboda P, Döring TF, Hamer M (2022) Remineralizing soils? The agricultural usage of siliate rock powders: a review. Sci Total Environ 807:150976. https://doi.org/10.1016/j.scitotenv.2021.150976

Toscani RGS, Campos JEG (2017) Uso de pó de basalto e rocha fosfatada como remineralizadores em solos altamente intemperizados. Geociências 36(2):259–274

Vienne A, Poblador S, Estrada MP, Ijiehon S, Wade P, Vicca S (2022) Enhanced weathering using basalt rock powder: carbon sequestration, co-benefits and risks in a mesocosm study with Solanum tuberosum. https://doi.org/10.3389/fclim.2022.869456

Viviani CA, Marchctti ME, Vitorino ACT, Novelin JO, Gonçalves MC (2010) Disponibilidade de fósforo em dois Latossolos argilosos e seu acúmulo em plantas de soja, em função do aumento do pH. Ciência e Agrotecnologia. https://doi.org/10.1590/S1413-70542010000100007

Wrege MS, Garrastazu MC, Soares MTS, Fritzsons E, Souza VA de, Aguiar AV de (2017) Principais fitofisionomias existentes no estado do Paraná e os novos cenários definidos pelas mudanças climáticas globais. Ambiencia. https://doi.org/10.5935/ambiencia.2017.03.05

Zhao X, Gao S, Lu D, Wang H, Chen X, Zhou J, Zhang L (2019) Can potassium silicate mineral products replace conventional potassium fertilizers in rice–wheat rotation?. Agron J 111(4):2075–83. https://doi.org/10.2134/agronj2019.01.0020

A Protocol for Assessing the Potential of Soil Acidity Correction by Employing Silicate Agrominerals Through Short-Term Incubation

Giuliano Marchi, Luise Lottici Krahl, Ewerton Gonçalves de Abrantes, Paula Caroline Ferreira Rodrigues, Elis Marina de Freitas, João Paulo Guimarães Soares, and Éder de Souza Martins

Abstract This chapter presents a protocol for evaluating the potential of soil acidity correction through the application of silicate agrominerals within a controlled short-term incubation framework. It explores theoretical underpinnings of soil acidity, including the dissolution mechanisms of silicate agrominerals, the role of oxidation processes in remineralizers containing iron, and the impact of incubation time on the efficacy of these amendments. By focusing on the mitigation of soil acidity using silicate agrominerals, the text offers a comprehensive methodology, emphasizing laboratory incubation of basic rocks to assess their reactivity and effectiveness in neutralizing acidic soils. The proposed protocol aims to bridge the gap between theory and practical application, providing a reliable foundation for future research and sustainable agricultural practices.

Keywords Soil neutralization · Soil reactions · Agromineral characterization

G. Marchi (✉) · L. L. Krahl · E. G. de Abrantes · P. C. F. Rodrigues · E. M. de Freitas · J. P. G. Soares · É. de Souza Martins
Agrogeological Laboratory, Empresa Brasileira de Pesquisa Agropecuária—Embrapa Cerrados, BR 020, km 18, Planaltina, Distrito Federal 73310970, Federal District, Brazil
e-mail: giuliano.marchi@embrapa.br

J. P. G. Soares
e-mail: jp.soares@embrapa.br

É. de Souza Martins
e-mail: eder.martins@embrapa.br

É. de Souza Martins and S. Huff Theodoro (eds.), *Soil Remineralizers and Silicate Fertilizers*, Frontier Studies in Soil Science,
https://doi.org/10.1007/978-3-032-14656-4_7

1 Introduction

Silicate agrominerals (ASi) are raw materials obtained from silicate rocks that contain bases and primary minerals for soil remineralizers (REM) and silicate fertilizers (FSi). Considering the chemical and mineralogical diversity of these inputs, it is essential to understand their elemental composition, constituent minerals, and granulometry for their optimal application in agriculture. It is possible to determine the capacity of each rock to provide essential nutrients and infer its resistance to dissolution by weathering processes after addition to soils (Ribeiro 2018). However, this data alone does not reveal its potential for correcting soil acidity.

This methodology was developed in response to the requirement for a straightforward adjustment in experiments conducted in pots within a greenhouse at Embrapa Cerrados: the pH adjustment of soils treated with silicate agrominerals. The addition of silicate minerals to soils often results in a change in pH, depending on the type of mineral and its category, such as potassium suppliers or acidity correctives. The initial research objective was complicated by the introduction of a new variable: soil pH variation caused by the addition of REM. This change meant that not only the release of potassium, but also other effects could either benefit or disadvantage an agromineral source. Thus, incubation curves were used to adjust soil pH with a 3:1 mix of $CaCO_3$ and $MgCO_3$, or dolomitic limestone.

In response to the need to evaluate new types of silicate agrominerals that are not potassium suppliers but rather belong to a category that corrects soil acidity and primarily provides calcium and magnesium as nutrients, a new experiment was conducted. This study aimed to assess the potential of metabasalt from the Carajás, Parauapebas—PA, Brazil, as a soil acidity mitigator in comparison to commercial limestone. A new incubation experiment was conducted in the laboratory to determine the optimal amount of agromineral required to achieve optimal pH values for agricultural use, without relying on limestone. The pH incubation method was successful, and researchers realized it could be applied to other silicate minerals where the proper usage or doses were unclear.

Some methods, such as the relative power of total neutralization (RPTN) and the calcium carbonate equivalent, were compared with the pH incubation method. All of them involve a degree of uncertainty due to the nature of agrominerals and the reactions of their components, which include carbonates of Ca and Mg, as well as simple and complex silicates of Ca and Mg. The kinetics of mineral dissolution in the soil, the release of FeII in the soil, and polymerization reactions of Si also affect the equilibrium constants of pH reactions in soils, altering the predictability.

The pH incubation method is a straightforward and rapid technique. Like the abrasion pH method, it provides data that enables the precise adjustment of silicate agrominerals dosage, which can be used to correct soil acidity in various types of soils. This method will enable experiments with unidentified acidity corrective materials to be evaluated within more logical dosage ranges that align closely with agricultural requirements. Additionally, it will assist field technicians in selecting the appropriate doses of acidity-corrective silicate agrominerals more accurately than current practices. This approach will also prevent soil pH from being raised to levels harmful to plants or administered in suboptimal doses.

2 Soil Acidity: Theory

Soil acidity, caused by H^+ ions from acidic components, influences the solubility of essential plant nutrients, thereby affecting soil fertility and plant growth. Soil correction neutralizes acidity with hydroxides (OH^-) and reduces toxic elements, such as Al, Mn, and Fe. It also increases the availability of Ca, Mg, P, B, and Mo and enhances microbial activity, including N fixation by legumes (Novais et al. 2007).

Soil solution, minerals, and soil exchange complex can reduce soil acidity. Soil minerals decrease soil acidity by consuming protons during chemical weathering reactions. Nevertheless, in the context of acid–base reactions in the soil solution and exchange complex, mineral weathering reactions are typically considered kinetically slow and are not often discussed or managed when addressing soil acidity. The interactions between the soil exchange complex and the solution components are both rapid and typically reversible, not restricted by kinetic constraints. Consequently, the composition of the exchange complex quickly adjusts in response to shifts in the soil solution's composition and can be effectively manipulated (Essington 2003).

Liming is commonly used to correct soil acidity in agriculture. However, several igneous, metamorphic, and sedimentary rocks have the potential to be used as soil acidity correctives. These rocks also contain essential macroelements (K, Ca, Mg, P) and micronutrients (Si, Fe, Zn, Mn, Cu) that are crucial for plant growth and development (Marchi et al. 2020; Martins et al. 2023). The effectiveness of soil acidity correction is influenced by the application rate, chemical and mineralogical composition, particle size characteristics, and soil properties. Certain silicate agromineral derivatives, including silicate agrominerals and silicate fertilizers (FSi), incorporate carbonates (calcite and magnesite) as well as calcium and magnesium silicates in their composition. These substances act as correctives for soil acidity. Calcite and magnesite exhibit greater reactivity than calcium and magnesium silicates and possess a higher neutralization capacity (Table 1).

Studies indicate that inputs derived from ultramafic rocks (Moretti et al. 2021), calcissilicate (Rodrigues et al. 2021), basic (Conceição et al. 2022), alkaline (Nogueira et al. 2021b), and ultramafic alkaline rocks (Ferreira et al. 2020) have the potential to aid in soil correction and provide nutrients.

2.1 Silicate Agromineral Dissolution

The dissolution of $CaCO_3$ in the soil consumes H^+ ions and releases CO_2 into the atmosphere, which is a gas that affects the planet's temperature.

Table 1 Neutralization capacity of various neutralizing agents relative to calcium carbonate ($CaCO_3$)

Constituents	Neutralization capacity relative to $CaCO_3$ ($ECaCO_3$)
Calcium carbonate	1.00
Magnesium carbonate	1.19
Calcium hydroxide	1.35
Magnesium hydroxide	1.72
Calcium oxide	1.79
Magnesium oxide	2.48
Calcium silicate	0.86
Magnesium silicate	1.00

[1] *Source* Alcarde (2005)

$$3CaCO_3(s) + 6H^+(aq) \rightarrow 3Ca^{2+}(aq) + 3CO_2(g) + 3H_2O(l) \quad (1)$$

The impact of ASi on the soil occurs continuously and can be divided into two phases. Silicates with finer granulometry and simpler structures, along with minerals from the carbonate group, are dissolved first. In many cases, ASi may not supply all the necessary nutrients and acidity correction required for crop growth in the short term. Over time, after fine particles are consumed, the dissolution process of coarser particles and those that undergo partial (incongruent) dissolution of minerals begins. This leads to the release of nutrients and the formation of new mineral phases (Basak 2018; Krahl et al. 2022), potentially generating both permanent and variable charges in the soil and reducing soil acidity. The second stage involves biological processes known as bioweathering (Krahl et al. 2022). During this process, silicon is released from amorphous silica (ASi) through a combination of hydrolysis and oxidation, thereby introducing new minerals into the system. Hydrolysis involves the interaction of water and protons (H^+), resulting in alterations to the chemical structure of the elements (Essington 2003).

The hydrolysis of a nesosilicate can be described by the following reaction:

$$\text{Primary silicate} + H_2O + H_2CO_3 \rightarrow \text{Basic cation} + HCO_3 + H_4SiO_4 + \text{New mineral phases} \quad (2)$$

Neutralization reactions happen in two stages. Initially, Ca or Mg reacts with H^+ within the exchange complex, leading to the replacement of H^+ by Ca^{2+} or Mg^{2+} at the exchange sites, which results in the formation of HCO_3. In the second stage, HCO_3 reacts with H^+ to form CO_2 and H_2O, resulting in an increased pH of the solution. This reaction involves consuming H_2O and H^+, releasing silicic acid (H_4SiO_4) and basic cations (Mg^{2+}, Ca^{2+}), and forming accessory minerals.

2.2 *Oxidation of Fe in Silicate Agrominerals*

Oxidation is a process of chemical weathering in which an element in the crystal structure loses an electron, resulting in the acidification of the medium. The most frequently oxidized element during the weathering of primary minerals in soil is FeII, with MnII being oxidized to a lesser extent. These elements are consistently found in the reduced state within the structure of primary minerals. Oxidation occurs through two primary mechanisms: (1) structural oxidation of Fe^{2+}, which initiates the decomposition of the mineral, and (2) the oxidation of Fe^{2+} following its release into the soil solution via hydrolysis. The oxidation of Fe^{2+} to Fe^{3+} within the mineral's structure disrupts its electrostatic neutrality. Consequently, cations exit the crystal to preserve electroneutrality (Essington 2003).

Initial reaction of olivine with water and carbon dioxide:

$$(Mg, Fe)_2SiO_4 + 4CO_2 + 4H_2O \rightarrow 2\left(Mg^{2+}, Fe^{2+}\right) + 4HCO_3^- + H_4SiO_4 \quad (3)$$

Formation of magnesium and iron carbonates:

$$Mg^{2+} + 2HCO_3^- \rightarrow MgCO_3 + CO_2 + H_2O \quad (4)$$

$$Fe^{2+} + 2HCO_3^- \rightarrow FeCO_3 + CO_2 + H_2O \quad (5)$$

These carbonates are important because they store carbon stably in temperate climate soils. However, in acidic tropical soils, these carbonates are not stable. In tropical soils with a pH below 6.37, the most prevalent form is carbonic acid (H_2CO_3), as described by the equilibrium reaction $H_2CO_3 \rightleftharpoons H^+ + HCO_3^-$,, where the dissociation constant (pKa) is 6.37. In soils with lower pH, carbon sequestration is limited due to less carbonate formation. Above pH 6, it increases because bicarbonate becomes highly soluble.

Precipitation of silica as quartz or other amorphous forms:

In aqueous medium, silica primarily exists as monomeric silicic acid (H_4SiO_4), which can undergo two sequential deprotonations:

$$H_4SiO_4 \rightleftarrows H_3SiO_4^- + H^+ \quad pK_{a1} \approx 9,8 \quad (6)$$

$$H_3SiO_4^- \rightleftarrows H_2SiO_4^{2-} + H^+ \quad pK_{a2} \approx 12,7 \quad (7)$$

Consequently, at pH levels below approximately 9.8, neutral H_4SiO_4 becomes unstable in solution.

2.2.1 *Oxidation of Fe^{2+} and the Release of Protons*

After its release from minerals, the oxidation of ferrous iron (Fe^{2+}) to ferric iron (Fe^{3+}) can initially take place without producing hydrogen ions, particularly in neutral or basic conditions. However, significant acidification occurs in the subsequent reaction when Fe^{3+} interacts with water.

Hydrolysis of Fe^{3+} and acidity generation

Fe^{3+} ions (as well as Al^{3+}) readily hydrate and form hydroxide complexes. The hydrolysis of ferric iron releases H^+ ions into the medium, increasing its acidity.

$$Fe^{3+} + H_2O \rightarrow Fe(OH)_2 + H^+ \quad (8)$$

$$Fe(OH)^{2+} + H_2O \rightarrow Fe(OH)_2^+ + H^+ \quad (9)$$

$$Fe(OH)_2^+ + H_2O \rightarrow Fe(OH)_3 + H^+ \quad (10)$$

2.3 *Incubation Time*

The mineralogical and chemical composition of ASi will lead to reactions that are similar to, but more complex than, those resulting from the dissolution of carbonates. The speed of these reactions can vary significantly. The solubility of pure calcium silicates ($CaSiO_3 = 0.095$ g dm^{-3}) in water is up to 6.8 times higher than that of calcium carbonate ($CaCO_3 = 0.014$ g dm^{-3}; Nolla et al. 2013). Calcium silicates also exhibit greater mobility in deeper soil horizons. Calcium and magnesium silicates may prove to be more effective than limestone in enhancing phosphorus availability and mitigating toxic Al levels (Lima Filho 2011). The rapid dissolution discussed here does not accurately represent the dissolution process for most agrominerals, which possess a more complex structure that incorporates aluminum. These minerals rely on both solubilization in water and bioweathering processes (Krahl et al. 2020).

Another factor affecting ASi dissolution rates is the type of mineral. The complexity of ASi dissolution increases with the presence of different minerals. The formation of aqueous species is often considered an immediate process. However, the equilibrium between a mineral and an aqueous phase can be influenced by kinetic factors. A solid in its standard state is pure, crystalline, and has a negligible specific surface area. However, precipitates that form within the soil environment generally do not meet these standard state criteria. The equilibrium solubility (logK sp) of amorphous minerals is higher than that of their crystalline counterparts, which is associated with an increased specific surface area. The mineral value log K sp for amorphous or microcrystalline minerals shows a positive deviation compared to crystalline ones, which increases with specific surface area (Essington 2003).

Even though, in a six-month leaching column simulation with 2750 mm of rainfall, pH, CEC, bases before treatment, and H^+ Al were measured in seven highly weathered soils treated with basalt doses (Gillman et al. 2001). The study indicated that the pH levels measured before and after the incubation period in the columns exhibited a strong similarity, particularly at higher basalt concentrations. The research conducted by Shanshuddin and Fauziah (2010) indicated that over an incubation period of 4–14 months, the pH levels of soil treated with varying dosages of basalt exhibited minimal change. Therefore, the swift balance between fine basalt particles, which effectively react within a short duration, and the soil suspension serves as a reliable parameter to reasonably estimate the necessary basalt dosage in numerous cases. Using longer incubation times may improve dose estimates, but seeking the ideal time can be impractical due to the material nature, soil biology dependence, and oxy-reduction reactions that may compromise the method.

Bioweathering contributes to the dissolution of minerals in soil. However, extended incubation periods can cause microorganisms to grow in moist, enclosed soil, potentially resulting in inaccurate pH readings. In a reducing environment, the pH of soil can naturally increase within a few hours without the addition of any additives (Gillman et al. 2001). Adding components that inhibit microbial growth can also affect the rate of mineral dissolution by preventing bioweathering, even in a short period.

2.4 *Mitigating Soil Acidity with Silicate Agrominerals*

Methods for mitigating acidity in tropical soils should focus on alleviating Al toxicity, given the sensitivity of the target crops. Additionally, addressing other soil limitations, such as calcium (Ca) and magnesium (Mg) deficiencies, is necessary (Gillman et al. 2001). Exchangeable acidity decreases exponentially with increasing soil pH. Soils with a pH of 5.1 or lower exhibit high levels of exchangeable acidity. However, not all low-pH soils possess this characteristic. Exchangeable acidity diminishes to nearly zero at a pH level of 5.5 and is virtually absent in soils with a pH above 6 (Abreu et al. 2003). Consequently, maintaining a target pH of 5.5 is adequate to address most issues related to aluminum toxicity.

Small amounts of most REM (usually 1–3 t ha^{-1}) will increase the pH of agricultural soil by a certain proportion. An elevation of 0.1 pH unit within the range of 4.5–5.5 is sufficient to precipitate part of the Al^{3+} in its hydroxide form (Gillman et al. 2001).

In acidic tropical soils, crop yields may not be affected by low pH levels, such as 5.0, depending on the plant species and other soil chemical properties (Zhang et al. 2023). The pH requirements for each crop vary, influenced by the specific soil conditions (Goulding 2016). Therefore, the target pH should be adjusted locally to suit the specific crop conditions and needs (Teixeira et al. 2020).

As illustrated in the data presented within this document (Table 5), silicate agrominerals with Neutralizing Power (NP) values and concentrations of CaO and

MgO (EDTA) that are nearly identical, such as metabasalt, basalt, and amphibole gneiss, exhibit significantly different pH reactivity in soil when assessed using the pH incubation method. The pH reactivity is also an indicator of REM dissolution (Marchi et al. 2024), although it is not straightforward due to the presence of some antagonistic effects. For example, the basalt MG hinders pH increase through the Fe oxidation process. The metabasalt demonstrates substantial reactivity, whereas the amphibole gneiss is almost inert with respect to soil pH correction. Therefore, the pH incubation curve method is essential for comprehending the reactivity of the mineral assemblages contained in agrominerals and determining doses of REM to alleviate soil acidity to suitable pH levels. Simultaneously, the levels of exchangeable Ca, Mg, and Al can be determined following pH measurement by introducing a spike of KCl solution as an extractant into the suspension, achieving a final concentration of 1 mol L^{-1}. This method may offer valuable insights into the availability of Ca and Mg, as well as Al toxicity.

For measuring exchangeable Ca, Mg, and Al in the soil after pH measurements, 10 g of soil can be placed into 125 mL Erlenmeyer tubes. Treatments such as control, REM, or limestone can be applied. Add 25 mL of water, then measure pH at 1, 24, and 72 h. Next, a solution with 99.41 g L^{-1} KCl can be used to bring the volume up to 100 mL. This will result in a final concentration of 1 mol L^{-1} KCl. Shake for 5 min using a circular horizontal shaker. After shaking, allow the mixture to rest overnight. Using a pipette, transfer three aliquots of 25 mL from the supernatant portion of the extractor solution into separate 125 mL Erlenmeyer flasks. This method will be used for determining the exchangeable Ca, Mg, and Al concentrations, following the specific soil analysis methodology outlined by Teixeira et al. (2017).

An additional significant consequence of the dissolution of basic silicates is their potential to buffer soil pH over an extended period (Marini 2006). This is attributed to the slower dissolution rate of silicates, which supports a more prolonged and sustained process compared to the rapid dissolution of carbonates. Furthermore, the dissolution rate of basic silicates is influenced by the pH level. Specifically, the rate increases in more acidic environments and decreases as the pH level rises, demonstrating its buffering capacity (White and Brantley 2003).

2.5 *Why Use Silicate Agrominerals Instead of Liming?*

Most waste rock in Brazil originates from mining activities and is considered an environmental liability if not properly managed (Lefebvre et al. 2019; Ramos et al. 2015). When selecting locally available rock materials for specific soil types, these materials have been shown to impact local agriculture positively (Ciceri and Allanore 2019; van Straaten 2006). Using silicate agrominerals in agriculture reduces quarry waste and environmental risks, promotes the Circular Economy, supports sustainability, and encourages the recycling of materials that would otherwise contribute to environmental contamination (Camara et al. 2021).

Some nutrients required for plant growth, such as K, Ca, Mg, and S, as well as micronutrients like Mn, B, Zn, Ni, Cu, and Fe, are abundant in either the structure or interlayer sites of minerals present in these rocks. Basic rocks, rich in Fe, Si, Ca, and Mg, such as basalt, are globally available in large quantities and exhibit a relatively rapid weathering rate (Anda et al. 2015; Lefebvre et al. 2019; Krahl et al. 2020).

The direct effects of silicate agrominerals in these soils include an increase in soil pH, nutrient supply, and the formation of new mineral phases that benefit the physicochemical characteristics of agricultural soils (Anda et al. 2015; Basak et al. 2018; Burghelea et al. 2018; Ciceri et al. 2019; Krahl et al. 2022). These include a range of clay minerals, such as vermiculite and smectite (Hinsinger et al. 1992; Norouzi and Khademi 2010), as well as amorphous phases and low-crystallinity minerals (Colombo et al. 2014; Yu et al. 2017), characterized by a large specific surface area (SSA) and higher cation exchange capacity (CEC). The weathering process, along with the amounts applied over time, can modify soil properties. Most soils under tropical conditions exhibit pH-dependent charges and losses of nutrient cations through leaching (Fontes and Alleoni 2006). Besides, the moisture conditions and the high temperatures favor the oxidation of soil organic matter (SOM), which is responsible for most of the cation retention in these soils.

An additional significant consequence of the dissolution of basic silicates is their potential to buffer soil pH over an extended period (Marini 2006). This buffering is attributed to the slower dissolution rate of silicates, which supports a more prolonged and sustained process compared to the rapid dissolution of carbonates. This gradual dissolution not only buffers soil acidity but also enhances carbon sequestration potential by forming stable mineral phases that lock carbon within the soil matrix. In contrast to traditional liming, the use of silicate agrominerals offers notable climate change mitigation benefits. The use of limestone for soil acidity correction releases CO_2 into the atmosphere as a by-product of its reaction with soil acids (Eq. 1). This reaction between lime and protons generates bicarbonate, which decomposes into water and carbon dioxide, contributing to greenhouse gas emissions. Thus, while liming provides a rapid pH correction, it also has environmental implications, as it increases atmospheric CO_2 levels.

As the application rate is common in the range of many tons per hectare, silicate agrominerals should be applied regionally (Martins et al. 2023) to reduce their lifecycle impact. The use of silicate agrominerals has a regional reach, determined by the economic radius of logistics, considering the volumes required by crops and the characteristics of the available rocks. Regional geology, soils, and crops necessitate specific recommendations for the use of remineralizers, aiming to achieve long-term benefits and facilitate structural soil improvement, while supplementing and complementing conventional fertility management practices.

Table 2 Chemical analysis of the Ferralsol (FR) and the Arenosol (AR)

Soil	pH[a]	P	K	Ca	Mg	Al	Na	H+Al	SB	CEC[b]
		($mg\ kg^{-1}$)		($cmol_c\ dm^{-3}$)						
FR	4.6	13	0.11	0.4	0.2	1.2	0.02	5.4	0.7	6.1
AR	4.8	13.2	0.10	0.4	0.2	0.5	0.02	3.7	0.7	4.4

Soil	BS	m	OM	C[c]	B	Cu	Fe	Mn	Zn
	(%)		($g\ kg^{-1}$)		($mg\ dm^{-3}$)				
FR	12	62	22.2	12.9	0.26	1.32	86.8	3.6	2.04
AR	16	41	7.1	4.1	0.31	1.16	144.8	2.3	1.59

[a]pH in water; [b]CEC at pH 7,0; [c]Organic carbon

3 Methodology

3.1 Soil Characterization

Two soils with differing textures, low pH values, and high levels of exchangeable Al were selected from native forest areas. Soil samples were collected using a loader from a depth of 0 to 20 cm and were thoroughly homogenized before the experiment. The first soil, a Ferralsol—FR (IUSS Working Group WRB 2022)/Dystrophic Red Latosol (Santos et al. 2018) with a clayey texture, was located at coordinates 15° 35′ 38.3″ S 47° 44′ 12.9″ W in Planaltina, DF, Brazil. The second soil, an Arenosol—AR (IUSS Working Group WRB, 2022)/Quartzarenic Neosol (Santos et al. 2018) with a sandy texture, was located at coordinates 15° 36′ 45.9″ S 47°44′22.5"W (Table 2). The soils were collected under forest conditions to ensure that the acidity correction method could be tested on soil that had not previously been treated with any type of corrective agents (Table 2).

3.2 Characterization of Silicate Agrominerals Used

Various silicate agrominerals with potential for acidity amendment were used in the study. These include basalt from Uberaba, MG (Famelli et al. 2021); metabasalt, a basic volcanic rock that have undergone regional or hydrothermal metamorphism processes from Carajás, PA (Ribeiro 2023); basalt from Balsas, MA (Nogueira et al. 2021a); wollastonite, composed of calcite formed in ultrahigh-temperature (UHT) metamorphism from Anicuns, GO (Moraes et al. 2007); amphibole gneiss, a metamorphic basic rock from Salvador, BA (Barbosa and Sabaté 2004); serpentinite from Nova Lima, MG (Viana et al. 2021); and calc-schist from Coromandel, MG (Tables 3 and 4).

The doses were determined based on the RPTN of each material (Table 5). For limestone (Table 6) and wollastonite, lower doses were administered, up to

Table 3 Mineralogical composition of silicate agrominerals using X-ray diffraction (DRX) and Rietveld refinement

	LCM[1]	Cc[2]	Dm[3]	Px[4]	Pl[5]	FK[6]	Bt[7]	Fl[8]	Mu[9]	Ct[10]	Am[11]	St[12]
Basalt MG	18.3	0	0	27.07	32.12	2.84	0	0	0	0	0	4.43
Metabasalt*	0	11.2	0	0	11.3	7.4	2.4	0	9.5	30.9	0	0
Basalt MA	15.2	0	0	30.42	21.68	2.54	0	0	0	0	0	2.15
Wollastonite	0	74.82	0	9.4	0	3.02	0	0	0.93	0.8	0	0
Amphibole Gneiss	0	0	0	0	42.78	13.05	14.82	0	0	4.02	3.69	0
Serpentinite	0	0	3.67	2.77	0	0	0	0	0	22.81	4.31	0
Calc-schist	0	8.31	24.21	0	6.98	0	0	2.95	21.26	8.74	0	0

	Vm[13]	Fp[14]	Qz[15]	Im[16]	Mt[17]	Hm[18]	Sp[19]	Ze[20]	Il[21]	Tc[22]	Total
Basalt MG	0	1.16	0	2.08	2.67	0	0	2.04	6.58	0	98.78
Metabasalt*	0	0	21.6	0	0	4.8	0	0	0	0	99.1
Basalt MA	0	2.78	0	0	0	1.69	0	15.39	6.61		98.46
Wollastonite	0	0	10.95	0	0	0	0	0	0	0	99.92
Amphibole Gneiss	0	0	16.65	0	2.68	0	0	0	0	1.03	98.72
Serpentinite	1.54	0	0	0	2.28	0	37.23	0	0	28.12	99.96
Calc-schist	0	0	26.56	0	0	0	0	0	0.79	0	100

Mineralogical analyses were performed by CRTI (https://crti.ufg.br). The materials analyzed included low-crystallinity minerals (**LCM[1]**), calcite (**Cc[2]**) $CaCO_3$, dolomite (**Dm[3]**) $CaMg(CO_3)_2$, pyroxene (**Px[4]**) $(Ca, Na)(Mg, Fe, Al, Ti)(Si, Al)_2O_6$, plagioclase (**Pl[5]**) $(Na, Ca)(Si, Al)_4O_8$, potassium feldspar (**FK[6]**) $KAlSi_3O_8$, biotite (**Bt[7]**) $K(Mg, Fe^{2+})_3[AlSi_3O_{10}(OH, F)_2]$, phlogopite (**Fl[8]**) $KMgSi_3O_{10}$ $(OH, F)_2$, muscovite (**Mu[9]**) $KAl_2(Si_3Al)O_{10}(OH, F)_{22}$, chlorite (**Ct[10]**) $(Mg, Fe^{2+})_5Al(Si_3Al)O_{10}(OH)_8$, amphibole (**Am[11]**) $A_{0-1}B_2C_5T_8O_{22}$ $(OH, F)_2$ (where: A = Na^+, K^+; B = Ca^{2+}, Na^+, Mg^{2+}, Fe^{2+}; Mn^{2+}, Li^+; C = Mg^{2+}, Fe^{2+}, Al^{3+}, Fe^{3+}; Mn^{2+}, Zn^{2+}, Cr^{3+}, Li^+, or Ti^{4+}; T = Si, Al), smectite (**St[12]**) $(Na, Ca)_{0.3}(Al, Mg)_2Si_4O_{10}(OH)_2{\bullet}n(H_2O)$, vermiculite (**Vm[13]**) $(Mg, Fe^{2+}, Al)_3(Al, Si)_4O_{10}(OH)_2{\bullet}4(H_2O)$, feldspathoid (**Fp[14]**) $(Na, K)AlSiO_4$, quartz (**Qz[15]**) SiO_2, ilmenite (**Im[16]**) $Fe^{2+}TiO_3$, magnetite (**Mt[17]**) $Fe^{2+}Fe^{3+}{}_2O_4$, hematite (**Hm[18]**) Fe_2O_3, serpentine (**Sp[19]**) $(Mg, Fe)_3(Si_2O_5)(OH)_4$, zeolite (**Ze[20]**) $M_{x/n}$ $[(AlO_2)_x$ $(SiO_2)_y].mH_2O$, illite (**Il[21]**) $(K, H_3O)(Al, Mg, Fe)_2(Si, Al)_4O_{10}[(OH)_2, (H_2O)]$, and talc (**Tc[22]**) $Mg_3Si_4O_{10}(OH)_2$. *Modified from Ribeiro (2023)

50 mg 10 g^{-1}. Higher doses were employed for materials with lower RPTN values. Consequently, pH curves were established according to the applied rock doses, simulating the mineral dissolution process—particularly of fine-sized Ca- and Mg-rich carbonates and silicates.

The reactivity of silicate agrominerals was assessed using a modified abrasion pH method (Table 7), which involved mixing 10 g of rock or mineral powder with 25 mL of water for 1 h. The pH of the suspension was then recorded after remixing. Magnetic susceptibility measurements were conducted using a Bartington MS2B magnetic susceptibility meter on 10 g soil samples (Table 7).

Table 4 Contents of chemical elements in silicate minerals

	SiO_2	TiO_2	Al_2O_3	Fe_2O_3	MnO	MgO	CaO	Na_2O	K_2O	P_2O_5	SO_3	LOI*	Total
	(%)												
Basalt MG[1]	49.04	3.25	12.76	15.55	0.21	5.74	8.66	1.94	1.15	0.33	0.32	1.67	100.6
Metabasalt[2]	49	0.8	12.5	14	0.4	5.4	5.5	1.1	2.6	0.1	0.28	5.6	97.28
Basalt MA[1]	51.01	1.12	14.21	11.05	0.18	6.58	9.01	1.91	1.01	< LQ	< LQ	4.55	100.6
Wollastonite[1]	18.02	< LQ	1.23	0.9	< LQ	0.87	45.92	< LQ	0.22	0.1	0.56	31.94	99.7
Amphibole Gneiss[1]	60.45	0.77	15.6	7.28	0.11	3.98	4.81	3.03	3.01	0.24	0.27	1.02	100.5
Serpentinite[1]	44	< LQ	1.95	9.06	0.13	32.96	1.02	< LQ	< LQ	< LQ	0.26	9.98	99.37
Calc-schist[1]	44.69	0.59	10.22	4.86	0.13	7.41	12.11	0.56	2.81	0.12	0.29	16.58	100.3

[1]Analyses were performed by CRTI (https://crti.ufg.br). [2]Modified from Ribeiro (2023). *Loss on ignition

Table 5 Relative power of total neutralization (RPTN) from selected silicate agrominerals

Material	CaO	MgO	NP	RPTN*	Passing through sieve (mesh %)		
		(%)			10	20	50
Basalt MG	1.7	4.4	9.0	8.8	100	99	95
Metabasalt	2.3	4.4	8.0	7.5	100	95	89
Basalt MA	2.3	3.6	8.5	6.4	100	93	45
Wollastonite	45.8	0.4	84.0	74.9	100	96	77
Amphibole Gneiss	2.3	4.4	9.0	7.2	100	94	56
Serpentinite	1.1	8.8	10.0	9.8	100	100	96
Calc-schist	10.3	6.8	10.4	8.8	100	90	71

NP: Neutralization potential

Table 6 Composition, reactivity, and weathering potential of the limestone

	CaO	$CaCO_3$	MgO	$MgCO_3$	NP	RE[1]	RPTN[2]	Passing through sieve (mesh %)			Classification
								10	20	50	
Limestone	29.7	52.9	18.7	39.3	117.5	81.3	95.5	98.9	89.6	64.3	Dolomitic

[1]Relative Efficiency; [2]Relative Power of Total Neutralization

Table 7 Abrasion pH of soil silicate agrominerals

	Abrasion pH	Magnetic susceptibility (m^3 kg^{-1})
Basalt MG	9.03	677
Metabasalt	9.05	104
Basalt MA	9.25	370
Wollastonite	8.99	6
Amphibole gneiss	9.27	1307
Serpentinite	9.11	1848
Calc-schist	9.53	13
Biotite schist	8.99	13
Limestone	9.48	3

3.3 *Incubation of Basic Silicate Agrominerals in the Laboratory*

A laboratory incubation experiment was conducted to assess the potential of selected silicate agrominerals for mitigating soil acidity.

In 50 mL containers, 10 g of soil was measured, and varying doses of agromineral were added to establish pH curves, based on estimates derived from their RPTN.

Three replicates were included for each amount, considering that the quantity of silicate agromineral added to the soil was measured for both practicality and precision, given the very small quantities involved. Afterward, 25 mL of water was introduced into each container holding the samples, followed by shaking the mixtures. After one hour, the suspensions were shaken again, and the pH was measured. This procedure was repeated 24 and 72 h after the start of incubation.

The results were modeled to derive the most straightforward equation, ideally a linear or first-order equation (Viana et al. 2023—exponential rise to the maximum—as found in SigmaPlot 14.5), that accurately represents the behavior of the data in relation to the doses, considering the inherent characteristics of the data. Based on the models adjusted for each soil and incubation period, doses of silicate minerals necessary to reach the desired pH were estimated.

$$\mathrm{pH} = \mathrm{a} + \mathrm{b(ASi)} \text{ or } pH = \mathrm{pH}^0 + \mathrm{a}\left(1 - e^{-\mathrm{b}ASi}\right) \quad (11)$$

where: pH is the desired pH value; a and b are components of the equation; ASi represents the amount of silicate agromineral in milligrams per cup.

The amount to be applied to the soil may be calculated from the following equation:

$$\mathrm{ASi}\left(\mathrm{t\ ha}^{-1}\right) = \mathrm{ASi(mg)} \times \mathrm{Ds} \times \mathrm{Soil\ depth(m)} \quad (12)$$

where: ASi (t ha^{-1}) is the amount of silicate agromineral for soil application; Ds = soil density (g cm^{-3}).

3.3.1 Comparative Analysis of Methods

Silicate Agromineral pH Incubation Curves

The curves resulting from the incubation of the agrominerals over time are shown in Figs. 1, 2, 3, 4, 5, 6, 7, and 8. The silicate agrominerals that promoted the increase of pH approaching a plateau, meaning that they were at equilibrium, were: metabasalt, basalt MA, wollastonite, calc-schist, and limestone (Figs. 2, 3, 4, 7, and 8). Application of other materials such as basalt MG, amphibole gneiss, and serpentinite at quantities exceeding 500 mg g^{-1} (equivalent to 100 t ha^{-1}, considering a soil depth of 20 cm) may lead to an increase in soil pH values (Figs. 1, 5, and 6).

It is expected that, as the materials are not promptly dissolved, with an increase in time, the latter results at 72 h will lead to a higher soil pH after the treatment. In certain instances, particularly with the basalt MG and metabasalt (Figs. 1 and 2), mainly in the Ferralsol, the values were lower than those obtained using the 1-h curve. The amphibole gneiss and serpentinite (Figs. 5 and 6) exhibited pH values at 72 h higher than those at 24 and 1 h. However, the overall pH remained low, preventing significant separation of the curves or achievement of higher pH values.

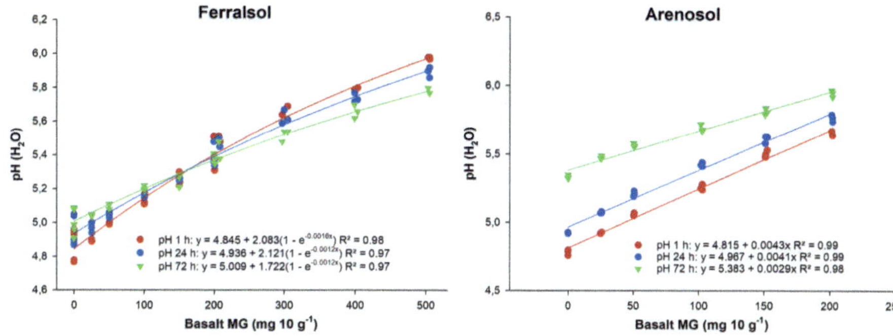

Fig. 1 Basalt MG pH incubation curves in the lab in Ferralsol (left) and Arenosol (right), after 1, 24, and 72 h

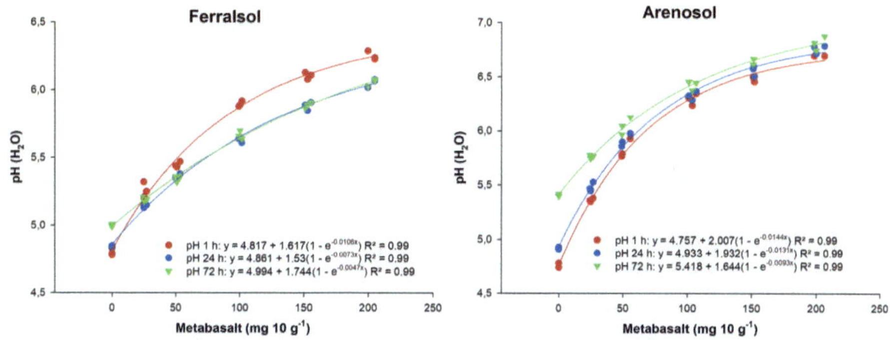

Fig. 2 Metabasalt pH incubation curves in the lab in Ferralsol (left) and Arenosol (right), after 1, 24, and 72 h

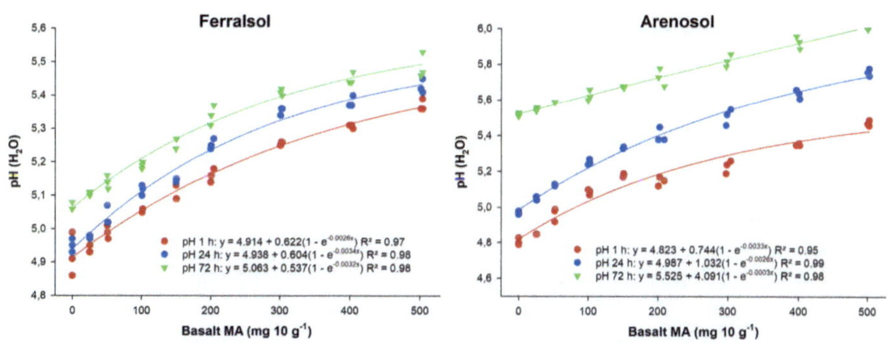

Fig. 3 Basalt MA pH incubation curves in the lab in Ferralsol (left) and Arenosol (right), after 1, 24, and 72 h

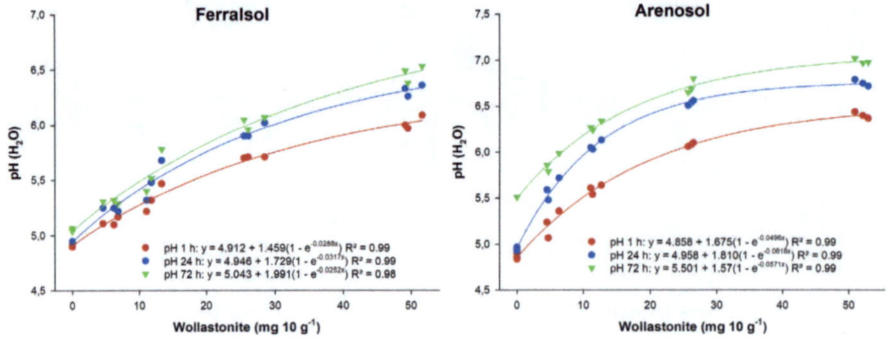

Fig. 4 Wollastonite pH incubation curves in the lab in Ferralsol (left) and Arenosol (right), after 1, 24, and 72-h

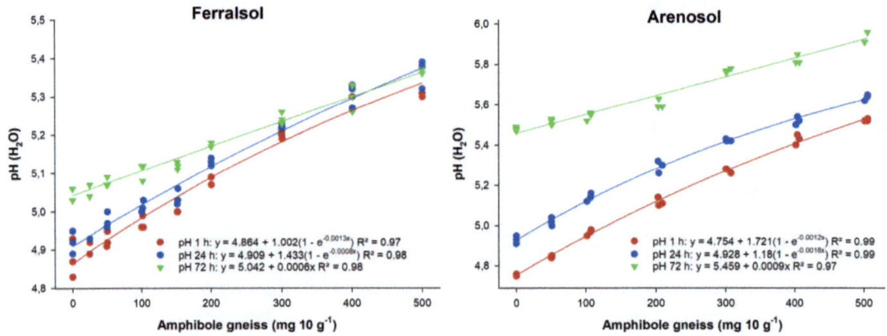

Fig. 5 Amphibole gneiss pH incubation curves in the lab in Ferralsol (left) and Arenosol (right), after 1, 24, and 72-h

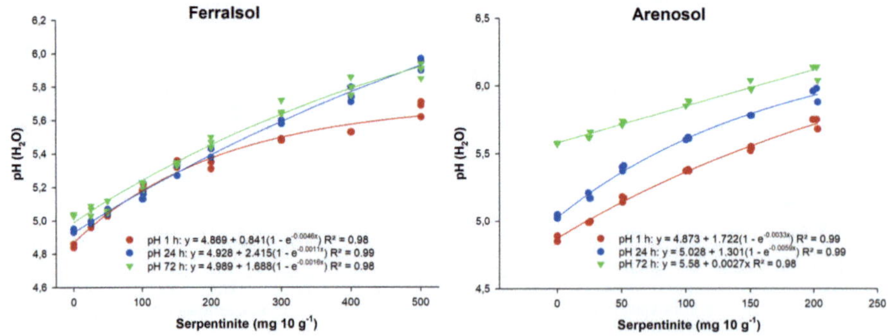

Fig. 6 Serpentinite pH incubation curves in the lab in Ferralsol (left) and Arenosol (right), after 1, 24, and 72 h

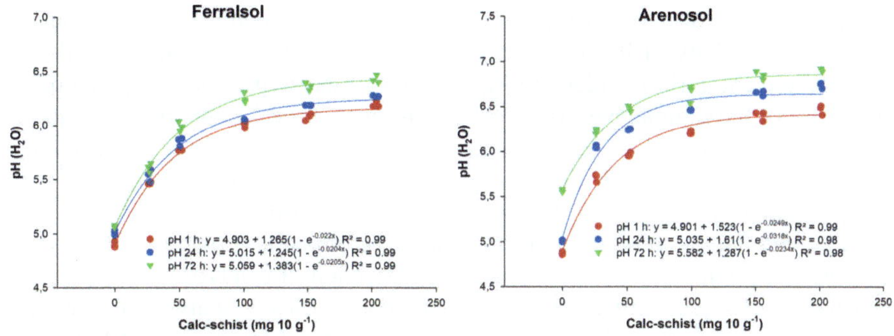

Fig. 7 Calc-schist pH incubation curves in the lab in Ferralsol (left) and Arenosol (right), after 1, 24, and 72 h

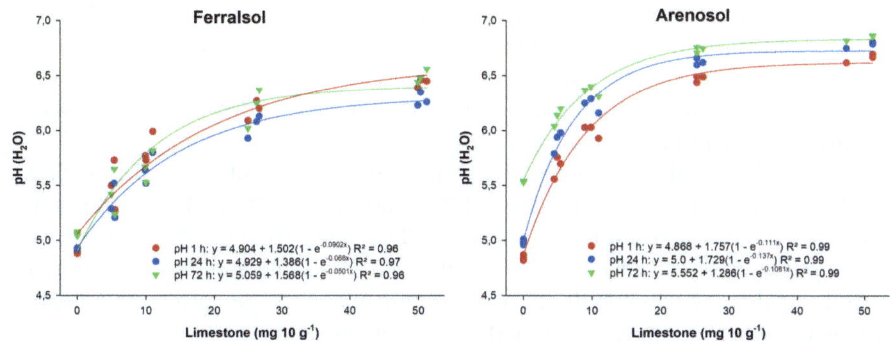

Fig. 8 Limestone pH incubation curves in the lab in Ferralsol (left) and Arenosol (right), after 1, 24, and 72-h

This is due to the Fe^{2+} content in these silicate agrominerals, which oxidizes upon mineral dissolution and causes a decrease in pH.

In fact, amphibole gneiss and serpentinite exhibit the highest magnetic susceptibility within the group, primarily due to the presence of magnetite, which accounts for 1307 and 1848 $m^3\ kg^{-1}$ (Table 7). Values of Fe^{2+} were sufficient to prevent the rise in pH during mineral dissolution. Basalt MG and metabasalt showed lower magnetic susceptibility, with values of 677 and 104 $m^3\ kg^{-1}$, respectively. Further studies are needed to measure redox potential in samples treated with these materials to explain the lower-than-expected pH values. This fact reinforces the need for the use of the pH incubation curve method, as the neutralization power (NP) measured by neutralizing the remaining HCl with NaOH (Embrapa 2017) may not reflect this effect, among others, such as Si polymerization and the dissolution of complex silicate minerals.

During NP measurement, neutralization with NaOH induces a reaction with Fe and Al, thereby reducing the NP of the materials. This contributes to the low RPTN of the materials, along with the low reactivity of silicates with HCl.

Serpentinite, in turn, is considered to have the highest potential for correction due to the high reactivity of its constituent minerals, particularly magnesium silicates (lizardite and antigorite), which are Mg silicates of the serpentine group and make up 37.23% of the sample. Nevertheless, the presence of Fe^{2+} in the rock, due to the presence of magnetite, chlorite, and enstatite, likely reduced this corrective capability. Consequently, the pH incubation curve can effectively distinguish materials with nearly identical RPTN values, demonstrating a unique pH correction potential for each material individually. When using silicate agrominerals containing Fe and Al that may reduce the effectiveness of acidity correction, it is advisable to apply the chosen dose of silicate mineral to the soil and conduct a soil analysis after at least 90 days. This analysis will determine if there is a need for additional limestone supplementation based on standard liming calculation methods.

Metabasalt has a higher neutralizing capacity (Fig. 2) than basalts due to its carbonate content, as calcite (11.2%). MG and MA basalts mainly consist of Ca silicates like labradorite (Plagioclase Group), associated with minerals with Fe, such as augite and pigeonite (Pyroxene Group), and provide a more minor increase in soil pH due to their lower content of carbonate minerals (Escosteguy and Klamt 1998). The amorphous minerals found in basalts may include iron due to the alteration of original iron-rich minerals.

Wollastonite exhibits high reactivity due to its composition, which includes 74.82% calcite, along with calcium and magnesium silicates (6.16% wollastonite and 3.24% diopside), resulting in a notable pH change during incubation (Fig. 4).

The presence of carbonates in calc-schist (8.31% calcite and 24.41% dolomite) as well as calcite in metabasalt and wollastonite enhances the correction potential, particularly when compared to those that contain only silicates.

The initial pH of the sandy soil at 0 mg g^{-1} concentration of silicate agrominerals showed noticeable changes over time. The rapid reductive effect of particulate organic matter under anoxic conditions likely contributed to the increase in pH levels observed in the sandy soil after 72 h. The soil was recently collected from the superficial strata of the forest (Cerrado). These are not soils originating from agricultural fields. Thus, they are susceptible to redox potential. The impact of low values of soil cation exchangeable capacity and low acidity buffering in Arenosol causes the effect of redox potential on pH to be more pronounced than in Ferralsol.

It is crucial to note that, due to the sensitivity to variations in redox potential, this method deviates from the actual conditions of the soil, and this deviation can increase with prolonged incubation time. Therefore, it is not possible to extend the duration of these curves beyond 72 h. However, other methods of recommending limestone for these two soils are also not ideal for correcting acidity. In this scenario, the 24-h curves should be used to select which of the three curves will determine the dose of silicate agromineral to be applied to the soil. In agricultural soils with low particulate organic matter content, the 72-h curve is likely to be the most appropriate for assessing the effect of silicate agrominerals, depending on their Fe release characteristics. The presence of Fe can lower the final soil pH, necessitating a higher application rate of the material. It is recommended to select a curve with intervals of 24–72 h that provides the highest dose of silicate agromineral for accurate dosing, ensuring that

Table 8 Calculated doses of silicate agrominerals to increase soil (Ferralsol) pH to 5.5 and 6.0, considering soil depth from 0 to 20 cm

	Time	Equation	pH 5.5	pH 6.0	pH 5.5	pH 6.0
			(mg 10 g^{-1})		(t ha^{-1})	
Basalt MG	24	$Y = 4.936 + 2.121(1 - \exp(-0.0012x))$	257.6	580.4*	51.5	–
	72	$Y = 5.009 + 1.722(1 - \exp(-0.0012x))$	279.72	714.02*	55.9	–
Metabasalt	24	$Y = 4.861 + 1.53(1 - \exp(-0.0073x))$	74.07	186.89	14.8	37.4
	72	$Y = 4.994 + 1.744(1 - \exp(-0.0047x))$	72.91	182.98	14.6	36.6
Basalt MA	24	$Y = 4.938 + 0.604(1 - \exp(-0.0034x))$	784.09*	–	–	–
	72	$Y = 5.036 + 0.537(1 - \exp(-0.0032x))$	525.26*	–	–	–
Wollastonite	24	$Y = 4.946 + 1.729(1 - \exp(-0.0317x))$	12.19	29.67	2.4	5.9
	72	$Y = 5.043 + 1.991(1 - \exp(-0.0252x))$	10.35	26.00	2.1	5.2
Amphibole gneiss	24	$Y = 4.909 + 1.433(1 - \exp(-0.0008x))$	664.68*	1790.89*	–	–
	72	$Y = 5.042 + 0.0006x$	133.33	966.67*	26.7	–
Serpentinite	24	$Y = 4.928 + 2.415(1 - \exp(-0.0011x))$	245.73	533.45*	49.14	–
	72	$Y = 4.989 + 1.688(1 - \exp(-0.0016x))$	225.36	571.02*	45.1	–
Calc-schist	24	$Y = 5.015 + 1.245(1 - \exp(-0.0204x))$	24.19	76.77	4.8	15.4
	72	$Y = 5.059 + 1.383(1 - \exp(-0.0205x))$	18.73	55.64	3.7	11.1
Limestone	24	$Y = 4.929 + 1.386(1 - \exp(-0.068x))$	7.81	21.79	1.56	4.4
	72	$Y = 5.059 + 1.568(1 - \exp(-0.0501x))$	6.59	18.30	1.3	3.7

*Value predicted outside data range

the dosage remains reasonable. Excessively high doses may occur depending on the material. Understanding the reactions of each material in advance can assist in selecting an appropriate time frame.

The pH incubation curves (Figs. 1, 2, 3, 4, 5, 6, 7, and 8) yielded predicted values of silicate agromineral doses to be applied to the soils to increase their pH to 5.5 or 6.0 (Tables 8 and 9). Some predicted values of doses did not achieve the desired pH, and some exceeded the maximum value in the data range. Therefore, these values were excluded from the table.

Relative Power of Total Neutralization (RPTN)

The chemical composition of limestone is characterized by its Ca and Mg carbonate content, which is quantified by its neutralizing power (NP). The NP of an acidity corrective is determined in the laboratory by dissolving a known amount of sample in excess of hydrochloric acid solution (HCl 0.5 mol L^{-1}). The corrective neutralizes the acid, and the excess acid is measured by titrating with NaOH. This process indirectly determines, by difference, the amount of acid that was neutralized, which corresponds to the quantity of the neutralizing component in the corrective sample.

Table 9 Calculated doses of silicate agrominerals to increase soil (Arenosol) pH to 5.5. and 6.0 considering soil depth from 0 to 20 cm

	Time	Equation	pH 5.5	pH 6.0	pH 5.5	pH 6.0
			($mg\ 10\ g^{-1}$)		($t\ ha^{-1}$)	
Basalt MG	24	$Y = 4.967 + 0.0041x$	130.00	251.95*	26.0	–
	72	$Y = 5.383 + 0.0029x$	40.34	212.76*	8.1	–
Metabasalt	24	$Y = 4.933 + 1.932(1 - \exp(-0.0131x))$	26.52	61.34	5.3	12.3
	72	$Y = 5.418 + 1.644(1 - \exp(-0.0093x))$	5.50	46.99	1.1	9.4
Basalt MA	24	$Y = 4.987 + 1.032(1 - \exp(-0.0026x))$	264.37	1536.47*	52.9	–
	72	$Y = 5.525 + 4.091(1 - \exp(-0.0003x))$	– 20.31*	411.40	–	81.2
Wollastonite	24	$Y = 4.958 + 1.81(1 - \exp(-0.0818x))$	4.35	10.48	0.9	2.1
	72	$Y = 5.501 + 1.57(1 - \exp(-0.0571x))$	– 0.01*	6.70	–	1.3
Amphibole gneiss	24	$Y = 4.928 + 1.18(1 - \exp(-0.0018x))$	368.39	1328.41*	73.7	–
	72	$Y = 5.459 + 0.0009x$	45.56	601.11*	9.1	–
Serpentinite	24	$Y = 5.028 + 1.301(1 - \exp(-0.0059x))$	76.38	233.02	15.3	–
	72	$Y = 5.58 + 0.0027x$	– 29.63*	155.56	–	31.1
Calc-schist	24	$Y = 5.035 + 1.61(1 - \exp(-0.0318x))$	10.72	28.77	2.1	5.8
	72	$Y = 5.582 + 1.287(1 - \exp(-0.0234x))$	– 0.26*	1.68	–	0.3
Limestone	24	$Y = 5.00 + 1.729(1 - \exp(-0.137x))$	2.50	6.32	0.5	1.2
	72	$Y = 5.552 + 1.286(1 - \exp(-0.1081x))$	– 0.37*	3.96		0.80

*Value predicted outside data range

The principle of chemical equivalence states that the quantity of acid neutralized matches the quantity of the neutralizing constituent within the corrective agent, with $CaCO_3$ serving as the standard reference. The quantity of $CaCO_3$ in the sample is calculated and reported as an equivalent percentage of calcium carbonate (% $CaCO_3$). The NP of a silicate agromineral is determined by the content and chemical nature of its neutralizers. The neutralization capacity of each constituent is quantified and compared to the standard capacity of $CaCO_3$ (Table 1).

The physical quality of limestone is determined by its granulometry, which is expressed through the relative efficiency (RE). This efficiency is calculated based on the distribution of particle sizes across a series of sieves and indicates the reactivity, or reaction rate, of the limestone within the soil. The relative total neutralization power (RPTN), which forms the basis for determining the limestone dosage, integrates the two indices (Rein et al. 2025). Detailed information regarding limestones and other acidity correctives, along with their soil reactions and properties, is available in Quaggio (2000) and Alcarde (2005).

Below are two formulas for calculating the lime dose for irrigated fields in the Cerrado (standard methods for calculation of lime requirement for the Cerrado). The formula that yields the highest recommended dose can be used (Rein et al. 2025). The first is the base saturation method (Eq. 13) and the second is the Ca + Mg method (Eq. 14).

$$\text{Limestone dose}\left(\text{t ha}^{-1}\right) = (\text{Target V}-\text{Current V}) \times \text{T/RPTN}) \times 1.33 \tag{13}$$

$$\text{Limestone dose}\left(\text{t ha}^{-1}\right) = \left[3-(\text{Ca} + \text{Mg})\right] \times (100/\text{RPTN}) \times 1.33 \tag{14}$$

In the first expression, the target base saturation (V) is 60%; V current represents the base saturation of soil sampled from the 0–20 cm layer; and T denotes the CEC at pH 7.0. In the second expression, Ca + Mg is the sum of the current soil calcium and magnesium contents ($cmol_c\ dm^{-3}$). The final multiplication factor of 1.33 has not yet been established as a definitive recommendation. It remains a calculation used to adjust the proposed V% value to match the field observations made by researchers from Embrapa Cerrados (Rein et al. 2025). Consequently, calculations will be conducted with and without applying this constant (Tables 10; 11).

Calcium Carbonate Equivalent Percent (%ECaCO$_3$) Derived from CaO and MgO Content in Silicate Agrominerals

The percentage of calcium carbonate equivalent method provides a way to overcome the uncertainties inherent in silicate agrominerals—namely, the underestimation of relative neutralizing power (RPTN) caused by iron oxidation, silicon polymerization (Chan 1989), and incomplete dissolution of complex, variably composed silicate minerals. $ECaCO_3$ is derived from the concentrations of calcium and magnesium present in silicate agrominerals through chemical analysis, such as CaO and MgO

Table 10 Recommended doses of silicate agromineral or limestone according to the base saturation method (Eq. 13) and RPTN of each material[1]

	Arenosol		Ferralsol	
	NC t ha^{-1} × 1.33	NC t ha^{-1}	NC t ha^{-1} × 1.33	NC t ha^{-1}
Basalt MG	15.11	11.36	44.25	33.27
Metabasalt	34.33	25.81	51.92	39.04
Basalt MA	40.23	30.25	60.85	45.75
Wollastonite	3.43	2.58	5.20	3.91
Amphibole Gneiss	35.76	26.89	54.09	40.67
Serpentinite	26.27	19.75	39.74	29.88
Calc-schist	15.11	11.36	44.25	33.27
Limestone	2.70	2.03	4.07	3.06
$CaCO_3$	2.57	1.94	3.89	2.93

[1]Quaggio (2000), Alcarde (2005) and Rein et al (2025)

Table 11 Recommended doses of silicate agromineral or limestone according to the Ca + Mg method (Eq. 14) and RPTN of each material[1]

	Ferralsol/Arenosol	
	NC t ha^{-1} × 1.33	NC t ha^{-1}
Basalt MG	36.27	27.27
Metabasalt	42.56	32.00
Basalt MA	49.88	37.50
Wollastonite	4.26	3.20
Amphibole Gneiss	44.33	33.33
Serpentinite	32.57	24.49
Calc-schist	36.27	27.27
Limestone	3.34	2.51
$CaCO_3$	3.19	2.4

[1]Quaggio (2000), Alcarde (2005) and Rein et al (2025)

EDTA on HCl extracts (Table 4). According to this calculation, Ca and Mg, irrespective of whether they exist as silicates or carbonates, are converted uniformly to carbonates in a 1:1 ratio. The method presents an error in the neutralization capacity relative to $CaCO_3$ of calcium and magnesium silicates, which are considered as carbonates (Table 1). Nevertheless, it provides a straightforward method for calculating doses of silicate agrominerals using basic routine analysis data. The method is calculated following the equation:

$$\%ECaCO_3 = (\%CaO \times 1,79) + (\%\,MgO \times 2,48) \tag{15}$$

Estimates obtained using this method yielded values of doses that were below the pH incubation curves and base saturation methods (Table 12). Equation 15 can

Table 12 Doses estimated from calcium carbonate equivalent percent (%$ECaCO_3$; Eq. 15) derived from CaO and MgO content in silicate agrominerals, considering a liming requirement = 2.93 and 1.94 t ha^{-1} $CaCO_3$, respectively for Ferralsol and Arenosol

	$ECaCO_3$%	Ferralsol	Arenosol
		(t ha^{-1})	
Basalt MG	14.0	20.98	13.87
Metabasalt	15.0	19.48	12.88
Basalt MA	13.0	22.45	14.84
Wollastonite	83.0	3.53	2.33
Amphibole Gneiss	15.0	19.48	12.88
Serpentinite	23.8	12.31	8.14
Calc-schist	35.3	8.29	5.48
Limestone	99.5	2.94	1.95
$CaCO_3$	100	2.93	1.94

be extended using the neutralization capacity relative to $CaCO_3$ for all species in Table 1.

Comparing Method Results

The results obtained from all methods exhibited significant variability. Dividing the results of the "base saturation method × 1.33" by the results of the other methods reveals a close relationship between the Ca + Mg method, as both are multiplied by the RPTN of the agrominerals (Table 13). The relationship is expressed as (Target V—Current V)/(Ca + Mg), equal to 1.22 for Ferralsol and 0.81 for Arenosol, respectively, resulting in equal values for all agrominerals (Table 12). Both methods showed a correlation with the $ECaCO_3$ method, producing Pearson correlation coefficients of 0.89 for Ferralsol and 0.84 for Arenosol, respectively. However, differences as significant as 5.34 times were observed with the calc-schist.

When the base saturation method results were divided by the values obtained from the pH incubation curve (Table 13), the findings indicated that basalt MG and MA, as well as serpentinite, presented values closer to 1. This indicates that the methods prescribe similar dosages of these agrominerals. However, other materials showed significant differences, with variations as high as 7.63 times in the case of calc-schist in the Ferralsol at pH 6.0. Calc-schist exemplifies the variations among the assessed methods, demonstrating an increase in soil pH with minimal doses and stabilization at a pH above 6.0 with larger doses over periods of 24 and 72 h. Alternative methods to the pH incubation curve did not identify its capacity to increase soil pH. Nevertheless, the pH incubation curve indicated a dose of limestone that was 3.39 times lower compared to the other standard methods (base saturation and Ca + Mg method) used for liming in the Ferralsol. In the Arenosol, the measurement was overestimated by a factor of 1.63 (equal to 1/0.61—Table 12—base saturation divided by pH incubation curves at pH 6.0) compared to the standard methods.

Table 13 Relative quantities of the required agromineral per hectare are calculated using the base saturation × 1.33 and $ECaCO_3$ methods, considering incorporation within a soil depth of 0–20 cm. Data for Ferralsol and Arenosol soil types

	Base saturation × 1.33				$ECaCO_3$%	
	Divided by:					
	$ECaCO_3$%	Ca + Mg method	pH incubation curves			
		X 1.33	pH 5.5	pH 6.0	pH 5.5	pH 6.0
Ferralsol						
Basalt MG	2.11	1.22	0.79	–	0.38	–
Metabasalt	2.67	1.22	3.51	4.22	1.32	1.58
Basalt MA	2.71	1.22	–	0.75	–	0.28
Wollastonite	1.47	1.22	2.17	2.48	1.47	1.68
Amph. Gneiss	2.78	1.22	2.03	–	0.73	–
Serpentinite	3.23	1.22	0.81	1.28	0.25	0.40
Calc-schist	5.34	1.22	9.22	7.63	1.73	1.43
Limestone	1.38	1.22	2.61	3.39	1.89	2.45
Arenosol						
Basalt MG	2.11	0.81	1.13	–	0.53	–
Metabasalt	2.66	0.81	6.48	0.92	2.43	0.34
Basalt MA	2.71	0.81	0.76	–	0.28	–
Wollastonite	1.47	0.81	3.81	0.58	2.59	0.40
Amph. Gneiss	2.78	0.81	0.49	–	0.17	–
Serpentinite	3.23	0.81	1.72	–	0.53	–
Calc-schist	5.34	0.81	13.93	1.90	2.61	0.36
Limestone	1.39	0.81	5.40	0.61	3.89	0.44

It should be noted that the RPTN of basalt MG is higher than that of metabasalt. The pH incubation curve indicated that at pH 5.5, for the Ferralsol, the required dose of metabasalt is 3.51 times lower than the dose recommended by the base saturation × 1.33 method, whereas for basalt MG, it is 1.26 (equals to 1/0.79—Table 12—base saturation divided by pH incubation curves at pH 5.5) times higher. It demonstrates the pH incubation curve method's ability to determine suitable silicate agromineral dosages in relation to increases in soil pH.

The $ECaCO_3$ method demonstrates a distinct behavior compared to other methods, suggesting that its theoretical foundation, which relies on the equivalence of oxides, yields different estimates. The most significant discrepancy was observed in the limestone, with a recommendation of 2.45 times that of the pH incubation curve method.

This analysis emphasizes the need to consider the specific goals of liming or remineralization (such as correcting surface and subsurface acidity, supplying cations, and adjusting pH) when selecting the most suitable method for dose recommendation.

4 Conclusions

The chapter highlights the short- and long-term reactivity of silicate agrominerals (ASi) in soil acidity correction due to their diverse mineral composition and polymerization of silicon. It recognizes the difficulty in aligning the neutralization power (NP) of ASi with calcium carbonate due to variability influenced by pH, particle size, oxidation, and bioweathering reactions, which differ substantially from simple acid attacks.

Fine particles of ASi demonstrate rapid dissolution and short-term correction potential, which can be assessed through a simple laboratory protocol analogous to the abrasion pH analysis of ground rocks. A reaction time of 24 h is sufficient for fine particle dissolution, enabling effective soil correction while minimizing the underestimation of rock doses. Larger particles dissolve more slowly, contributing to long-term effects by enhancing soil resilience and nutrient supply through bioweathering over time.

Experimental findings, based on the pH correction curve of limestone, confirmed that higher doses of silicate rocks achieved pH correction efficiencies comparable to limestone at its desired pH value. These results establish the potential of silicate agrominerals for both short- and longer-term soil acidity correction.

Acknowledgements The authors gratefully acknowledge the financial support provided by Embrapa and by FINEP—Fundo Setorial Mineral through the project 'Rede de PD&I sobre remineralizadores e agrominerais (REMAg)' (Project No. 1541/2022), which made possible the development of this book chapter.

Author's Contribution Giuliano Marchi conceived the study, designed the methodology, and led the interpretation of the results and the writing of the chapter. Luise Lottici Krahl contributed to methodology development, data analysis, and manuscript writing. Ewerton Gonçalves de Abrantes and Paula Caroline Ferreira Rodrigues contributed to methodology development, conducted the laboratory analyses, participated in data interpretation, and contributed to the writing. Elis Marina de Freitas performed the statistical analyses and contributed to data interpretation. João Paulo Guimarães Soares and Éder de Souza Martins contributed to the interpretation of the results and critically revised the final version of the chapter.

Declaration of Competing Interest The authors declare that they have no known competing financial interests or personal relationships that could have appeared to influence the work reported in this paper.

References

Abreu CH Jr, Muraoka T, Lavorante AF (2003) Relationship between acidity and chemical properties of Brazilian soils. Scientia Agricola 60:337–343

Alcarde JC (2005) Corretivos da acidez dos solos: Características e interpretações técnicas. São Paulo: Associação Nacional para Difusão de Adubos. 21 p. (ANDA. Boletim Técnico, 6). Disponível em: https://anda.org.br/wp-content/uploads/2019/03/boletim_06.pdf. Acesso em: 31 ago. 2024

Anda M, Shamshuddin J, Fauziah CI (2015) Improving chemical properties of a highly weathered soil using finely ground basalt rocks. Catena 124:147–161. https://doi.org/10.1016/j.catena.2014.09.012

Barbosa JSF, Sabaté P (2004) Archean and Paleoproterozoic crust of the São Francisco Craton, Bahia, Brazil: geodynamic features. Precambrian Res 133(1–2):1–27

Basak BB (2018) Waste mica as alternative source of plant-available potassium: evaluation of agronomic potential through chemical and biological methods. Nat Resour J 27:1–13. https://doi.org/10.1007/s11053-018-9430-3

Basak BB, Sarkar B, Sanderson P, Naidu R (2018) Waste mineral powder supplies plant available potassium: evaluation of chemical and biological interventions. J Geochem Explor 186:114–120. https://doi.org/10.1016/j.gexplo.2017.11.023

Burghelea CI, Dontsova K, Zaharescu DG, Maier RM, Huxman T, Amistadi MK, Hunt E, Chorover J (2018) Trace element mobilization during incipient bioweathering of four rock types. Geochim Cosmochim Acta 234:98–114. https://doi.org/10.1016/j.gca.2018.05.011

Camara GDR, Faitanin BX, Silveira LLLD, Chiodi Filho C, Santos ESD (2021) Utilização de rochas ornamentais ricas em minerais potássicos como fonte alternativa de insumo agrícola via rochagem: Parte I. CETEM/MCTI, Rio de Janeiro, p 75, (Série Rochas e Minerais Industriais; 30)

Chan SH (1989) A review on solubility and polymerization of silica. Geothermics 18(1–2):49–56

Ciceri D, Allanore A (2019) Local fertilizers to achieve food self-sufficiency in Africa. Sci Total Environ 648:669–680. https://doi.org/10.1016/j.scitotenv.2018.08.154

Ciceri D, Close TC, Barker AV, Allanore A (2019) Fertilizing properties of potassium feldspar altered hydrothermally. Commun Soil Sci Plant Anal 50:482–491. https://doi.org/10.1080/00103624.2019.1566922

Colombo C, Palumbo G, He J-Z, Pinton R, Cesco S (2014) Review on iron availability in soil: interaction of Fe minerals, plants, and microbes. J Soils Sed 14:538–548. https://doi.org/10.1007/s11368-013-0814-z

Conceição JM et al (2022) Basic rocks as a resource for soil fertility enhancement. Geochem Int 59(2):101–120

Embrapa (2017) Manual de métodos de análise de solo. Centro Nacional de Pesquisa de Solos, Rio de Janeiro

Escosteguy PAV, Klamt E (1998) Basalto moído como fonte de nutrientes. Rev Bras Ciênc Solo 22:11–20. https://doi.org/10.1590/S0100-06831998000100002

Essington ME (2003) Soil and water chemistry: an integrative approach. CRC PRESS, Boca Raton. https://doi.org/10.1201/b12397

Famelli N, Lima EF, Carmo IO (2021) Lithostratigraphy of the Serra Geral Formation in the northern portion of the Paraná-Etendeka Igneous Province: a tool for tracking Early Cretaceous paleoenvironmental changes. J Volcanol Geotherm Res 410. https://doi.org/10.1016/j.jvolgeores.2020.107152

Ferreira RL et al (2020) Ultramafic alkaline rocks: implications for sustainable agriculture. Environ Geochem Health 42(5):1101–1115

Fontes MPF, Alleoni LRF (2006) Electrochemical attributes and availability of nutrients, toxic elements, and heavy metals in tropical soils. Scientia Agricola 63:589–608

Gillman GP, Burkett DC, Coventry RJ (2001) A laboratory study of application of basalt dust to highly weathered soils: effect on soil cation chemistry. Soil Res 39(4):799–811

Goulding KWT (2016) Soil acidification and the importance of liming agricultural soils with particular reference to the United Kingdom. Soil Use Manage 32(3):390–399

Hinsinger P, Jaillard B, Dufey JE (1992) Rapid weathering of a trioctahedral mica by the roots of ryegrass. Soil Sci Soc Am J 56:977–982. https://doi.org/10.2136/sssaj1992.03615995005600030049x

Krahl LL, Paz SPA, Angélica RS, Valadares LF, Souza-Silva JC, Marchi G, Martins ÉS (2020) Successive offtake of elements by maize grown in pure basalt powder. Afr J Agric Res V15(2):229–239

Krahl LL, Marchi G, Paz SPA, Angélica RS, Sousa-Silva JC, Valadares LF, Martins ÉS (2022) Increase in cation exchange capacity by the action of maize rhizosphere on Mg or Fe biotite-rich rocks. Pesquisa Agropecuária Trop 52. https://doi.org/10.1590/1983-40632022v5272376

Lefebvre D, Goglio P, Williams A, Manning DAC, Azevedo AC, Bergmann M, Meersmans J, Smith P (2019) Assessing the potential of soil carbonation and enhanced weathering through life cycle assessment: a case study for Sao Paulo State. Brazil J Clean Prod 233:468–481. https://doi.org/10.1016/j.jclepro.2019.06.099

Lima Filho OF (2011) Modificação de características químicas em solos de Mato Grosso do Sul incubados com escórias de siderurgia. Embrapa Agropecuária Oeste, Dourados, MS, p 35

Marchi G, Guelfi-Silva DR, Malaquias JV, Guilherme LRG, Spehar CR, Martins ÉS (2020) Solubility and availability of micronutrients extracted from silicate agrominerals. Pesq Agrop Brasileira 55:e00807

Marchi G, Sousa-Silva JC, Soares JPG, Abrantes EG, Martins ÉS (2024) Growth of urochloa grass in an oxisol treated with powdered silicate materials. Peer Rev 6(2):248–272

Marini L (2006) Geological sequestration of carbon dioxide: thermodynamics, kinetics, and reaction path modeling, vol 11. Elsevier

Martins ÉS, Martins ES, Hardoim PR (2023) Princípios geoquímicos, mineralógicos e biológicos do manejo de remineralizadores de solos. Informe Agropecuário 44(321):26–39

Moraes RD, Fuck RA, Brown M, Piccoli P, Baldwin J, Dantas EL, Laux JH, Junges SL (2007) Wollastonite-scapolite-clinopyroxene marble of the Anápolis-Itauçu Complex, Goiás: more evidence of ultrahigh-temperature metamorphism. Revista Brasileira de Geociências 37(4 supl):11–17

Moretti LG et al (2021) Potential of ultramafic rocks in soil correction and nutrient supply. J Soil Sci Plant Nutr 21(3):345–360

Nogueira AFR, Rabelo CEN, Góes AM, Cardoso AR, Bandeira J, Rezende GL, Santos RF, Truckenbrodt W (2021a) Evolution of Jurassic intertrap deposits in the Parnaíba Basin, northern Brazil: the last sediment-lava interaction linked to the CAMP in West Gondwana. Palaeogeogr Palaeoclim Palaeoecol 572. https://doi.org/10.1016/j.palaeo.2021.110370

Nogueira TAR, Miranda BG, Jalal A, Lessa LGF, Filho MCMT, Marcante NC, Abreu-Junior CH, Jani AD, Capra GF, Moreira A, Martins ÉD, Martins, ÉS (2021b) Nepheline syenite and phonolite as alternative potassium sources for maize. Agronomy 11(7):1385

Nolla A, Korndörfer GH, Silva CAT, Silva TRB, Zucarelli V, Silva MA (2013) Correcting soil acidity with the use of slags. Afr J Agric Res 8(41):5174–5180

Norouzi S, Khademi H (2010) Ability of alfalfa (Medicago sativa L.) to take up potassium from different micaceous minerals and consequent vermiculitization. Plant Soil 328:83–93. https://doi.org/10.1007/s11104-009-0084-0

Novais RF, Alvarez VHV, Barros NF, Fontes RLF, Cantarutti RB, Neves JC (2007) Fertilidade do Solo. Sociedade Brasileira de Ciência do Solo, Viçosa, 1 edn., p 1017. ISBN: 978-85-86504-08-2

Quaggio J (2000) Acidez e calagem em solos tropicais. Instituto Agronômico, Campinas, p 111

Ramos CG, Querol X, Oliveira MLS, Pires K, Kautzmann RM, Oliveira LFS (2015) A preliminary evaluation of volcanic rock powder for application in agriculture as soil a remineralizer. Sci Total Environ 512–513:371–380. https://doi.org/10.1016/j.scitotenv.2014.12.070

Rein TA, Santos Junior JDG, Souza KW, Marchi G (2025) Correção da fertilidade dos solos para culturas anuais irrigadas no Cerrado. In: Rodrigues LN (ed) Agricultura Irrigada no Cerrado: subsídios para o desenvolvimento sustentável, p 633

Ribeiro D (2023) Análise das garantias geoquímicas e mineralógicas de coproduto de rochas máficas extraídas como estéreis nas minas de ferro de Carajás. Relatório Inédito, VALE

Ribeiro GM (2018) Caracterização de pós de rochas silicáticas, avaliação da solubilidade em ácidos orgânicos e potencial de liberação de nutrientes como remineralizadores de solos agrícolas. Doctoral dissertation, Tese de Doutorado, CAV/UDESC

Rodrigues M, Vahl LC, Silveira CAP, Salé MM, Nanni MR, Batista MA (2021) Co-products from the limestone mining as sources of calcium, magnesium and sulphur. Environ Nanotechnol Monit Manage 15:100446

Santos HG, Jacomine PKT, Anjos LHC, Oliveira VA, Lumbreras JF, Coelho MR, Almeida JA, Araújo Filho JC, Oliveira JB, Cunha TJF (2018) Sistema Brasileiro de Classificação de Solos. 5. ed., rev. e ampl.Brasília: Embrapa. 356p

Shamshuddin J, Fauziah CI (2010) Alleviating acid soil infertility constraints using basalt, ground magnesium limestone and gypsum in a tropical environment. Malays J Soil Sci 14:1–13

Souza FNS (2014) O potencial de agrominerais silicáticas como fonte de nutrientes na agricultura tropical. Brasília, p 144. Tese (Doutorado em Geociências). Instituto de Geociências, Universidade de Brasília

Teixeira PC, Donagemma GK, Fontana A, Teixeira WG (2017) Manual de métodos de análise de solo. Embrapa, Brasília, DF, 577p

Teixeira WG, Alvarez VH, Neves JCL (2020) New methods for estimating lime requirement to attain desirable pH values in Brazilian soils. Rev Bras Ciênc Solo 44:e0200008

van Straaten P (2006) Farming with rocks and minerals: challenges and opportunities. An Acad Bras Ciênc 78:731–747. https://doi.org/10.1590/S0001-37652006000400009

Viana JHM, Coelho AM, Thomazini A, Carvalho MPF (2021) Evaluation of the agricultural potential of the serpentinite rock as a soil remineralizer. An Acad Bras Cienc 93(3):e20201614

Viana JH, Albuquerque Filho MRD, Santos FCD, Ladeira DDA (2023) Nonlinear modeling of liming reaction and extractable base curves. Revista Brasileira De Engenharia Agrícola e Ambiental 27(10):820–827

White AF, Brantley SL (2003) The effect of time on the weathering of silicate minerals: why do weathering rates differ in the laboratory and field? Chem Geol 202:479–506. https://doi.org/10.1016/j.chemgeo.2003.03.001

Yu G, Xiao J, Hu S, Polizzotto ML, Zhao F-J, Mcgrath SP, Li H, Ran W, Shen Q (2017) Mineral availability as a key regulator of soil carbon storage. Environ Sci Technol 51:4960–4969. https://doi.org/10.1021/acs.est.7b00305

Zhang S, Zhu Q, de Vries W, Ros GH, Chen X, Muneer MA, Zhang F,. Wu L (2023) Effects of soil amendments on soil acidity and crop yields in acidic soils: a world-wide meta-analysis. J Environ Manage 345:118531

Assessment of Soil Acidity Correction Potential Using Metabasalt

Luise Lottici Krahl, Giuliano Marchi, Elis Marina de Freitas, and Éder de Souza Martins

Abstract This chapter examines the application of metabasalt as a soil remineralizer in the Cerrado region, where soil acidity presents a significant challenge to agriculture. It is based on research that investigates the potential of metabasalt powder, rich in essential nutrients such as calcium and magnesium, to enhance the chemical properties of acidic soils. This study aims to evaluate the effects of metabasalt on soil pH and nutrient availability, as well as its impact on the productivity of maize (*Zea mays* L.) and soybean [*Glycine max* (L.) Merrill.]. The objective is to determine the feasibility of using metabasalt as an alternative to traditional liming materials. This determination will be achieved by employing an innovative method that adjusts the results of pH obtained from increasing doses of metabasalt mixed with soil over 24 h. The results indicate that metabasalt can effectively correct soil acidity, enhance nutrient availability, and improve crop yields, offering a promising solution for sustainable agriculture in the Cerrado biome.

Keywords Soil quality · Soil acidity · Basic silicates

L. L. Krahl (✉) · G. Marchi · E. M. de Freitas · É. de Souza Martins
Agrogeological Laboratory, Empresa Brasileira de Pesquisa Agropecuária—Embrapa Cerrados, BR 020, km 18, Planaltina, Distrito Federal 73310970, Federal District, Brazil
e-mail: luisekrahl82@gmail.com

G. Marchi
e-mail: giuliano.marchi@embrapa.br

E. M. de Freitas
e-mail: elismarina1.freitas@gmail.com

É. de Souza Martins
e-mail: eder.martins@embrapa.br

É. de Souza Martins and S. Huff Theodoro (eds.), *Soil Remineralizers and Silicate Fertilizers*, Frontier Studies in Soil Science,
https://doi.org/10.1007/978-3-032-14656-4_8

1 Introduction

The Cerrado, a tropical savanna-like biome, spans more than 2,000,000 km^2, accounting for approximately 23% of Brazil's territory. It is marked by strong seasonality, with dry winters and rainy summers, and harbors one of the richest floras among the world's savannas (Sano et al. 2008). Since the late twentieth century, extensive areas of the biome have been converted into major grain-producing regions through the adoption of advanced soil management technologies (Goedert, 1983; Rada, 2013). Ferralsols are the predominant soil type in the Cerrado. Agriculture in the Cerrado has become feasible through extensive studies that have evaluated the primary limiting factors related to soil fertility in this region. The inherently acidic nature of these soils necessitates the use of appropriate liming as an initial agronomic practice for most crops (Lopes and Guilherme 2016).

Although liming is the usual method of soil correction, several igneous, metamorphic, and sedimentary silicate rocks have the potential to be used as agrominerals to correct soil acidity, also containing macro- and micronutrients important for plant growth and development (Marchi et al. 2020; Martins et al. 2024).

Basic rocks, rich in Fe, Si, Ca, and Mg, such as basalts, are globally available in large quantities and exhibit a relatively rapid weathering rate (Nunes et al. 2014; Anda et al. 2015; Lefebvre et al. 2019). Basalt powders are typical agrominerals, which are weathered in the soil on the agronomic time scale (Hinsinger et al. 2001; Krahl et al. 2020), present a balance in the proportion of bases and micronutrients (Theodoro et al. 2021), promote the correction of exchangeable soil aluminum (Melo et al. 2012; Conceição et al. 2022), make silicon available (Batista 2013), and other micronutrients (Marchi et al. 2020). Basalt powders also have the potential to improve the productivity of various crops, including grains (Luchese et al. 2021), pastures (Korchagin et al. 2019, 2022), and trees (Shafar et al. 2017). Research shows that the incubation of basalt powder in acidic soils promotes improvements in pH, base, and phosphorus availability (Gillman et al. 2002; Melo et al. 2012; Batista 2013; Rodrigues et al. 2024).

Silicate agrominerals address soil acidity through the presence of calcium and magnesium carbonates, as well as by dissolving silicate minerals. In some cases, the oxidation of iron also occurs in silicate agrominerals used for acidity correction. During dissolution, the pH may initially rise and then decrease to some extent. Iron (FeII) is commonly oxidized during the weathering of primary minerals in the soil, with manganese (MnII) also being oxidized, but to a lesser degree. Oxidation occurs through two mechanisms: (1) structural oxidation of FeII, which begins the decomposition of the mineral, and (2) oxidation of Fe^{2+} after its release into the soil solution via hydrolysis. The oxidation of FeII to FeIII within the mineral structure disrupts electrostatic neutrality, leading to the release of cations from the crystal to maintain electroneutrality (Essington 2005).

The reactions involving agrominerals vary in complexity based on their mineralogical and chemical composition. Reaction rates can differ significantly, with some calcium silicates ($CaSiO_3 = 0.095$ g dm^{-3}) showcasing solubility in water that is up to 6.8 times faster than that of calcium carbonate ($CaCO_3 = 0.014$ g dm^{-3}) (Alcarde and Rodella 2003; Castro et al. 2016; Amoakwah et al. 2023), resulting in greater mobility to deeper soil horizons. Consequently, certain silicate agrominerals can increase the pH of agricultural soils. However, the extent of this pH elevation is influenced by various factors, including the rock's chemical and mineralogical characteristics, soil properties, application rate, particle size, and others.

During measurement of the neutralization potential (NP) of silicate agrominerals, reactions involving Fe and Al during NaOH neutralization can consume hydroxide through dissolution or complexation of amphoteric Fe/Al phases, which reduces the measured NP. This effect, together with the silicates, generally slow dissolution kinetics and limited stoichiometric reactivity with HCl and NaOH, helps explain their low measured NP. Variability in mineral composition and the degree of Si polymerization (Alves et al. 2017) further complicate correlating the NP of silicate agrominerals with the more predictable neutralization behaviour of carbonates. As a result, particles with slower dissolution rates may react differently than they would in cultivated soil, where bioweathering reactions differ from simple acid attack, including pH reduction, oxidation (Essington 2005), and complexation with various substances produced by rhizosphere microorganisms, as well as the buffering effect of silicates (White and Brantley 2003).

Although labor-intensive, the use of pH incubation curves with silicate agrominerals in soils, along with concentrations of Ca and Mg, could be an accurate method for estimating doses for each soil type until more practical methods are developed for each material. Metabasalts are among the silicate agrominerals that are notably reactive for acidity correction and are categorized as calcium-magnesium-bearing silicate agrominerals (ASiCaMg) (Martins et al. 2024).

Metabasalts are basic volcanic rocks that have undergone regional metamorphism processes (Ribeiro 2023). The mineral composition differs from that of the original basalt due to metamorphic processes, including pressure and temperature, superimposed by hydrothermal activity, resulting in a distinct mineralogical makeup. Calcium, formerly found in plagioclase, is now mainly present as carbonate (calcite), making the rock highly reactive. These rocks release calcium more efficiently than magnesium because magnesium is contained in chlorite, a magnesium silicate that is more resistant to weathering. Agronomic studies of metabasalts have shown results like those found with basalts (Korchagin et al. 2019, 2022; Zuffo et al. 2022). Therefore, as metabasalt powder contains essential nutrients for plants, it has potential for use in agriculture (Korchagin et al. 2019; Zuffo et al. 2022). Some of these rocks contain carbonates (calcite and magnesite) and Ca- and Mg-silicates in their structure, which function as soil acidity correctors. Calcite and magnesite are more reactive than Ca and Mg silicates.

Assessing the neutralization potential of these calcium- and magnesium-rich basic rocks through simple measurements is crucial for understanding their reactivity and effectiveness in altering soil chemical properties, their role in correcting soil acidity, and determining suitable application guidelines. Consequently, an experiment was conducted to examine the feasibility of using a metabasalt from Paraupebas—PA, Brazil, as a soil remineralizer. This study evaluates the changes in the chemical properties of two Cerrado soils under maize (*Zea mays* L.) and soybean [*Glycine max* (L.) Merrill.] cultivation, as well as the resultant agronomic effects.

2 Metabasalt Classification and Characterization

Silicate agrominerals are raw materials composed of silicate rocks used in the development of inputs for managing soil fertility (Martins et al. 2010, 2014, 2024). These materials represent one of the strategic approaches for sustainable agricultural development (van Straaten 2006). Soil remineralizers (REM) are derived from silicate agrominerals that have only undergone particle size reduction processes (Brazil 2013) and comply with the criteria defined by Normative Instruction 5/2016 (IN 5) of the Ministry of Agriculture, Livestock, and Supply (MAPA) (Brazil 2016).

The registration criteria for a soil remineralizer, according to IN 5/2016 (Brazil 2016), include: a sum of total bases ($CaO + MgO + K_2O$) of at least 9%; a minimum total K_2O content of 1%; a maximum free silica (quartz) content of 25%; and maximum concentrations of potentially toxic elements (arsenic—As, < 15 mg kg^{-1}; cadmium—Cd, < 10 mg kg^{-1}; mercury—Hg, < 0.1 mg kg^{-1}; lead—Pb, < 200 mg kg^{-1}). Additionally, an agronomic test must demonstrate the rock's efficacy in modifying soil fertility properties and promoting plant development. Beyond these minimum criteria, NI 05/2016 (Brazil 2016) also mandates the declaration of the abrasion pH of rock dust and requires that its granulometry is classified into one of three categories: bran, powder, or filler.

Silicate rocks with potential for adoption as REM should be organized into classes to indicate their primary and secondary effects on soil and establish the usage guidelines for these rocks. The following chemical classification of agrominerals is based on the bases in the form of oxides: CaO, MgO, and K_2O. The sum of these oxides ($CaO + MgO + K_2O$) must exceed 9%, which is the minimum criterion for REM in Brazilian regulation (Martins et al. 2024). When the sum of bases in the form of oxides is between 9 and 20%, with less than 3% K_2O, the agromineral is classified into calcium-magnesium silicate agrominerals (ASiCaMg) (See Chapter 3). These are derived from basic rocks (e.g., basalts, diabases, gabbros, andesites, amphibolites, basic granulites) and have a neutralizing power of around 30%.

The metabasalt originates from mine spoil generated by iron ore production activities from Carajás Complex—PA, PA, Brazil (Tables 1, 2, and 3). The material tested is a representative sample of the mean composition of metabasalt from mine waste, evaluated for its potential as an agricultural product (Ribeiro 2023). The total bases in oxide form ($CaO + MgO + K_2O$) constitute 13.4%, with a K_2O content of 2.6%, which exceeds the minimum requirements set by NI 5/2016 at 9% and 1%, respectively. The levels of potentially toxic elements are below the maximum limits specified by NI 5/2016 (Table 2).

The mineralogical composition of the metabasalt (Table 1) is predominantly chlorite and quartz, with additional presence of albite, orthoclase, muscovite, biotite, and notably calcite (11.2%). Consequently, the free silica content (21.6% quartz) is within

Table 1 Chemical contents by mineral phase and mineralogical proportion from mineralogical analyses of metabasalt using X-ray diffraction (XRD) and Rietveld Refinement.[1]

Mineral[1]	SiO_2	Al_2O_3	FeO	Fe_2O_3	CaO	MgO	TiO_2	P_2O_5	Na_2O	K_2O	CO_2	LOI*	XRD
	(%)												
Hematite	–	–	–	100	–	–	–	–	–	–	–	–	4.8
Orthoclase	64.8	18.3	–	–	–	–	–	–	–	16.9	–	–	7.4
Quartz	100	–	–	–	–	–	–	–	–	–	–	–	21.6
Albite	67.4	20.4	–	–	1.1	–	–	–	11.2	–	–	1.1	11.3
Chlorite	30.3	17.1	15.1	–	–	25.4	–	–	–	–	–	12.1	30.9
Calcite	–	–	–	–	56	–	–	–	–	–	44	44	11.2
Muscovite	45.2	38.4	–	–	–	–	–	–	–	11.8	–	4.1	9.5
Biotite	41.6	11.8	8.3	–	–	23.2	–	–	–	10.9	–	3.6	2.4

[1]*Source* Ribeiro (2023). *Loss on ignition

Table 2 Chemical contents of metabasalt major elements in the form of oxides[1]

SiO_2	Al_2O_3	Fe_2O_3	CaO	MgO	TiO_2	P_2O_5	Na_2O	K_2O	MnO	LOI*
(%)										
49.0	12.5	14.0	5.5	5.4	0.8	0.1	1.1	2.6	0.4	5.6

[1]Modified from Ribeiro (2023). *Loss on ignition

Table 3 Chemical contents of potentially toxic trace elements in metabasalt[1]

Cd	Pb	Hg	As
(mg kg^{-1})			
1.28	30	< 0.01	< 1

[1]Modified from Ribeiro (2023)

the acceptable limit of 25% as required by IN 5/2016. The granulometric composition reveals that the material is predominantly powder, with 100% passing through a 2 mm sieve and 80% passing through a 0.3 mm sieve, as reported by Ribeiro (2023).

3 Methods

The experiment consisted of three phases: a lab pH incubation to assess acidity correction, greenhouse pot incubation, and a final stage involving maize and soybean plants.

3.1 Soil Characterization

Two soils with differing textures, low pH values, and high levels of exchangeable aluminum were selected from native forest areas. Soil samples were collected using a loader from a depth of 0–20 cm and were thoroughly homogenized before the experiment. The first soil, a Ferralsol (IUSS Working Group WRB 2022)/Dystrophic Red Latosol (Santos et al. 2018a, b), with a clayey texture, was located at coordinates – 15° 35′ 38.3″ S, – 47° 44′ 12.9″ W in Planaltina—DF, Brazil. The second soil, an Arenosol (IUSS Working Group WRB 2022)/Quartzarenic Neosol (Santos et al. 2018a, b), with a sandy texture, was located at coordinates – 15° 36′ 45.9″ S, – 47° 44′ 22.5″ W (Table 4). The soils were collected under forest conditions to ensure that the acidity correction method could be tested on soil that had not previously been treated with any type of corrective agents.

Table 4 Chemical analysis of the Ferralsol (FR) and the Arenosol (AR)*

Soil	pH[a]	P	K	Ca	Mg	Al	Na	H + Al	SB	CEC[b]	BS	m	OM	C[c]	B	Cu	Fe	Mn	Zn
		mg kg	$cmol_c\ dm^{-3}$							%			$g\ kg^{-1}$		$mg\ dm^{-3}$				
FR	4.6	13	0.11	0.4	0.2	1.2	0.02	5.4	0.7	6.1	12	62	22.2	12.9	0.26	1.32	86.8	3.6	2.04
AR	4.8	13.2	0.10	0.4	0.2	0.5	0.02	3.7	0.7	4.4	16	41	7.1	4.1	0.31	1.16	144.8	2.3	1.59

*Analyses performed by Soloquímica Laboratory. [a]pH in water; [b]CEC at pH 7,0; [c]Organic carbon

Table 5 Doses of metabasalt and limestone (mg 10 g^{-1} of soil) and their equivalent in t ha^{-1}

Metabasalt		Limestone	
mg 10 g^{-1}	t ha^{-1}	mg 10 g^{-1}	t ha^{-1}
0	0	0	0
2.5	0.25	2.5	0.25
5	0.5	5	0.5
7.5	0.75	7.5	0.75
10	1	10	1
20	2	15	1.5
40	4	20	2
60	6	30	3
80	8	40	4
100	10	50	5

3.2 Laboratory Incubation

Metabasalt reactivity was measured using a modified abrasion pH method, in which 10 g of rock or mineral powder was mixed with 25 mL of water for 1 h. After mixing again, the pH of the suspension was recorded. This method is useful for assessing the weathering potential of minerals and rocks. The procedure involved adding increasing doses of crushed rock to soil over a set reaction time (Table 5). In 50 mL cups, 10 g of soil were combined with varying amounts of metabasalt or limestone, followed by 25 mL of water. After 24 h, the mixture was stirred, and the pH of the suspension was measured. Three replicates were used for each dose, including a control. Agronomic doses of lime were also established for reference (see Table 6).

3.3 Greenhouse Incubation

To perform the incubation process in a greenhouse, soils FR and AR were thoroughly homogenized, placed in 6 kg pots, and maintained in a greenhouse for a 90-day incubation period. The pots received varying doses of metabasalt, depicted in Table 6, to assess the dissolution and reactivity of the rock in each soil. The control treatment consisted of pots containing soil without metabasalt.

During the incubation of the metabasalt in a greenhouse, the pots were kept consistently moistened to maintain approximately 70% of field capacity until the end of the 90-day incubation cycle. During the final 30 days, an additional lime treatment was applied (Table 3) at the recommended agronomic dose for each soil type, aiming to increase base saturation for soybean and maize cultivation to 60% (van

Table 6 Composition, reactivity and weathering potential of the limestone and metabasalt

	CaO	$CaCO_3$	MgO	$MgCO_3$	PN	RE[1]	PRNT[2]	Abrasion pH	Passing through sieve (mesh %)			Classification
									10	20	50	
Metabasalt	2.3		4.4		8.0		7.5	9.05	100	95	89	ASiCaMg[3]
Limestone	29.7	52.9	18.7	39.3	117.5	81.3	95.5		98.9	89.6	64.3	Dolomitic

[1]Relative Efficiency; [2]Relative Power of Total Neutralization; [3]Estimated by calculation; [3]According to Martins et al. (2024)

Raij 1997). Consequently, the pots with FR received 9.75 g of limestone (equivalent to 3.25 t ha^{1}), and those with AR received 6.45 g (equivalent to 2.15 t ha^{1}).

At the conclusion of the incubation period, soil samples were collected from each treatment to analyze the following parameters: pH(H_2O), exchangeable aluminum (Al), potassium (K), phosphorus (P), calcium (Ca), magnesium (Mg), sum of bases (SB), base saturation (BS%), and effective cation exchange capacity (CEC).

3.4 Experiment with Plants in a Greenhouse

After the incubation period, the pots were supplied with macro and micronutrients appropriate for soybean and maize crops, thereby initiating the cultivation stage. The maize pots received fertilization with 3.951 g of KH_2PO_4 and 4.247 g of $(NH_4)_2SO_4$. The soybean pots were inoculated with *Bradyrhizobium* and *Azospirillum* (1×10^9 CFU mL^{1}) (Barbosa et al. 2021) and subsequently fertilized with 3.398 g of $Ca(H_2PO_4)_2$ and 2.006 g of K_2SO_4. In addition, all pots were supplemented with 0.3 g of FTE-BR10 (comprising B 2.5%, Co 0.1%, Cu 1%, Fe 4%, Mn 4%, Mo 0.1%, and Zn 7%).

Maize (*Zea mays* L., KWS 7510 VIP3) and soybean (*Glycine max* L., BRS 7582) seeds were sown in the soil samples, maintaining two plants per pot. After 45 days of cultivation, the shoots were harvested and oven-dried at 65 °C until a constant weight was achieved. The dried shoots were then weighed and milled for tissue nutrient analysis. The dry biomass was analyzed for K, Ca, and Mg, which were extracted using a digestion block with HNO_3:$HClO_4$ and determined by inductively coupled plasma-optical emission spectrometry (ICP-OES) (Embrapa 2017).

3.5 Statistical Analysis

The means of the soil parameters for each dose of metabasalt were modeled, and the equations were selected based on the analysis of variance (ANOVA) for the coefficients: $^{**}p < 0.01$, $^{*}p < 0.05$. Subsequently, soil and plant tissue data were tested for normality using the Shapiro–Wilk test and for homoscedasticity, with significant p-values considered to be < 0.05. These analyses were conducted using Sigma Plot software, version 12.0.

The Scott-Knott test at a 5% significance level was used to compare the means of soil and plant tissue attributes using R software, version 4.3.0.

4 Results and Discussion

4.1 Laboratory Incubation

Based on the pH values obtained in the laboratory, equations representing the pH change as a function of the doses of rock and limestone were determined (Fig. 1). These equations were used to define the doses of metabasalt to be applied per pot during greenhouse incubation, according to the target pH value (Table 7).

Although the current pH incubation method does not account for metabasalt as a source of calcium (Ca) and magnesium (Mg) for dose determination, its objective is to provide an accurate approximation of the final soil pH. Future advancements should aim to integrate the values of Ca and Mg, as well as analyze low-crystallinity silicon (Si) and iron (Fe), to determine the most appropriate dosage of silicate minerals with potential acidity correction. This would consider the oxidation reactions involving Fe and the polymerization of Si. Consequently, it would be possible to estimate suitable dosages for each soil type, like the process used for limestone.

The selection of a 24-h incubation period was based on initial evaluations. Laboratory tests with various silicate agrominerals indicated that a 72-h incubation period generally produced slightly higher reaction rates compared to 24 h. However, for some rocks such as dunite and kamafugite, the opposite effect was observed, likely due to the presence of significant amounts of low-crystallinity iron (Fe). It was determined that a 24-h incubation period would be preferable for silicate agrominerals

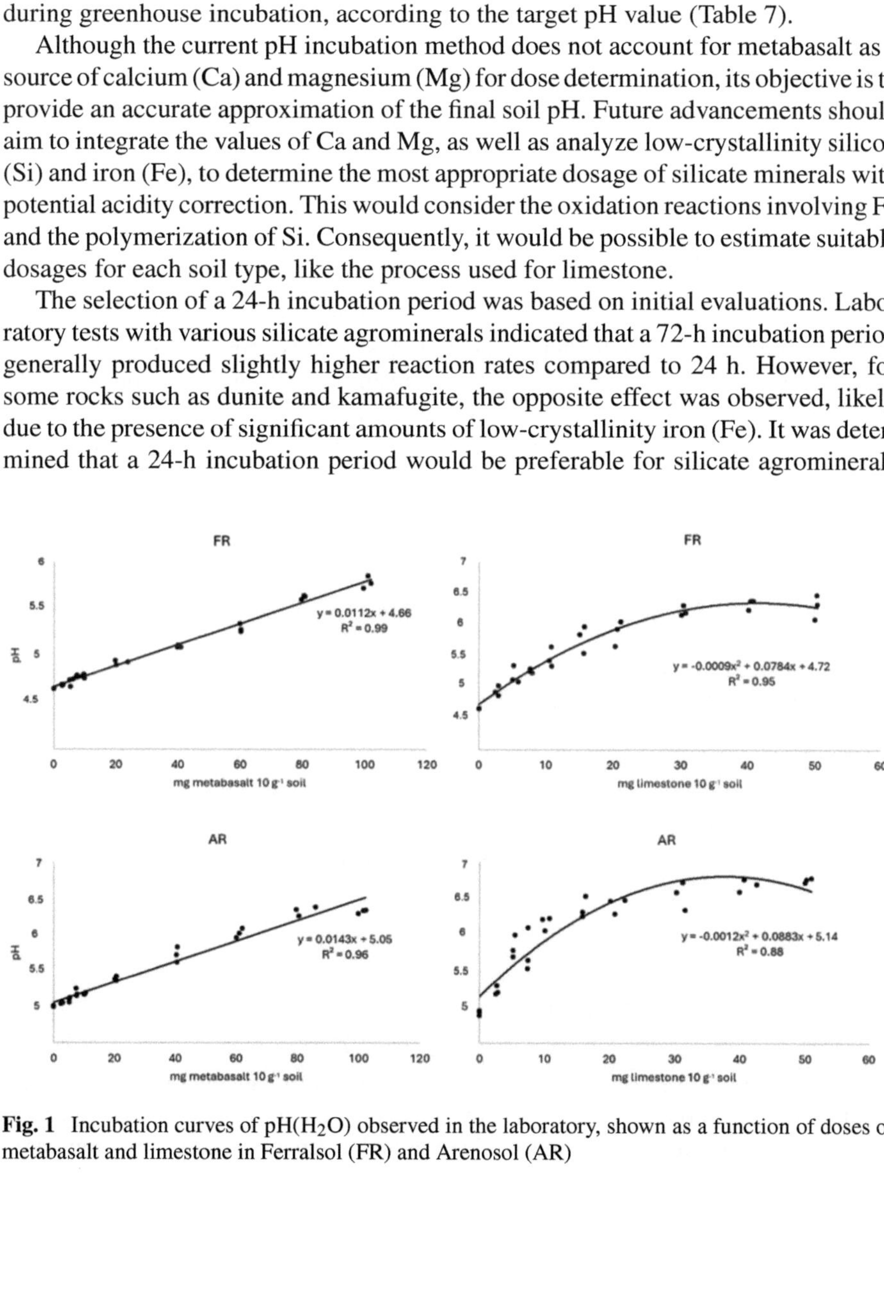

Fig. 1 Incubation curves of pH(H_2O) observed in the laboratory, shown as a function of doses of metabasalt and limestone in Ferralsol (FR) and Arenosol (AR)

Table 7 Metabasalt application rates determined from laboratory-generated pH incubation curves for Ferralsol (FR) and Arenosol (AR)

Soil	Equation	pH[1]	mg 10 g^{-1} soil	g metab pot^{-1*}	t $ha^{-1\#}$	Identification
FR	y = 0.0112x + 4.66	4.75	0	0	0	D0
		5	30	18	3	D1
		5.5	75	45	7.5	D2
		6	100	60	10	D3
		6.5	150	90	15	D4
AR	y = 0.0143x + 5.05	5	0	0	0	D0
		5.25	10	6	1	D1
		5.5	30	18	3	D2
		5.75	45	27	4.5	D3
		6	60	36	6	D4

[1] Target pH; *For pots with 6 kg of soil; #t ha^{-1} considering 0–10 cm depth

with low magnetic susceptibility, like metabasalt, as it would provide data supporting the calculation of slightly higher application doses. This approach helps to ensure that the doses applied to the soil are not underestimated. A 24-h duration was sufficient for the reaction of fine particles, so that the increase in this time reflects a small increase in the pH values. The shorter reaction time also reduces the chance of underestimating the dose of rock to be applied, increasing the supply of Ca and Mg to the soil. With this, it was possible to establish the short-term potential of correcting soil acidity with remineralizers in a simple laboratory analysis protocol, analogous to the abrasion pH of ground rocks.

The pH curves were obtained at different doses of the rock, simulating the dissolution process of minerals, especially carbonates and silicates rich in calcium (Ca) and magnesium (Mg), of small size. The initial rapid reaction observed reflects the dissolution of fine particles of ASi (Basak 2018; Krahl et al. 2022), and a short-term potential for rock correction can be considered.

The long-term potential is relative to larger particles and slower-dissolving minerals, such as chlorite $(Mg, Fe^{2+})_5Al(Si_3Al)O_{10}(OH)_8$. These particles persist in the soil and interact over time based on their physical, chemical, and mineralogical properties, as well as the biological characteristics influenced by agroecosystem management practices.

4.2 *Soils After Incubation*

Both soils from native Cerrado underwent oxidation during incubation. The initially stable particulate organic matter was exposed to oxygen during the soil homogenization stage due to tillage, leading to oxidation and a general acidification compared to the initial pH values of the soils (4.6 for FR and 4.8 for AR, Table 5). This acidification was observed even in the control group, with final pH values of 4.0 for FR and 3.6 for AR (Fig. 2). Acidification was observed in all treatments, including those with lime application, resulting in only a marginal change in the final pH. Luchese et al. (2021) noted similar acidification in their pot experiments involving basalt, although the underlying causes were not fully explained. This phenomenon may be attributed to the oxidation of the soil organic matter or the oxidation of iron released from the basalt during its dissolution. Nonetheless, the applied treatments improved soil acidity components, evidenced by increases in pH, Ca^{2+}, Mg^{2+}, sum of bases (SB), and base saturation (BS%), along with a reduction in exchangeable Al levels (Fig. 2).

Under the conditions of the incubation test, an increase in metabasalt doses led to a rise in soil pH according to the dose applied to each soil (Fig. 2). The presence of CaO in the rock's chemical composition (Table 1) can react and promote an increase in

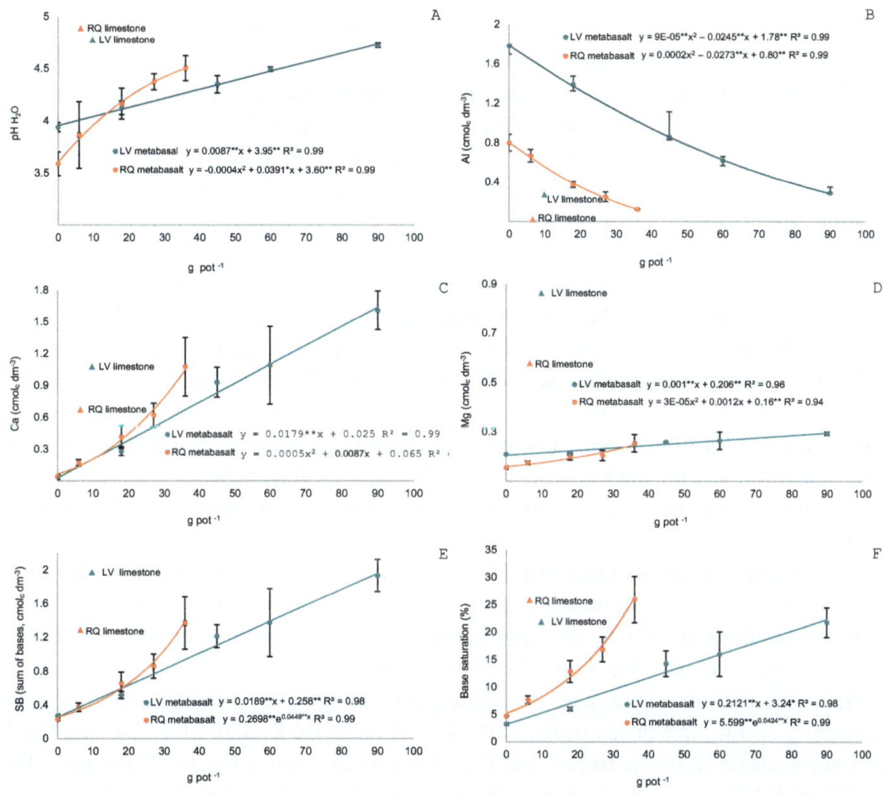

Fig. 2 Chemical attributes of Ferralsol (FR) and Arenosol (AR) under increasing doses of metabasalt (D0, D1, D2, D3, and D4) and a fixed, recommended dose of limestone, incubated for 90 days in a greenhouse. $^{**}p < 0.01$; $^{*}p < 0.05$. **a** pH in water; **b** exchangeable Al; **c** exchangeable Ca; **d** exchangeable Mg; **e** sum of bases; **f** base saturation

the soil's pH value. Calcite is primarily found in sedimentary rocks and metamorphic limestones, but it also exists in various igneous rocks as both a major mineral and a secondary mineral. The pH increase associated with the use of silicate rocks has been documented in other studies (Toscani and Campos 2017; Luchese et al. 2021, 2023). This effect occurs due to the presence of calcium oxides in the material and the consumption of H^+ during reactions, such as orthoclase reacting to form $HAlSi_3O_8$ and K^+ (Curi et al. 2005; Luchese et al, 2021), which increases the soil pH.

The results indicated a significant increase in pH and calcium levels, coupled with a reduction in exchangeable aluminum, as the metabasalt doses increased. The D4 doses demonstrated similar efficiency to lime treatment, except for magnesium supply, where limestone provided significantly higher values (Fig. 2).

The higher extraction of Ca in metabasalt treatments can be attributed to the presence of calcite in the rock composition (Table 1). These results indicate that metabasalt gradually releases this nutrient, leading to increased concentrations in the soil. The clayey soil exhibited higher calcium contents at all doses compared to the sandy soil, which is explained by the higher doses applied in the FR. Recommended calcium values in soil are above 1.5 $cmol_c\ dm^{-3}$ (Sousa and Lobato 2004); this value was achieved by metabasalt at D4 (90 g pot^{-1}, equivalent to 15 t ha^{-1}) of the FR.

The Mg^{2+} content in soils exhibited minimal changes following the application of metabasalt in the soils. Previous studies have documented slight increases in Mg^{2+} content following the application of various types of rock materials (Theodoro et al. 2020; Luchese et al. 2021). The ideal magnesium levels in soil range from 0.5 to 2 $cmol_c\ dm^{-3}$ (Sousa and Lobato 2004), and the 90-day period of incubation was insufficient to reach this range (Fig. 1). However, in the AR treatment, there was an increasing release of Mg, corresponding to the dosages applied, with significant differentiation from the control. The limestone treatment released more Mg than metabasalt, which is expected since the limestone used is dolomitic, containing high levels of magnesium carbonate ($MgCO_3$). A significant distinction between the dissolution mechanisms of multi-oxides lies in the numerous potential reaction pathways available for dissolution. There is a preferential release of elements from rock materials into the soil: $CaO > MgO >> Al(OH)_3 > Fe(OH)_3 > SiO_2$. This results in a nonlinear rate of nutrient release from rock materials (Luchese et al. 2021; Oelkers 2001).

In both soils, metabasalt increased the values of base sum (SB) and base saturation (BS%) compared to the control samples (Fig. 2). The highest dose of metabasalt in each soil resulted in a 603% increase in SB in VT and a 495% increase in AR. For BS%, the increase was 563% in FR and 446% in AR.

After incubation, there was no significant change in K and P contents with increasing metabasalt doses in the soil: $K = 10.6 \pm 1.64^s$, and 14 ± 2.4^s mg kg^{-1} for the FR and AR, respectively; and $P = 3.3 \pm 0.25^s$, and 3.4 ± 0.12^s mg kg^{-1} for the FR and AR, respectively. For soils with limestone, the values were: $K = 11 \pm 1.4^s$, and 13 ± 1.4^s mg kg^{-1} for the FR and AR, respectively; and $P = 3.2 \pm 0.02^s$, and 3.3 ± 0.01^s mg kg^{-1} for the FR and AR, respectively.

4.3 Soils After Maize and Soybean Cultivation

The pH and exchangeable aluminum levels in FR and AR after 45 days of maize cultivation, which is a nutrient-demanding crop, remained consistent with those observed post-incubation (Fig. 3). However, pH levels were significantly lower than those found in the limestone treatment, likely due to slower metabasalt dissolution kinetics. During soybean cultivation, the acidification observed post-incubation stabilized at a higher pH value, consequently reducing exchangeable Al values.

In the two soils cultivated with soybean, an increase in pH was observed to correlate with the incremental application of the previously incubated metabasalt dose. Notably, the sandy soil achieved a desirable pH level, comparable to the liming treatment, at dosage D3, and the loamy soil reached similar levels at dosage D4.

The application of metabasalt in soils cultivated with the soybean crop resulted in a continuous reduction of exchangeable Al values throughout the cultivation cycle, which continued from the dosage applied from the incubation phase. In both soybean-cultivated soils, a significant decrease in exchangeable Al was observed even at the lowest doses of metabasalt. Conversely, in soils cultivated with maize, there was only a slight reduction in Al values compared to the initial incubation levels.

In pots cultivated with maize under the higher dose of metabasalt (D4), the increase in exchangeable Ca^{2+} from the final incubation (Fig. 2) to the final cultivation (Fig. 4)

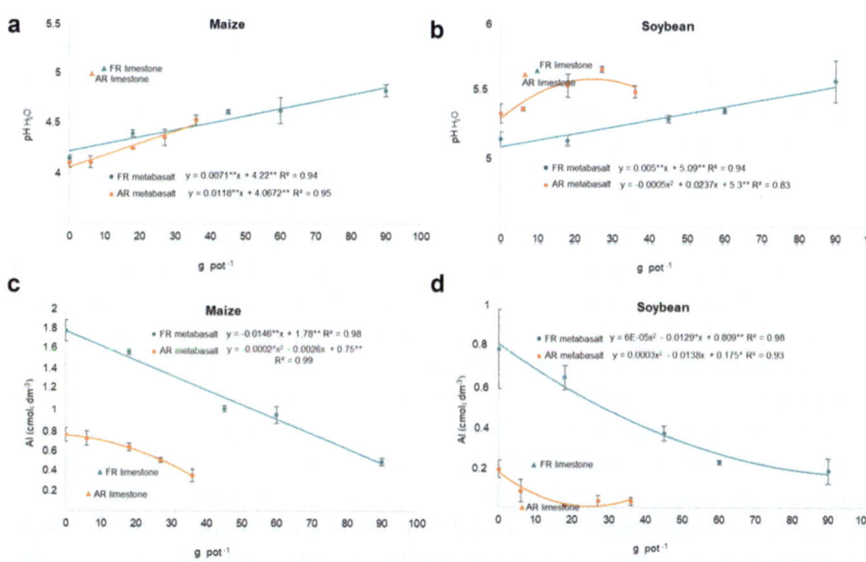

Fig. 3 pH values in water and exchangeable aluminum obtained in Ferralsol (FR) and Arenosol (AR) under increasing doses of metabasalt and a fixed dose of limestone after maize and soybean cultivation for 45 days in a greenhouse. $^*p < 0.01$; $p < 0.05$. **a** pH in water after maize crop; **b** pH in water after soybean crop; **c** exchangeable Al after maize crop; **d** exchangeable Al after soybean crop.

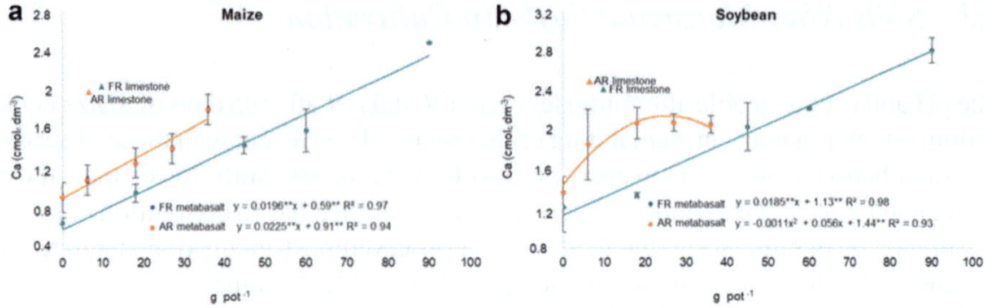

Fig. 4 Exchangeable calcium values obtained in Ferralsol (FR) and Arenosol (AR) under increasing doses of metabasalt and a fixed dose of limestone after maize and soybean cultivation for 45 days in a greenhouse. $*p < 0.01$; $p < 0.05$. **a** Exchangeable Ca after maize crop; **b** exchangeable Ca after soybean crop

was 54% in the FR soil (from 1.61 to 2.49 $cmol_c\ dm^{-3}$) and 69% in the AR soil (from 1.07 to 1.82 $cmol_c\ dm^{-3}$), indicating continuous calcium release during incubation and throughout the cultivation period. At the end of the cultivation period, the D4 dose for maize showed a higher calcium content compared to the liming treatment. This suggests that metabasalt dissolution and the subsequent calcium supply occur at a slower rate over a more extended period. Values of Ca in pots cultivated with soybean in both soils at all doses were higher than at the end of incubation at the highest dose (D4), with an increase of 74% in FR soil (from 1.61 to 2.81 $cmol_c\ dm^{-3}$) and 91% in AR soil (from 1.07 to 2.06 $cmol_c\ dm^{-3}$). However, in the AR soil cultivated with soybeans, regardless of the dose from D2 onwards, there was a tendency for the Ca content in the soil to stabilize (18 g pot^{-1}, equivalent to 3 t ha^{-1}), with values around 2.07 ± 0.01 $cmol_c\ dm^{-3}$ (Fig. 4).

After cultivating maize and soybeans, the exchangeable calcium (Fig. 4), as well as the sum of bases (SB), base saturation percentage (BS%), and effective cation exchange capacity (CEC) (Fig. 5), increased. Studies indicate that plant roots enhance the dissolution rates of calcium and magnesium silicates in basaltic minerals (Anda et al. 2015; Silva et al. 2017; Krahl et al. 2020). Metabasalt doses in maize and soybean soils did not significantly affect potassium and phosphorus (Table 8).

Analyzing the D4 dose post-incubation (Fig. 2) and post-crop (Fig. 5), there was an observed increase in BS by 49% in the FR soil cultivated with maize and by 96% in the FR soil cultivated with soybean. For AR, the increase in BS was 62% for pots grown with soybean and 200% for those grown with maize. In terms of BS%, the increases were 115% in FR soil cultivated with maize and 155% in FR soil cultivated with soybeans; 11% in AR soil cultivated with maize and 57% in AR soil cultivated with soybeans. Limestone applied at the recommended doses resulted in significantly higher values of SB, BS%, and effective CEC, except for soybean cultivation in FR, where these values were equivalent. After maize cultivation, the pH of treatments with metabasalt was significantly lower than that of limestone, resulting in a reduction in the effective CEC compared to limestone. The sum of bases and base saturation were significantly influenced by the higher Mg content in limestone

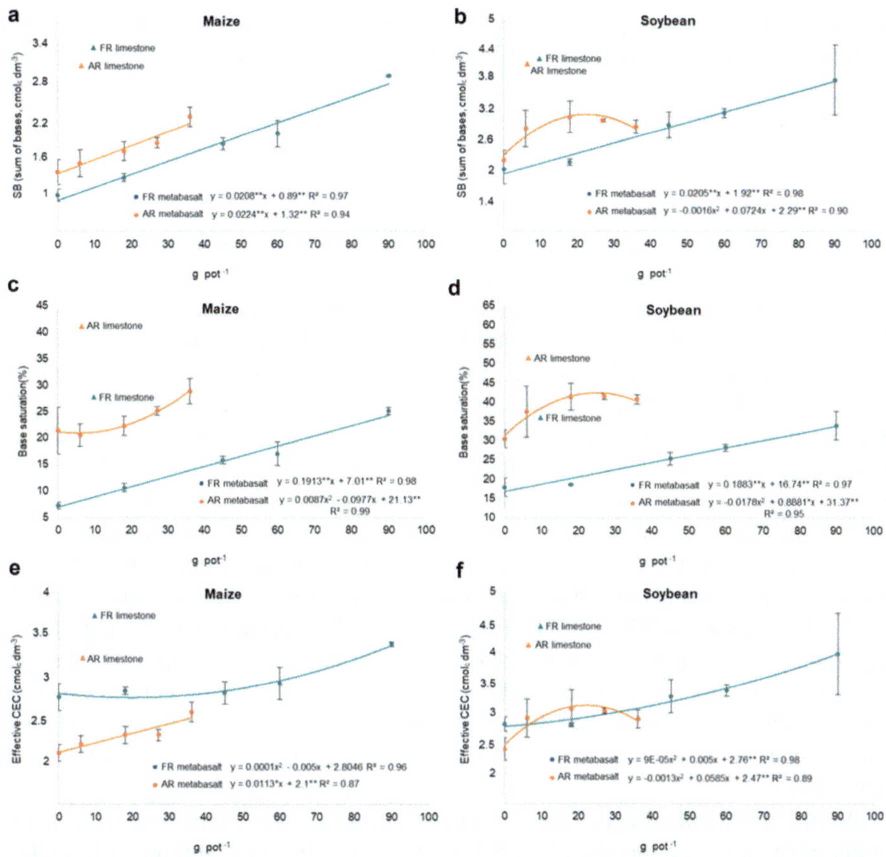

Fig. 5 Values of sum of bases (SB), base saturation (BS%), and effective cation exchange capacity (effective CEC) obtained in Ferralsol (FR) and Arenosol (AR) under increasing doses of metabasalt and a fixed, recommended dose of limestone after maize and soybean cultivation for 45 days in a greenhouse. $*p < 0.01$; $p < 0.05$. **a** Sum of bases after maize crop; **b** sum of bases after soybean crop; **c** base saturation after maize crop; **d** base saturation after soybean crop; **e** effective CEC after maize crop; **f** effective CEC after soybean crop

Table 8 Exchangeable K and P in soils after maize and soybean cultivation

Agromineral	Soil	K		P	
		mg kg^{-1}			
		Maize	Soybean	Maize	Soybean
Metabasalt	FR	85.33 ± 7.54[s]	277.38 ± 10.83	21.67 ± 1.45	21.74 ± 1.28
	AR	108.3 ± 6.7	299.7 ± 32.52	32.16 ± 2.77	33.52 ± 1.73
Limestone	FR	94.1 ± 2.5	274.49 ± 27.5	23.06 ± 1.21	18.6 ± 1.5
	AR	112.35 ± 1.6	326.6 ± 44.7	38.11 ± 2	39.7 ± 2.33

[s]Standard deviation

compared to metabasalt, given that K levels were similar and the exchangeable Ca content of metabasalt was significantly higher than that of limestone.

4.3.1 Dry Matter Production and Nutrient Extraction by Maize and Soybean

The dry matter values in the shoots and roots of maize within the FR group were comparable between metabasalt and limestone treatments (Fig. 6), with the highest production observed at the maximum dose (D4, 90 g pot^{-1}, equivalent to 15 t ha^{-1}). This was significantly higher compared to the control treatment (D0). In the AR group, dry matter production in the shoots was similar from dose D2 (18 g pot^{-1}, equivalent to 3 t ha^{-1}) onwards and was consistent in the lime treatment, with no significant differences observed between doses above 3 g pot^{-1}. In soybean cultivation in FR, no statistically significant differences were observed in shoot dry matter production across various metabasalt treatments. However, in AR, the dry matter mean values of soybean shoots were notably higher at doses D4 and liming.

In all treatments for the two soils (metabasalt doses and limestone), the adequate K concentrations in shoots of maize were exceeded (33.83 ± 3.29^s and 30.53 ± 2.18^s g kg^{-1} of K for FR and AR, respectively). The optimal growth range for K requirement in vegetative parts of maize is between 17.5 and 22.5 g kg^{-1} (Malavolta et al. 1997). In this study, soybean shoots were close to the recommended range in both soils (17.04 ± 1.06^s and 17.59 ± 1.2^s g kg^{-1} of K for FR and AR, respectively). The adequate foliar K levels in soybeans range from 17 to 25 g kg^{-1} of K (Malavolta et al. 1997).

Despite the standard application of potassium fertilization across all treatments, comparable doses of metabasalt and limestone resulted in a significantly higher accumulation (plant element concentration x dry mass) of K in the plant tissue of maize in FR compared to D0 (Fig. 7). In AR, potassium accumulation increased up to D3 and was comparable to that of limestone. In contrast, potassium accumulation in soybeans showed no significant differences between the treatments in either soil type.

The increase in exchangeable Ca^{2+} in the soils resulted in a direct increase in the Ca content in plant shoots in most treatments with the application of metabasalt and liming. The Ca requirement for maize growth was met, as the concentration in maize leaves in metabasalt treatments ranged from 8.1 (D0) to 13.6 (D4), and 7.9 (D0) to 14.3 (D4) g kg^{-1} of Ca in the FR and AR, respectively, as compared to the recommended range for optimum growth of 2.5 to 8.0 g kg^{-1} of Ca (Malavolta et al. 1997). For limestone application, the Ca values were 12.33 ± 2.56^s for FR and 9.97 $\pm 1.65^s$ for AR.

For soybeans, the Ca content in plant shoots increased linearly with the metabasalt dose, starting from 11.3 (D0) and reaching 13.83 (D4) g kg^{-1} of Ca at the highest dose in AR, within the recommended sufficiency range (3.5–20 g kg^{-1} of Ca), according to Embrapa (2010). The Ca content in soybean plants grown in FR did not increase significantly with metabasalt treatments, but it remained within the sufficiency range

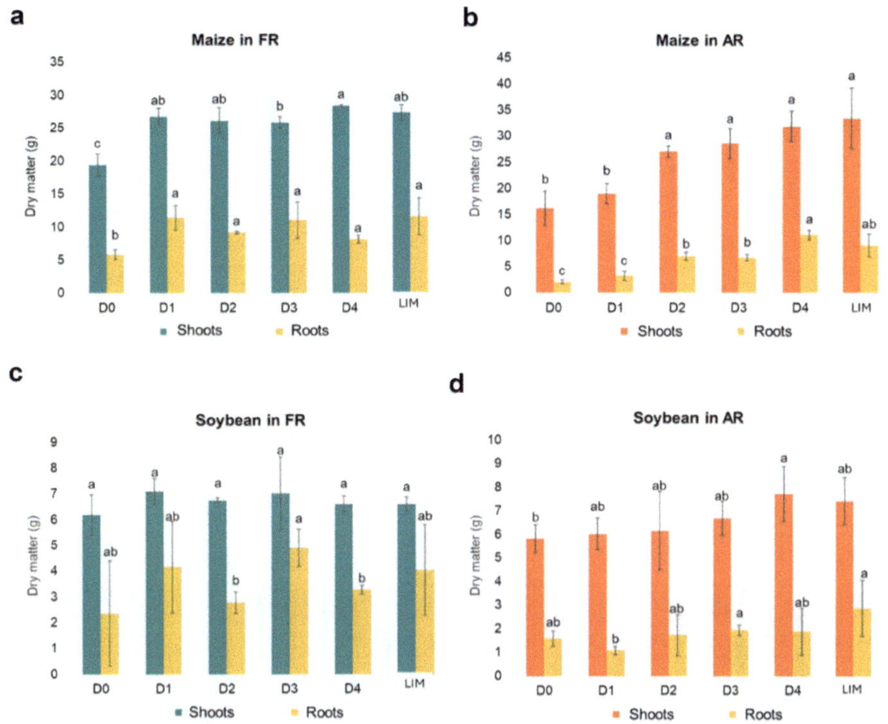

Fig. 6 Dry matter production of shoot and roots for maize and soybean cultivated for 45 days in a greenhouse in Ferralsol (FR) and Arenosol (AR) under increasing doses of metabasalt and a fixed dose of limestone; Scott-Knott test at 5% probability. **a** Dry matter production of maize cultivated in FR; **b** dry matter production of maize cultivated in AR; **c** dry matter production of soybean cultivated in FR; **d** dry matter production of soybean cultivated in AR

(12.54 ± 2.5^{s} g kg^{-1} of Ca). For limestone application, the Ca values were 18.34 ± 1.4^{s} for FR and 14.27 ± 2.2^{s} for AR.

The application of metabasalt and lime resulted in increased Ca content in the soils and higher Ca concentration in the plants, promoting greater Ca accumulation (Fig. 7). Although plant Ca levels were similar between the two crops, K accumulation in maize was significantly higher than in soybeans. In maize, Ca accumulation increased proportionally with higher doses of metabasalt and lime across both soil types. For soybeans grown in FR soil, there was no significant difference in Ca accumulation between the doses. However, in AR soil, all doses of metabasalt led to higher Ca accumulation compared to the treatment without metabasalt, with doses D4 and lime treatment showing the most substantial increase.

Despite the soil contents being significantly lower than those treated with lime, maize exhibited comparable magnesium accumulation in the D3, D4, and limestone treatments for clayey soil, as well as in the D2 treatment for sandy soil.

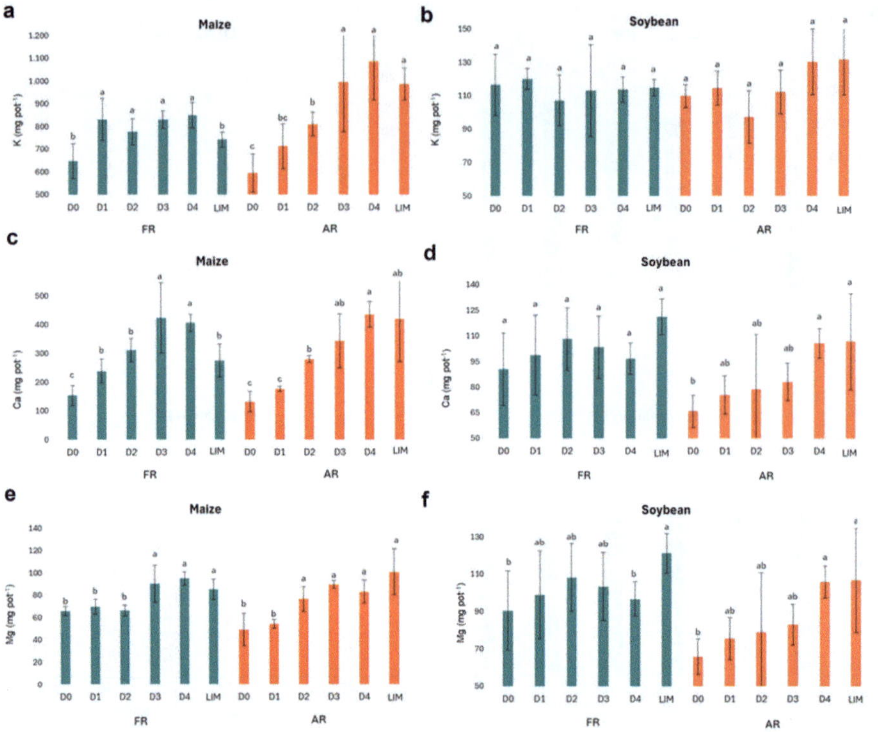

Fig. 7 Accumulation of K, Ca, and Mg in the dry matter of maize and soybean cultivated for 45 days in a greenhouse in a Ferralsol (FR) and Arenosol (AR) for increasing doses of metabasalt and doses of limestone; Scott Knott test at 5% probability. **a** K accumulation in maize; **b** K accumulation in soybean; **c** accumulation of Ca in maize; **d** accumulation of Ca in soybean; **e** accumulation of Mg in maize; **f** accumulation of Mg in soybean. Accumulation = plant element concentration x dry mass

5 Conclusions

The application of metabasalt demonstrated significant improvements in the soil's chemical properties, which, in turn, positively influenced the growth and nutrient uptake of maize and soybean. Metabasalt proved to be a viable alternative to traditional lime in correcting soil acidity and enhancing the availability of essential nutrients, particularly calcium and magnesium, which are vital for plant health. The study found that metabasalt increased the pH levels in both Ferralsol and Arenosol soils, reduced exchangeable aluminum, and enhanced the accumulation of essential nutrients, including potassium, calcium, and magnesium, in plant tissues. These findings suggest that metabasalt can play a crucial role in sustainable agriculture by promoting healthier soil and more productive crops.

Metabasalt behaves similarly to basalts, increasing soil pH and base saturation and boosting crop productivity as application rate increases. However, the presence of calcium, primarily as carbonate in the metabasalt and as plagioclase in basalts, enhances mineral reactivity and the soil liming effects, especially for the metabasalt.

6 Declaration of Competing Interest

The authors declare that they have no known competing financial interests or personal relationships that could have appeared to influence the work reported in this publication.

Acknowledgements The authors gratefully acknowledge the financial support provided by Embrapa, by FINEP — Fundo Setorial Mineral through the project 'Rede de PD&I sobre remineralizadores e agrominerais (REMAg)' (Project No. 1541/2022), and by VALE-Embrapa Cerrados partnership through the project 'Tecnologia de manejo de estéreis de mina como insumos minerais para a fertilidade do solo agrícola (AgroGeol)' which made possible the development of this book chapter.

Author's Contribution Luise Lottici Krahl conducted the experiments and was responsible for data collection, data analysis, and manuscript writing. Giuliano Marchi and Éder de Souza Martins conceived and designed the study; both contributed to the writing and critically revised the final version of the chapter. Elis Marina de Freitas performed the statistical analyses and contributed to the interpretation of the results.

References

Alcarde JA, Rodella AA (2003) Qualidade e legislação de fertilizantes e corretivos. Tópicos em Ciência do Solo. Sociedade Brasileira de Ciência do Solo, Viçosa, pp 291–334

Alves EEN, Siebeneichler EA, da Costa LM, Vergütz L, da Fonseca TC, Fontes RLF (2017) Efeito da polimerização do silício na diminuição da eficiência na correção da acidez do solo pelo silicato de cálcio. In IV Simpósio Mineiro de Ciência do Solo. Sociedade Brasileira de Ciência do Solo, Viçosa

Amoakwah E, Shim J, Kim S, Lee Y, Kwon S, Sangho J, Park S (2023) Impact of silicate and lime application on soil fertility and temporal changes in soil properties and carbon stocks in a temperate ecosystem. Geoderma 433:116431. https://doi.org/10.1016/j.geoderma.2023.116431

Anda M, Shamshuddin J, Fauziah CI (2015) Improving chemical properties of a highly weathered soil using finely ground basalt rocks. Catena 124:147–161. https://doi.org/10.1016/j.catena.2014.09.012

Barbosa JZ, Hungria M, Sena JVS, Poggere G, Reis AR, Corrêa RS (2021) Meta-analysis reveals benefits of co-inoculation of soybean with *Azospirillum brasilense* and *Bradyrhizobium* spp. in Brazil. Appl Soil Ecol 163:e103913

Basak BB (2018) Waste mica as alternative source of plant-available potassium: evaluation of agronomic potential through chemical and biological methods. Nat Resour J 27:1–13. https://doi.org/10.1007/s11053-018-9430-3

Batista NTF (2013) Atributos químicos do solo e componentes agronômicos na cultura da soja pelo uso da rochagem. Dissertação de Mestrado, UFG, Campus de Jataí. https://files.cercomp.ufg.br/weby/up/217/o/DISSERTAÇÃO_VERSÃO_FINAL_1_Nayra.pdf

Brazil (2016) Instrução Normativa No 5, de 10 de março de 2016. Ministério da Agricultura, Pecuária e Abastecimento (MAPA). Gabinete da Ministra, 2016. http://www.agricultura.gov.br/assuntos/insumos-agropecuarios/insumos-agricolas/fertilizantes/legislacao/in-5-de-10-3-16-remineralizadores-e-substratos-para-plantas.pdf

Brazil (2013) Lei No 12.890, de 10 de dezembro de 2013, que altera a Lei No 6.894 de 16 de dezembro de 1980, para incluir os remineralizadores como uma categoria de insumo destinado à agricultura. Presidência da República, Casa Civil, Subchefia para Assuntos Jurídicos. http://www.planalto.gov.br/ccivil_03/_Ato2011-2014/2013/Lei/L12890.htm

Carvalho AM, Burle ML, Pereira J, Silva A (1999) Manejo de adubos verdes no cerrado, vol 4. Circular Técnica, Brasília, Embrapa Cerrados. https://www.infoteca.cnptia.embrapa.br/bitstream/doc/546470/1/cirtec04.pdf

Castro GSA, Crusciol CAC, da Costa CHM, Ferrari Neto J, Mancuso MAC (2016) Surface application of limestone and calcium-magnesium silicate in a tropical no-tillage system. J Soil Sci Plant Nutr 16:362–379

Conceição LT, Silva GN, Holsback HMS, de Figueiredo OC, Marcante NC, Martins ÉS, Santos EF (2022) Potential of basalt dust to improve soil fertility and crop nutrition. J Agric Food Res 10:100443. https://doi.org/10.1016/j.jafr.2022.100443

Curi N, Kämpf N, Marques JJ (2005) Mineralogia e formas de potássio em solos do Brasil. In: Yamada T, Roberts TL (eds.), Potássio na agricultura brasileira. Associação Brasileira para a Pesquisa da Potassa e do Fosfato. Piracicaba, SP, pp 71–92

Embrapa (2017) Manual de métodos de análise de solo. Embrapa Solos, Rio de Janeiro, RJ

Embrapa (2008) Tecnologias de produção de soja—Região Central do Brasil 2009 e 2010 (Sistemas de Produção, 13). Embrapa Soja, Londrina, PR

Essington ME (2005) Soil and water chemistry: an integrative approach. CRC Press, Boca Raton.

Fageria NK, Stone LF (1999) Manejo da acidez dos solos de cerrado e de várzea do Brasil. Santo Antônio de Goiás, Embrapa-CNPAF. Documentos, p 92

Gillman GP, Burkett DC, Coventry RJ (2002) Amending highly weathered soils with finely ground basalt rock. Appl Geochem 17(8):987–1001. https://doi.org/10.1016/S0883-2927(02)00078-1

Goedert WJ (1983) Management of the Cerrado soils of Brazil: a review. J Soil Sci 34:405–428. https://doi.org/10.1111/j.1365-2389.1983.tb01045.x

Hinsinger P, Fernandes Barros ON, Benedetti MF, Noack Y, Callot G (2001) Plant-induced weathering of a basaltic rock: experimental evidence. Geochimica et Cosmochimica Acta 65(1):137–152. https://doi.org/10.1016/S0016-7037(00)00524-X

IUSS Working Group WRB (2022) World reference base for soil resources. International soil classification system for naming soils and creating legends for soil maps, 4th edn. International Union of Soil Sciences (IUSS), Vienna, Austria

Korchagin J, Caner L, Bortoluzzi EC (2019) Variability of amethyst mining waste: a mineralogical and geochemical approach to evaluate the potential use in agriculture. J Clean Prod 210:749–758. https://doi.org/10.1016/j.jclepro.2018.11.039

Korchagin J, Escosteguy PAV, Bortoluzzi EC (2022) Nutrient transfer in rangelands under rock powder amendment. J Plant Nutr Soil Sci 185(5):656–667. https://doi.org/10.1002/jpln.202200059

Krahl LL, Marchi G, Paz SPA, Angélica RS, Sousa-Silva JC, Valadares LF, Martins ÉS, (2022) Increase in cation exchange capacity by the action of maize rhizosphere on Mg or Fe biotite-rich rocks. Pesquisa Agropecuária Trop 52. https://doi.org/10.1590/1983-40632022v5272376

Krahl LL, Paz SPA, Angélica RS, Valadares LF, Sousa-Silva JC, Marchi G, Martins ES (2020) Successive offtake of elements by maize grown in pure basalt powder. Afr J Agric Res 15(2):229–239. https://doi.org/10.5897/AJAR2019.14546

Lefebvre D, Goglio P, Williams A, Manning DAC, Azevedo AC, Bergmann M, Meersmans J, Smith P (2019) Assessing the potential of soil carbonation and enhanced weathering through life cycle assessment: a case study for Sao Paulo State. Braz J Clean Prod 233:468–481. https://doi.org/10.1016/j.jclepro.2019.06.099

Lopes AS (2016) Guilherme LRG (2016) A career perspective on soil management in the Cerrado region of Brazil. Adv Agron 137:1–72. https://doi.org/10.1016/bs.agron.2015.12.004

Luchese AV, Leite IJGC, Giaretta APS, Alves ML, Pivetta LA, Missio RF (2023) Use of quarry waste basalt rock powder as a soil remineralizer to grow soybean and maize. Heliyon 9:e14050. https://doi.org/10.1016/j.heliyon.2023.e14050

Luchese AV, Pivetta LA, Batista MA, Steiner F, Giaretta APS, Curtis JCD (2021) Agronomic feasibility of using basalt powder as soil nutrient remineralizer. Afr J Agric Res 17(3):487–497. https://doi.org/10.5897/AJAR2020.15234

Malavolta E, Vitt, GC, Oliveira SA (1997) Avaliação do estado nutricional das plantas:princípios e aplicações. Associação Brasileira de Potassa e do Fósforo, Piracicaba

Marchi G, Guelfi-Silva DR, Malaquias JV, Guilherme LRG, Spehar CR, Martins ÉS (2020) Solubility and availability of micronutrients extracted from silicate agrominerals. Pesquisa Agropecuária Brasileira, Themat Issue—Reg Agrominerals 55:e00807. https://doi.org/10.1590/s1678-3921.pab2020.v55.00807

Martins ÉS, Martins ES, Hardoim PR (2021) Princípios geoquímicos, mineralógicos e biológicos do manejo de remineralizadores de solos. Informe Agropecuário. EPAMIG 44(321):26–39. https://ainfo.cnptia.embrapa.br/digital/bitstream/doc/1156783/1/Eder-principios-geoquimicos-mineralogicos.pdf

Martins ÉS, Resende AV, Oliveira CG, Furtini Neto AE (2010) Materiais silicáticos como fontes regionais de nutrientes e condicionadores de solos. In: Fernandes FRC, da Luz AB, Castilhos ZC (eds) Agrominerais para o Brasil. CETEM/MCT, pp. 89–104. http://mineralis.cetem.gov.br/bitstream/cetem/907/1/cap5.pdf

Martins ÉS, Silveira CAP, Bamberg AL, Martinazzo R, Bergmann M, Angélica RS (2014) Silicate agrominerals as nutrient sources and as soil conditioners for tropical agriculture. In: International scientific centre of fertilizers (ed) World Fertilizer Congress of CIEC, 16. Technological innovation for a sustainable tropical agriculture: proceedings. Rio de Janeiro

Martins ÉS, Theodoro SH, Bernardez FFG, Luchese AV, Bergmann M, Siqueira DS, Teixeira A, Azevedo A, Curtis JCD (2024) Produção Brasileira de Remineralizadores e Fertilizantes Naturais: 2019 a 2023. Revista Novosolo Ano 3, Edição 6, ABREFEN

Melo VF, Uchôa SCP, Dias FO, Barbosa GF (2012) Doses de basalto moído nas propriedades químicas de um Latossolo Amarelo distrófico da savana de Roraima. Acta Amazon 42(4):471–476

Miranda LN, Miranda JC, Rein TA, Gomes AC (2005) Utilização de calcário em plantio direto e convencional de soja e milho em Latossolo Vermelho. Pesq Agropec Bras 40:563–572. https://doi.org/10.1590/S0100-204X2005000600006

Nunes JMG, Kautzmann RM, Oliveira C (2014) Evaluation of the natural fertilizing potential of basalt dust wastes from the mining district of Nova Prata (Brazil). J Clean Prod 84:649–656. https://doi.org/10.1016/j.jclepro.2014.04.032

Oelkers E (2001) General kinetic description of multioxide silicate mineral and glass dissolution. Geochim Cosmochim Acta 65(21):3703–3719

Oliveira AKM, Pina JC, Pereira SR, Bono JAM, Matias R, Pires FF, Assis TE (2020) Effect of basalt rock powder associated with different substrates on the initial development of aroeira seedlings (*Myracrodruon urundeuva*). Res Soc Dev 9(12):e5591210790. https://doi.org/10.33448/rsd-v9i12.10790

Rada N (2013) Assessing Brazil's cerrado agricultural miracle. Food Policy 38:146–155. https://doi.org/10.1016/j.foodpol.2012.11.002

Resk DVS, Ferrreira EAB, Figueiredo CC, Zinn Y (2008) Dinâmica da matéria orgânica no Cerrado. In: Santos GA, Silva LS, Canellas LP, Camargo FAO (eds) Fundamentos da matéria orgânica do solo: ecossistemas tropicais e subtropicais. Editora Cinco Continentes, Porto Alegre, pp 359–417

Ribeiro D (2023) Análise das garantias geoquímicas e mineralógicas de coproduto de rochas máficas extraídas como estéreis nas minas de ferro de Carajás. Relatório Inédito, VALE

Rodrigues M, Silva Junges LF, Mozorovicz C, Ziemmer GS, Neto CK, Andrade EA, Melo Teixeira L (2024) Paraná basin basalt powder: a multinutrient soil amendment for enhancing soil chemistry and microbiology. J S Am Earth Sci 104957. https://doi.org/10.1016/j.jsames.2024.104957

Sano, SM, Almeida, SP, Ribeiro, JF (Eds.) (2008) Cerrado: ecologia e flora. Embrapa Informação Tecnológica, Brasília, DF 2 v. (1.279 p.)

Santos HG, Jacomine PKT, Anjos LHC, Oliveira VA, Lumbreras JF, Coelho MR, Almeida JÁ, Araújo Filho JC, Oliveira JB, Cunha TJF (2018) Sistema Brasileiro de Classificação de Solos. 5. ed., rev. e ampl. Brasília, Embrapa, p 356

Silva RCS, Cury ME, Ieda JJC, Sermarini RA, Azevedo AC (2017) Chemical attributes of a remineralized Oxisol. Ciência Rural 47:e20160982. https://doi.org/10.1590/0103-8478cr20160982

Shafar JM, Noordin WD, Zulkefly S, Shamshuddin J, Hanafi MM (2017) Improving soil chemical properties and growth performance of Hevea brasiliensis through basalt application. In: Proceedings of international rubber conference. Bali, Indonesia. https://doi.org/10.22302/ppk.procirc2017.v1i1.483

Soil Survey Staff (2022) Keys to soil taxonomy, 13th edn. USDA-Natural Resources Conservation Service

Sousa DMG, Lobato E (2004) Cerrado: correção do solo e adubação. Embrapa Cerrados: Embrapa Informação Tecnológica, Planaltina, DF

Theodoro SH, de Paula Medeiros F, Ianniruberto M, Baiocchi Jacobson TK (2020) Soil remineralization and recovery of degraded areas: an experience in the tropical region. J South Am Earth Sci 103:103014. https://doi.org/10.1016/j.jsames.2020.103014

Theodoro SH, Sander A, Burbano DFM, Almeida GR (2021) Rochas basálticas para rejuvenescer solos intemperizados. Revista Liberato 45–58. https://doi.org/10.31514/rliberato.2021v22n37

Toscani RGS, Campos JEG (2017) Uso de pó de basalto e rocha fosfatada como remineralizadores em solos altamente intemperizados. Geociências 36(2):259–274

van Raij B (1997) Recomendações de adubação e calagem para o Estado de São Paulo, vol 2. Instituto Agronômico, Campinas, p 285

van Straaten P (2006) Farming with rocks and minerals: challenges and opportunities. An Acad Bras Ciênc 78(4):731–747

White AF, Brantley SL (2003) The effect of time on the weathering of silicate minerals: why do weathering rates differ in the laboratory and field? Chem Geol 202(3–4):479–506

Zuffo AM, Dias de Morais KA, González Aguilera J, Felippe Ratke R, Steiner F (2022) Agronomic performance of maize crop in response to the application of metabasalt powder rates. Bulg J Agric Sci 28(3). https://www.agrojournal.org/28/03-17.pdf

Agronomic Characterization of Magnesium Soil Remineralizers and Silicate Fertilizers in Tropical Conditions

Carlos Alexandre Costa Crusciol, Sirlene Lopes Oliveira, Luiz Gustavo Moretti, and João William Bossolani

Abstract This chapter explores the physical, chemical, and microbiological factors influencing the solubilization of magnesium (Mg) silicates. In addition, studies of the effectiveness of Mg silicate-containing rocks, such as dunite and thermomagnesium (TM), as alternative fertilizers in tropical soils are discussed. Greenhouse and field experiments have demonstrated that these materials enhance Mg and silicon availability, improve photosynthetic efficiency, and increase crop yields in soybean and **corn** systems. Dunite application raises soil pH and provides steady release of Mg, while TM, especially in finer particle sizes, has superior nutrient release and agronomic performance. These findings highlight the potential of Mg-silicate agrominerals as eco-friendly fertilizers that can reduce reliance on conventional inputs and support sustainable agriculture in nutrient-depleted tropical soils.

Keywords Ultramafic rocks · Ferronickel slags · Sustainable fertilization

C. A. C. Crusciol (✉) · S. L. Oliveira · L. G. Moretti · J. W. Bossolani
Department of Crop Science, College of Agricultural Sciences, São Paulo State University (UNESP), Botucatu, São Paulo, Brazil
e-mail: carlos.crusciol@unesp.br

S. L. Oliveira
e-mail: sirlene.oliveira@ifmt.edu.br; sirlene.lopes@unesp.br

L. G. Moretti
e-mail: souzamoretti@gmail.com

J. W. Bossolani
e-mail: bossolani.agro@gmail.com

C. A. C. Crusciol
Center for Carbon Research in Tropical Agriculture (CCARBON), University of São Paulo (USP), Piracicaba, São Paulo, Brazil

S. L. Oliveira
Federal Institute of Mato Grosso (IFMT), Guarantã do Norte, Mato Grosso, Brazil

É. de Souza Martins and S. Huff Theodoro (eds.), *Soil Remineralizers and Silicate Fertilizers*, Frontier Studies in Soil Science,
https://doi.org/10.1007/978-3-032-14656-4_9

1 Introduction

Magnesium (Mg) is an essential plant nutrient that plays a crucial role in physiological and biochemical processes, including chlorophyll synthesis, photoassimilate production, transport, and utilization; enzyme activation; and protein synthesis (Ishfaq et al. 2022). Approximately 5 to 25% of a plant's magnesium (Mg) content is located in chloroplasts, reflecting the importance of these organelles in photon capture and energy transfer reactions of light-harvesting complexes (de Bang et al. 2021). Mg-deficient plants accumulate photoassimilates in their leaves, which culminates in excessive production of reactive oxygen species (ROS) in chloroplasts. The deficient translocation of photoassimilates to roots leads to significantly reduced root growth (Liu et al. 2017). The first visual symptoms of Mg deficiency are interveinal chlorosis, which develops gradually from the tips of old leaves and is eventually accompanied by purplish discoloration (de Bang et al. 2021)..

High temperatures and rainfall make Mg deficiency a common problem in highly weathered tropical soils (Schroth et al. 2002). These soils are typically nutrient-poor and have a low cation retention capacity, which favors magnesium leaching, especially in sandy soils (Pavan et al. 1982). High acidity reduces nutrient availability because Mg is easily replaced by other cations, such as aluminum (Al^{3+}) and hydrogen (H^+), making Mg unavailable to plants. Excess Al^{3+} is toxic to plants, causing oxidative stress that manifests as poor root development and reduced crop productivity (Pavan et al. 1982; Lisboa et al. 2021).

Liming is an effective method for neutralizing soil acidity and toxic aluminum ($A13^+$), precipitating it as insoluble hydroxides [$Al(OH)_3$] that are unavailable to plants (Auler et al. 2019; Nolla et al. 2020; Miotto et al. 2020). Furthermore, reducing acidity decreases competition between H^+, $A13^+$, and Mg^{2+} at soil exchange sites, freeing Mg for uptake by roots (Auler et al. 2019; Miotto et al. 2020). However, excessive application of calcitic limestone and gypsum and continuous applications of potassium (K) can lead to Mg deficiency via imbalanced nutrient absorption, as excess calcium (Ca^{2+}), K^+, manganese (Mn^{2+}), and zinc (Zn^{2+}) cause competitive inhibition (Moreira et al. 2003; Malavolta 2006; Cakmak and Yazici 2010). The application of dolomitic limestone ($CaMg(CO_3)_2$) directly provides MgO ($\sim$12% Mg [wt%]) to the soil solution that can be taken up by roots and enhance crop nutrition (Castro and Crusciol 2015). However, in cultivated tropical areas, other agrominerals, such as magnesium silicates, are increasingly used as fertilizers to replenish not only magnesium but also other essential nutrients, including potassium (K), silicon (Si), and iron (Fe), in a more balanced manner (Moretti et al. 2021).

The slow nutrient release and lasting impact of agrominerals necessitate extended experimental evaluations. Adopting agrominerals as fertilizers allows farmers to sustainably enhance soil health and crop yields while reducing conventional fertilizer use and mitigating associated environmental harm (Passos et al. 2024). Basalt (an igneous rock) and smectite-rich sedimentary rocks both contain silica (Fernandes et al. 2010; Van Ranst et al. 2020). In Brazil, basalt—comprising pyroxene and plagioclase minerals—is utilized as a natural fertilizer. Though it weathers easily, its high

Ca and Mg content makes it agriculturally beneficial (Ferreira et al. 2009; Hanisch et al. 2013). The diversity of rocks identified as potential sources of Mg for fertilizer is gradually increasing, and exploratory studies, geochemical analyses, method development, and systematic agronomic evaluations are needed to validate the efficiency of these materials as fertilizers. In this chapter, we review the key physical, chemical, and biological mechanisms governing magnesium silicate dissolution and examine their agronomic potential as alternative fertilizers in tropical soils. Special emphasis is placed on recent greenhouse and field experiments with dunite and thermomagnesium (TM), which provide new evidence of their efficiency in supplying magnesium and silicon, improving soil chemistry and crop performance. By combining a synthesis of dissolution processes with original agronomic trials, this chapter offers a critical review of current knowledge and new experimental insights to guide the development of sustainable magnesium silicate fertilization strategies.

2 Minerals Sources of Magnesium and Factors Affecting Their Dissolution

Mg is the 8th most abundant mineral element on Earth (Maguire and Cowan 2002). Mg in soil is derived from igneous/metamorphic and sedimentary rocks. Basic and ultrabasic rocks rich in Mg include basalts (rich in pyroxene and olivine), gabbro (pyroxene and amphiboles), peridotite (olivine and pyroxene), and dunite (nearly pure olivine) (Jaques et al. 1983; Hébert et al. 1983; Kelemen 1990). Dolomitic marble (metamorphosed dolomite) and serpentinite (formed by the metamorphic alteration of peridotites and dunites and rich in serpentinite) are examples of metamorphic rocks (Menzel et al. 2019; Martin et al. 2022). The main source of Mg for plants is sedimentary rocks, such as carbonates, dolomite ($CaMg(CO_3)_2$) and dolomitic limestone (mixture of calcite and dolomite), in addition to magnesian clays such as smectite (as montmorillonite) and hydrated silicates (which may contain serpentine formed by the alteration of ultrabasic rocks) (Sperber et al. 1984; Bakhshandeh et al. 2011; Mathews et al. 2024).

The Mg content in silicate rocks varies considerably (muscovite > biotite > hornblende > augite > olivine) due to substitutions. Mg^{2+} is often replaced by $Al3^{+}$ in silicate mineral structures, creating permanent charge deficits that critically influence soil physicochemical behavior (Gransee and Führs 2013). Table 1 shows the average Mg concentrations of the main mineral sources of Mg. The Mg content in the soil can vary between 0.05 and 0.5%, depending on the source material (Maguire and Cowan 2002). Clayey or silty soils typically have a higher Mg content than sandy soils, and fixation may occur in extreme cases where there are silicates poor in Mg, such as chlorites (Schachtschabel et al. 1976; Papenfuß and Schlichting 1979).

In silicates, Mg is bound in the interlayers, and its release is slow on an agronomic time scale, which can be a challenge for plant nutrition. The dissolution of Mg silicates is influenced by several factors, including mineral structure, pH, temperature,

Table 1 Mineral sources of magnesium

Mineral	Chemical formula	Class	Mg content (wt%)	Occurrence	References
Dolomite	$CaMg(CO_3)_2$	Carbonate	13.2	Sedimentary rocks	Nogueira et al. (2017)
Magnesite	$MgCO_3$	Carbonate	2.8	Metamorphic deposits	Correia (2001)
Brucite	$Mg(OH)_2$	Hydroxide	41.7	Peridotite alteration	Simandl et al. (2007)
Forsterite	Mg_2SiO_4	Nesosilicate	34.6	Earth's mantle	Simandl et al. (2007)
Serpentine	$Mg_3Si_2O_5(OH)_4$	Phyllosilicate	26.3	Ophiolites	Simandl et al. (2007)
Carnallite	$KMgCl_3.6(H_2O)$	Halide	8.8	Evaporite deposits	Simandl et al. (2007)
Olivine	$(Mg,Fe)_2SiO_4$	Nesosilicate	25.4	Ultramafic igneous rocks	Simandl et al. (2007)
Biotite	$K(Mg,Fe)_3AlSi_3O_{10}(F,OH)_2$	Mica	10	Granites/schists	Lebedev et al. (2018), Cao et al. (2022)
Phlogopite	$KMg_3AlSi_3O_{10}(F,OH)_2$	Mica	16	Ultramafic rocks	Kuwahara and Aoki (1995)
Chlorite	$(Mg,Fe)_5Al_2Si_3O_{10}(OH)_8$	Phyllosilicate	11.34	Hydrothermal alteration	Gailhanou et al. (2009)
Glauconite	$(K,Na)(Fe,Mg,Al)_2(Si,Al)_4O_{10}(OH)_2$	Phyllosilicate	< 1.8	Sedimentary rocks	Dillenburg et al. (2000)

CO_2 pressure, presence of organic and inorganic ligands, and chemical conditions of the environment (Luce et al. 1972; Wang and Giammar 2013; Farhang et al. 2019). Compared with multioxide minerals, olivine has a simplified dissolution process that requires the breaking of multiple types of metal–oxygen bonds (Oelkers 2001a). Olivine dissolution occurs initially by rapid reversible exchange of Mg for protons at the mineral surface, facilitated by the breaking of divalent metal–oxygen ionic bonds, which occurs more rapidly than the breaking of covalent bonds. This exchange is completed within a few minutes and is necessary to maintain charge equilibrium. Proton penetration into the olivine structure is limited due to the structural configuration of the mineral. At low pH, Mg depletion occurs at the olivine surface, whereas at basic pH, Mg is retained, and Si is preferentially released. Dissolution rates are non-stoichiometric in the early stages and become stoichiometric after the olivine surfaces equilibrate with the reactive fluids (Oelkers et al. 2018).

Closed-system experiments have revealed that biotite dissolution is highly pH-dependent and partly non-stoichiometric. At acidic pH (2.5), the release of Mg, Fe, and Al is close to stoichiometric, while that of Si is 2.5–3 times lower. With increasing pH, the concentrations of the elements decrease, and release becomes non-stoichiometric, with a preference for Mg at pH > 4. Smaller particles (25–53 μm) dissolve faster than larger ones (53–500 μm), except for Fe, whose release is independent of particle size, possibly due to the precipitation of secondary phases. The presence of organic ligands (citric acid, oxalic acid, and Desferrioxamine B increases the release of Fe and Al at pH > 3 but does not significantly affect Si-based dissolution rates. In an open system, dissolution rates reach a quasi-steady state after ~ 100 h, with rates consistent with those of a closed system after 12 h. Dissolution is dominated by the biotite rims, which are 45–120 times more reactive than basal surfaces, resulting in the preferential removal of Mg and Fe from the octahedral layers and Al and Si from the tetrahedral layers. The dissolution stoichiometry varies with pH and the presence of ligands; at neutral pH, Fe is retained due to the low solubility of secondary phases (Bray et al. 2015).

A study of the dissolution of layer-type Mg silicate minerals (brucite, antigorite, talc, and phlogopite) in water at 25 °C and 1 atm CO_2 pressure revealed incongruent dissolution; Mg from octahedral sheets is released faster than Si from tetrahedral sheets. The dissolution kinetics are controlled by the breakdown of tetrahedral silica sheets, with Mg release decreasing in the order brucite > antigorite > phlogopite > talc. This pattern reflects the ratio of octahedral to tetrahedral layers in each mineral. Brucite forms hydromagnesite as a secondary phase, while antigorite and talc reach silica saturation at different times. The pH of the solution increases with the dissolution of the minerals and varies according to the mineral: talc (4.8–5.2), phlogopite (5.1–5.4), antigorite (6–7), and brucite (7–9). Mineral hydrolysis consumes H^+ ions, so the more minerals that dissolve, the higher the final pH (Lin and Clemency 1981). CO_2 pressure also influences the rate of mineral dissolution, as it alters the pH conditions in aqueous media. The initial dissolution rate of forsterite generally increases with increasing CO_2 pressure because higher CO_2 pressure leads to a higher concentration of dissolved CO_2 in water, resulting in the formation of carbonic acid (H_2CO_3) and a decrease in the solution's pH (Wang and Giammar 2013).

Increasing temperatures favor the dissolution of Mg silicates, such as forsterite and serpentine (Wang and Giammar 2013; Farhang et al. 2019). The positive relationship between the dissolution rate of these minerals and temperature follows the Arrhenius equation (Blum and Lasaga 1988; Awad et al. 2000; Hänchen et al. 2006; Hu and Jun 2012; Wang and Giammar 2013; Olsen et al. 2015). The dissolution of forsterite accelerates with increasing temperature, especially under acidic pH conditions (Wang and Giammar 2013). A study of the dissolution of a biotite at different temperatures (35, 55, 75, and 95 °C) and the analysis of the ionic concentrations demonstrated preferential release of K^+ by ion exchange (Na^+–K^+) in which the kinetics are controlled by diffusive transport. By contrast, the structural ions (Mg, Fe, Al, Si) exhibit distinct behaviors according to temperature: at 95 and 75 °C, dissolution is congruent (all ions released proportionally), while at 55 and 35 °C, it becomes incongruent, with greater release of Mg, Fe, and Al than Si (Hu and Jun 2012).

In Mg silicates, Mg-O bonds break more rapidly than Si-O bonds, and the rate of breakage increases under acidic conditions and elevated temperatures (Oelkers and Schott 2001; Oelkers 2001b). In serpentine, the initial dissolution rates of Mg and Si increase with increasing temperature (Farhang et al. 2019). However, at higher temperatures, the formation of secondary phases, such as amorphous silica, occurs more rapidly, creating passivating layers on the mineral's surface that prevent further dissolution (Oelkers 2001b). This can result in a plateau of the dissolved Mg concentration at higher temperatures, whereas dissolution can continue for a more extended period at lower temperatures, leading to greater total Mg extraction (Farhang et al. 2019). Studies also suggest that forsterite forms a silica-rich layer, which may slow dissolution over time, even though the initial rate increases with temperature (Wang and Giammar 2013). Additionally, temperature influences the incorporation of water into the structure of minerals, such as pyroxenes. Under certain conditions of pressure and oxygen fugacity, the OH content in these minerals increases with increasing temperature (Liu and Yang 2020).

In addition to physical and chemical factors, organic and microbiological factors are important determinants of the solubilization of silicate minerals (Fig. 1). Plants and microorganisms in the rhizosphere produce a range of organic acids (OAs) that facilitate the solubilization of minerals and make nutrients more readily available for root absorption. OAs produced by rhizosphere microbes include acetic, citric, fumaric, glycolic, phenyl, 2-ketogluconic, lactic, oxalic, propionic, succinic, and tartaric acids (Macias-Benitez et al. 2020; Wen et al. 2020; Liu et al. 2024). For example, OAs mediate the dissolution of olivine (Mg_2SiO_4) by adsorbing on the mineral surface, where cations, especially Mg^{2+}, preferentially bind to the carbonyl oxygen of the carboxyl groups. This interaction causes charge transfer, orbital hybridization, and the breaking of the H–O bond of the carboxyl group, resulting in the hydroxylation of the surface and facilitating dissolution. The greater the number of carboxyl groups, the more intense the corrosion, with pitting. Additionally, the released Mg^{2+} forms stable complexes with the OAs, thereby favoring continued dissolution (Sun et al. 2023).

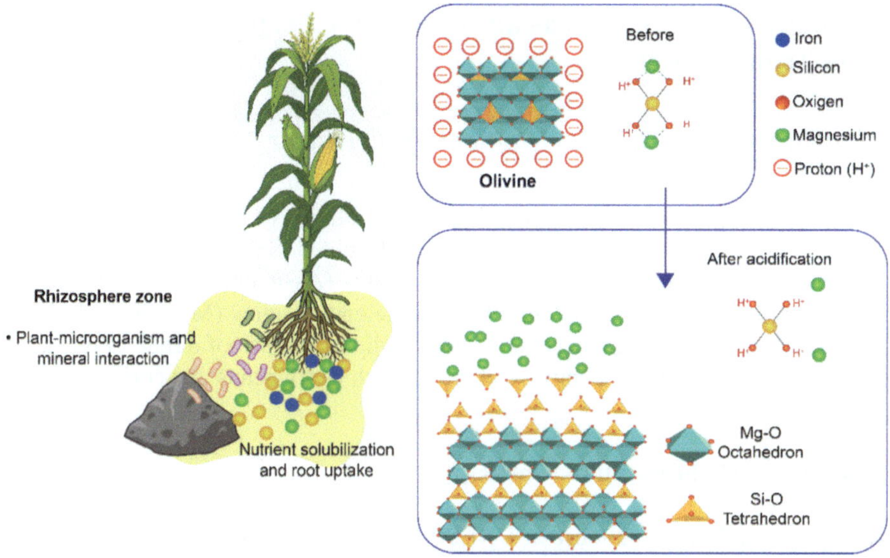

Fig. 1 Dynamics of solubilization of olivine from acidic solutions. The rhizosphere is an environment that favors the acidification of the medium, due to the exudation of organic acids from roots and microorganisms, promoting the solubilization of nutrients more easily. Adapted from Liu et al. (2024)

Microbial inoculation has proved efficient for Mg acquisition. Mycorrhizal hyphae of *Paxillus involutus*, in association with *Pinus sylvestris* roots, accelerate the alteration of biotite through localized acidification and extraction of elements such as K, Mg, Fe and Al. The weathering rate reached 0.04 µmol m^{-2} h^{-1}, demonstrating that mycorrhizae promote mineral weathering more effectively than abiotic processes by combining chemical and mechanical action (Bonneville et al. 2011). *Aspergillus niger* interacts differentially with the silicate minerals forsterite and talc; Mg is released from forsterite more effectively through the secretion of oxalic acid, resulting in the formation of the secondary mineral glushinskite. Furthermore, the size of the forsterite particles influences the morphology of the fungus (pellet-like or dispersed hyphae), affecting its metabolism and resulting in different crystalline forms of glushinskite (Cai et al. 2013). Combining olivine melilitite powder with beneficial bacteria (*Azospirillum brasilense, Bacillus subtilis, and B. megaterium*) enhances soil microbial activity and increases corn productivity, even with 25% less chemical fertilizer (Galina et al. 2024).

Understanding the solubilization mechanisms of Mg silicates is crucial for developing techniques that optimize the availability of this nutrient in the soil and enhance its absorption by plants. However, most studies have been conducted under controlled laboratory conditions, which limits the extrapolation of the results to real field conditions and makes it difficult to assess their impact on agricultural productivity accurately. Studies of the effects of applying Mg silicates on crop development are discussed next.

3 Magnesium Silicates as an Alternative Fertilizer in Tropical Soils

3.1 Dunite as a Silicon–Magnesium Fertilizer

Dunite is an ultramafic rock composed mainly of olivine (95−99%); the remainder (1−5%) is pyroxene, serpentine, chlorite, and spinel (Saif et al. 2023). Olivine contains approximately 40% MgO and 34% SiO_2 (Assis and Dias 2007). Dissolution (Moretti et al 2021) and agronomic efficiency studies (Moretti et al. 2019; Crusciol et al. 2019) have been conducted to validate the use of dunite (Table 2) as an alternative fertilizer for commercial crops, such as soybeans and corn.

3.1.1 Dunite Solubilization Kinetics

An experiment in a greenhouse at UNESP Botucatu evaluated the effect of dunite application (1.7 g kg^{-1}, particle size 50 mesh) to three Latosols (clayey, medium, and sandy) using a completely randomized design with two treatments (with and without dunite) and four replicates. The samples were incubated in 500-mL pots (300 g of soil) under controlled conditions (27–32 °C, 60–80% field capacity). For 180 days, 13 collections were carried out (with intervals of 10–30 days) to analyze pH, Mg (using the resin method), and Si (via spectrophotometry) (see Moretti et al. 2021, for more information).

Table 2 Trace elements, mineralogical composition, and main oxides of two dunite sources from Botucatu, São Paulo, Brazil (Moretti et al. 2021)

Category	Component	Symbol/Formula	Value	Unit
Trace elements	Arsenic	As^{3+}	3.79	mg kg^{-1}
	Cadmium	Cd^{2+}	0.8	mg kg^{-1}
	Lead	Pb^{2+}	2.5	mg kg^{-1}
	Mercury	Hg^{2+}	< 0.20	mg kg^{-1}
	Selenium	Se^{2-}	< 0.10	mg kg^{-1}
	Chromium	Cr^{3+}	0.037	%
	Nickel	Ni^{2+}	0.239	%
Mineralogical Phases	Lizardite	$Mg_3(Si_2O_5)(OH)_4$	48	%
	Magnetite	Fe_3O_4	9	%
	Brucite	$Mg(OH)_2$	26	%
	Magnesioferrite	$MgFe_2O_4$	17	%
Composition	Magnesium oxide	MgO	40	%
	Silicon dioxide	SiO_2	34	%

The results demonstrated that the application of dunite significantly increased the pH in the tropical acid soils, with distinct patterns depending on the soil texture: quadratic behavior in sandy and medium-textured soils and linear behavior in clayey soils due to greater buffering power (Fig. 2). In the soil solution, the dissolution and hydrolysis of magnesium silicate ($MgSiO_3$) releases Mg^{2+} and $SiO_3{}^{2-}$, which react with water and raise the pH through the formation of OH^- and orthosilicic acid (H_4SiO_4) (Alcarde 2005).

$$MgSiO_3 \rightleftharpoons Mg^{2+} + SiO_3^{2-} \quad (1)$$

$$SiO_3^{2-} + H_2O \rightleftharpoons HSiO_3^- + OH^- \quad (2)$$

$$HSiO_3^- + H_2O \rightleftharpoons H_2SiO_3 + OH^- \quad (3)$$

$$H_2SiO_3 + H_2O \rightleftharpoons H_4SiO_4 \quad (4)$$

Subsequently, the polymerization of H_4SiO_4 releases H^+ (Alves 2014) partially reducing the neutralizing effect:

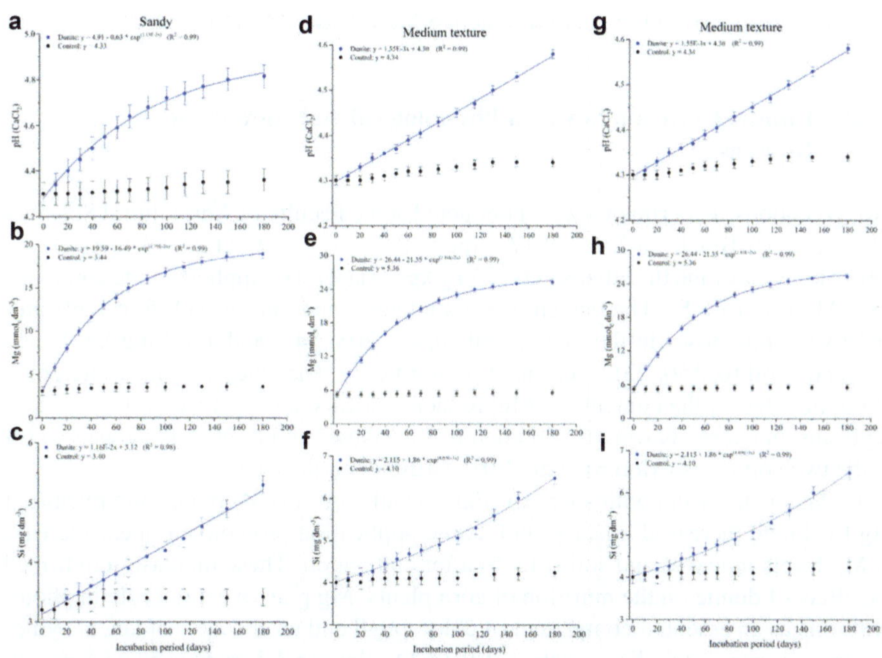

Fig. 2 pH (**a**, **d**, **g**), Mg content (**b**, **e**, **h**), and Si (**c**, **f**, **i**) content in soil with sandy, medium or clayey texture, respectively, as a function of dunite application and incubation period. Botucatu, São Paulo, Brazil (Moretti et al. 2021)

$$4H_4SiO_4 \rightleftharpoons H_6Si_4O_{12}^{2-} + 2H^+ + 4H_2O \tag{5}$$

Therefore, the total neutralization of acidity by $MgSiO_3$ is limited because 4 mol of the mineral neutralize only 6 mol of H^+ due to the generation of 2 mol of H^+ during polymerization:

$$4MgSiO_3 + 6H^+ \rightleftharpoons 4Mg^{2+} + H_6Si_4O_{12}^{2-} \tag{6}$$

This process explains the moderate and controlled increase in pH by dunite, which differs from the effects of correctives such as limestone and reduces the impact on soil microbiota (Imaizumi et al. 2018; Moretti et al. 2021).

The dunite exhibited high efficiency in releasing Mg^{2+} during the first 30 days, followed by stabilization, while Si was released more slowly and consistently (Moretti et al. 2021). Because the study was performed only with soil, i.e., in the absence of plants, the system reached an equilibrium between the solid phase of the mineral and the soil solution. In other words, as the dunite dissolved, the concentration of Mg in the solution increased until it reached a point where there was no longer "demand" to continue dissolution on a large scale. According to Carvalho (2012), this stabilization is natural and expected in closed systems (without cultivation) because, without the removal of nutrients by plants or leaching, the dissolution of minerals decreases. Under real field conditions, however, root uptake and other processes would extend the period of active Mg release (Moretti et al. 2021).

3.1.2 Dunite: Corn and Soybean Physiological and Agronomic Response

Two greenhouse experiments were conducted at the Faculty of Agricultural Sciences (Botucatu, SP, Brazil) in 2016–2017 using two dystrophic Red Latosols: one clayey (602 g kg^{-1} clay) and the other sandy (724 g kg^{-1} sand). The applied dunite contained 24% Mg and 16% Si. The design was completely randomized with four replicates and five dunite doses in the clayey soil (0, 42, 208, 542, and 1542 mg kg^{-1}) and the sandy soil (0, 150, 238, 411, and 933 mg kg^{-1}). The study evaluated the effect of dunite rates on the contents of Mg, reducing sugars, and sucrose in the leaves of soybeans and corn, as well as soil chemical attributes and corn and soybean yields in the two soil types (Moretti et al. 2019; Crusciol et al. 2019).

In this open system with soybean (Fig. 3) and corn (Fig. 4) plants, the release of Mg by dunite increased linearly, and dunite application promoted a linear increase in Mg levels in leaves and soil (Fig. 5) after cultivation. These increases confirmed the effect of dunite on the nutrition of corn plants. Mg plays a central role in photosynthesis as an essential component of chlorophyll and an activator of key enzymes in carbohydrate metabolism, such as RuBisCO (ribulose-1,5-bisphosphate carboxylase/oxygenase) and ATPases involved in sugar transport (Tränkner et al. 2018). The increase in reducing sugars (glucose + fructose) reflects greater carbon fixation due to the optimization of photosynthetic activity by Mg (Crusciol et al. 2019).

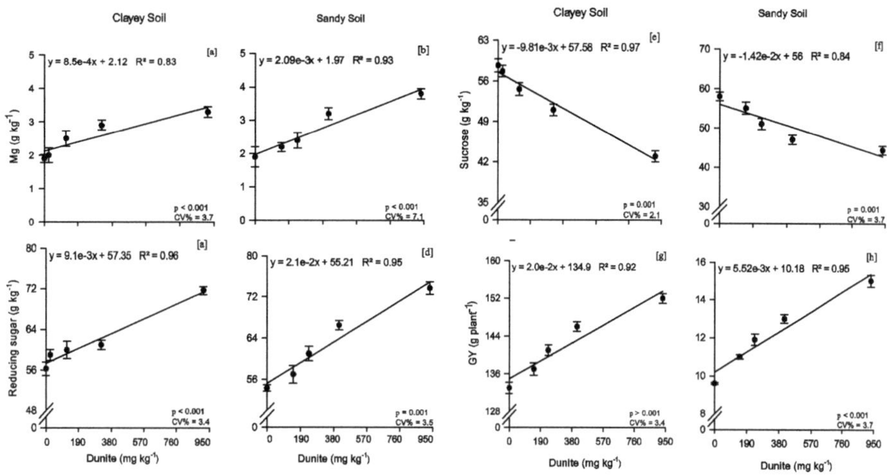

Fig. 3 Leaf contents of Mg (**a**, **b**), reducing sugars (**c**, **d**), and sucrose (**e**, **f**), and grain yield (**g**, **h**) of corn grown in clayey and sandy soil as a function of the dunite rate (Crusciol et al. 2019)

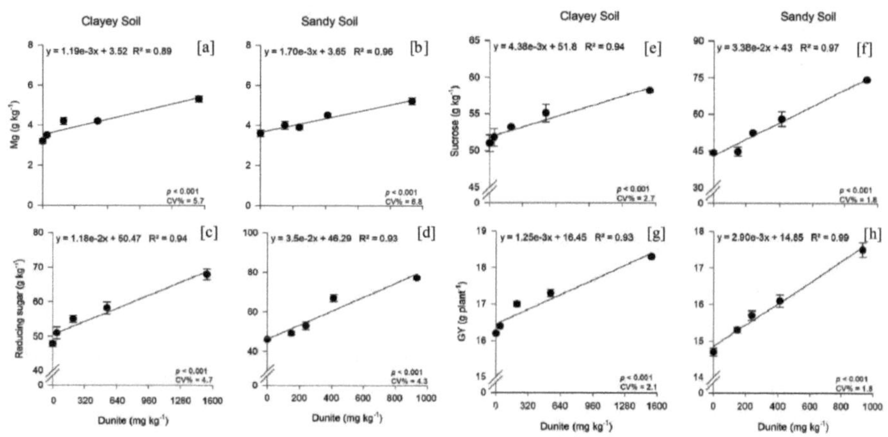

Fig. 4 Leaf contents of Mg (**a**, **b**), reducing sugars (**c**, **d**), and sucrose (**e**, **f**) and grain yield (**g**, **h**) of soybean grown in clayey and sandy soil as a function of the dunite rate (Crusciol et al. 2019)

At the same time, the reduction in leaf sucrose indicates a more efficient partitioning of photoassimilates, with greater translocation to the sink organs (grains). Mg activates H^+-ATPases in the plasma membrane, which generate the proton gradient necessary for the loading of sucrose into the phloem via sucrose transporters (Farhat et al. 2016). In addition, the decrease in leaf starch (vide Moretti et al. 2019; Crusciol et al. 2019) suggests that reserve carbohydrates were more rapidly mobilized to meet the energy demands of the forming grains, a process mediated by enzymes such

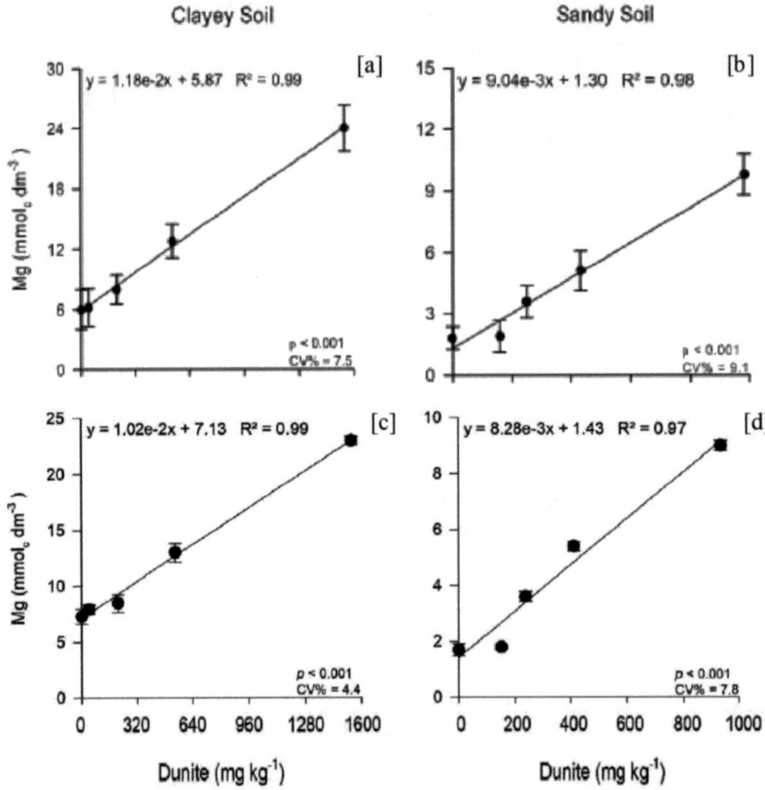

Fig. 5 Mg soil content after corn (**a, b**) and soybean (**c, d**) cultivation in clayey and sandy soil as a function of the dunite rate (Crusciol et al. 2019)

as sucrose-phosphate synthase and the starch-breaking enzyme β-amylase (Ceylan et al. 2016).

3.2 Thermomagnesium as a Fertilizer for a Soybean–Corn System

Thermomagnesium (TM, $MgSiO_3$) is a by-product generated during the thermal process of nickel (Ni) extraction. The ore is melted at high temperatures (1600 °C) in an electric furnace to produce an iron-nickel alloy (Fe-Ni) and a residue rich in magnesium oxide (MgO) and silicon dioxide (SiO_2) (Table 3). This by-product, after rapid cooling and processing (drying and sieving), has potential as an alternative fertilizer due to its high magnesium content (Bossolani et al. 2021; Crusciol et al. 2022; de Melo et al. 2022). Three studies have evaluated the effect of TM on soybean and corn crops in a succession system. First, Bossolani et al. (2021) evaluated the

Table 3 Chemical and mineralogical characteristics of thermomagnesium from nickel ore mining (de Melo et al. 2022)

Component	Value	Unit
Metal concentrations		
Arsenic (As)	< 5	$mg \cdot kg^{-1}$
Cadmium (Cd)	< 5	$mg \cdot kg^{-1}$
Lead (Pb)	< 5	$mg \cdot kg^{-1}$
Mercury (Hg)	< 0.20	$mg \cdot kg^{-1}$
Selenium (Se)	< 0.10	$mg \cdot kg^{-1}$
Chromium (Cr)	0.88	%
Nickel (Ni)	0.11	%
Mineralogical phases		
Amorphous phase	74	%
Forsterite ($Mg_{1.8}Fe_{0.2}SiO_4$)	24	%
Lizardite ($Mg_3Si_2(OH)_4O_5$)	2	%
Chemical composition		
Magnesium oxide (MgO)	28	%
Silicon dioxide (SiO_2)	48	%

effect of TM doses in clayey soil (0, 55, 273, 709, and 2018 mg kg^{-1}) and sandy soil (0, 293, 410, 645, and 1260 mg kg^{-1}) on the development and productivity of corn under greenhouse conditions. Second, Crusciol et al. (2022) evaluated the influence of different doses of TM (0, 350, 700, 1050, 1400, and 1750 kg ha^{-1}) on the productive performance of soybeans and corn grown in an Oxisol under field conditions. Finally, De Melo et al. (2022) evaluated the effect of different particle sizes (control, 20, and 100 mesh) of TM on the nutrition, physiology, and productivity of soybeans and corn in a succession system. under field conditions. Finally, De Melo et al. (2022) evaluated the effect of different particle sizes (control, 20, and 100 mesh) of TM on the nutrition, physiology, and productivity of soybeans and corn in a succession system.

Bossolani et al. (2021) demonstrated that under greenhouse conditions, TM is an efficient fertilizer for corn, with positive effects on several parameters. Increasing TM doses increased Mg and Si concentrations in both clayey and sandy soils and, consequently, in the leaves of corn plants. An increase in reducing sugars and decreases in sucrose and starch were also observed in the leaves, indicating a change in carbohydrate metabolism. These adjustments in metabolism were accompanied by greater growth, grain filling, and corn productivity, especially at higher TM doses in both soil types (Fig. 6). The results of this initial research in a controlled environment established the potential of TM as an alternative and sustainable source of Mg and Si for corn, providing a foundation for field-based research.

Crusciol et al. (2022) expanded the evaluation of TM application in a soybean–corn rotation system to field conditions during three consecutive harvests (Fig. 7). Similar to the findings of Bossolani et al. (2021), the field study revealed that

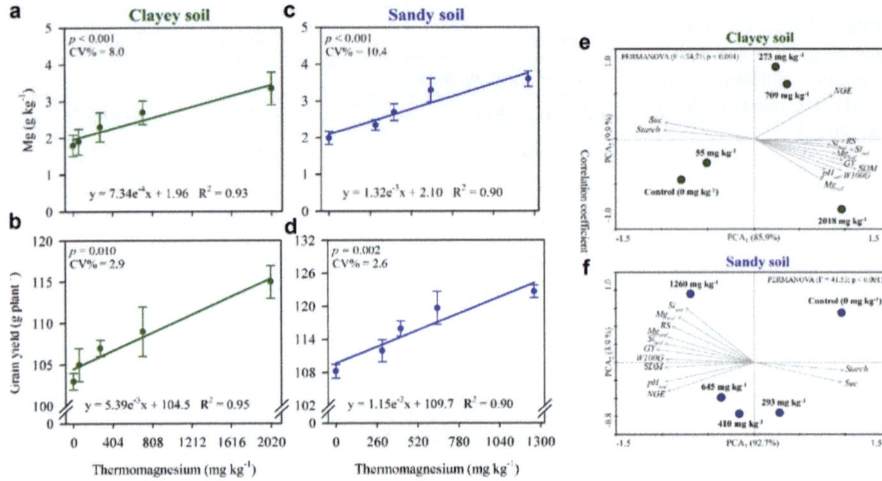

Fig. 6 Leaf Mg content (**a, c**) and grain yield (**b, d**) of corn plants grown in a greenhouse as a function of different doses of thermomagnesium in clayey and sandy soils. Principal component analysis (PCA) (**e, f**) of soil parameters and corn plants (Bossolani et al. 2021)

increasing doses of TM increased the concentrations of Mg and Si in the soil and, to a lesser extent, in the leaves of soybean and corn. Additionally, TM application contributed to an increase in soil pH. Positive impacts on photosynthetic pigments and gas exchange parameters, such as the photosynthetic rate, were also observed. In line with the greenhouse study, the leaf contents of reducing sugars and sucrose increased, whereas starch content decreased. More importantly, TM application resulted in significant increases in the 100-grain weight and yield of both soybean and corn, with the 1050 kg ha^{-1} dose producing the best results for both crops. This field study reinforced the viability of TM as a fertilizer, demonstrating its beneficial effects under real agricultural conditions and in a crop rotation system.

De Melo et al. (2022) investigated an additional factor in TM efficiency: particle size. The application of TM with particle sizes of 20 or 100 mesh was compared in a soybean–corn–soybean rotation system (Fig. 8). The results showed that smaller particles (100 mesh) generally led to higher Mg^{2+} concentrations in the soil after 24 months and higher Si concentrations in corn leaves. Consistent with the results of other studies, the application of TM, especially 100 mesh, increased photosynthetic pigments and the photosynthetic rate. Reducing sugar and sucrose concentrations were higher when 100-mesh TM was applied, while TM application decreased starch concentrations regardless of particle size. In terms of productivity, 100-mesh TM increased the weight of 100 soybean grains and the yields of the first soybean and corn crops, suggesting greater efficiency of smaller particles in the short term. Thus, this study demonstrated that TM particle size influences the nutrient release rate and, consequently, agronomic efficiency, especially in the initial stages of cultivation.

Fertilization with Mg silicate is efficient in providing magnesium (Mg) to crops, but it can also be an interesting source of nickel (Ni) and Selenium (Se) (Kierczak

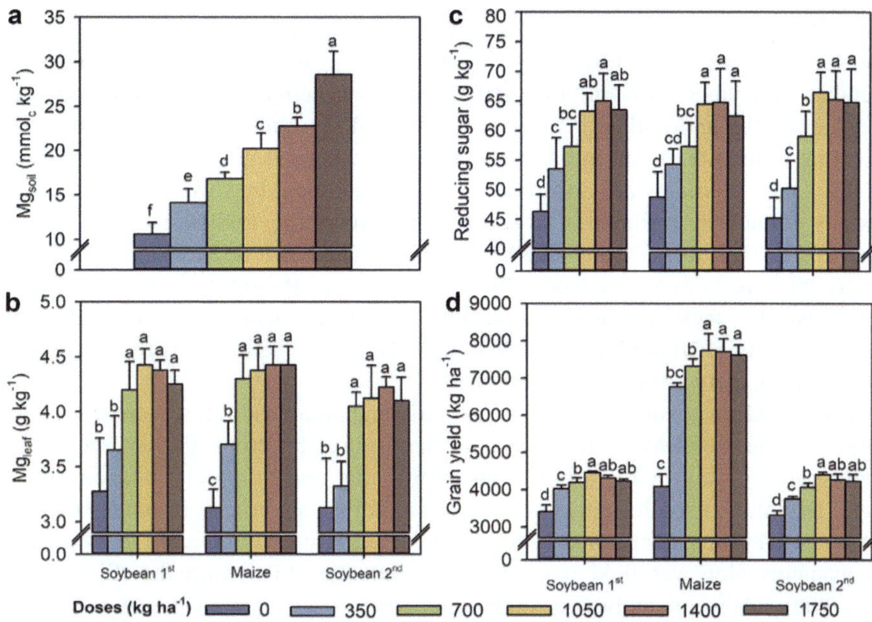

Fig. 7 Soil content of Mg (**a**), leaf contents of Mg (**b**) and reducing sugars (**c**), and grain yield (**d**) in soybean and corn plants in a succession system under different doses of thermomagnesium. Bars with different lowercase letters within the same crop season are significantly different according to Student's t-test at $p \leq 0.05$ (Crusciol et al. 2022)

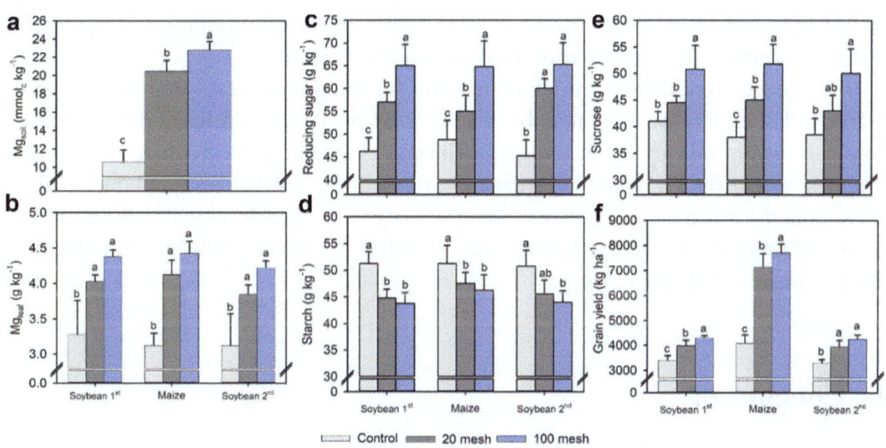

Fig. 8 Soil Mg content (**a**), leaf contents of Mg (**b**), reducing sugars (**c**), starch (**d**), and sucrose (**e**), and grain yield (**f**) of soybean and corn plants in a succession system as a function of the application of thermomagnesium of different particle sizes. Bars with different lowercase letters within the same crop season are significantly different according to Student's t-test at $p \leq 0.05$ (de Melo et al. 2022)

et al. 2021; Kaur et al. 2023). Ni is essential for the activity of enzymes such as urease, which participates in nitrogen metabolism, as well as other antioxidant and metabolic enzymes (Ameen et al. 2019). Se is a beneficial element for plants, as it enhances the antioxidant activity of plants and the human body, in addition to affecting immunity and hormonal regulation in humans (Liu et al. 2022; Bai et al. 2024). These elements are absorbed in small quantities; however, they can cause toxicity to plants if available in large quantities. According to these studies (Moretti et al. 2019, 2021; Bossolani et al. 2021; Crusciol et al. 2022; de Melo et al. 2022), no phytotoxic effects were observed, indicating that it is a safe source for plant nutrition. Si may be a key factor in this situation, since silicate can increase soil pH, decreasing the availability of Ni to plants, which would lead to a reduction in its toxic potential (Khaliq et al. 2016; Abd-Allah et al. 2019).

These studies, conducted under greenhouse and field conditions, were crucial for validating the use of Mg silicates as a nutrient source for crops, thereby increasing photosynthesis, improving carbohydrate partitioning, and ultimately enhancing grain yield. Dissolution studies under controlled conditions are crucial for understanding the potential utility of rocks as a nutrient source for agriculture and the associated mechanisms. However, field experiments are necessary to measure the efficiency of these materials when applied to commercial crops.

pH, temperature, moisture, and mineral structure are the main factors influencing the solubilization of silicate minerals; however, root activity may have a greater contribution than expected. Nutrient uptake creates a depletion zone that can induce the removal of nutrients from minerals (Tinker and Nye 2000). The role of root exudates under controlled conditions is relatively well understood. However, measuring their actual contributions under field conditions remains challenging due to the complexity of the soil–root–microorganism system.

Slow-release fertilizers are advantageous for deep and weathered soils, mainly sandy soils, as they prevent loss by leaching. However, this release must coincide with the agronomic cycle to avoid nutritional deficiency. Heat treatment of Mg-silicates is an effective way to increase their solubility and has proven beneficial for enhancing the nutrition and physiological activity of crops. Moreover, these materials are a sustainable source of nutrients and contribute to the economy by transforming mining waste into high-value.

4 Conclusion

The results presented in this chapter demonstrate that magnesium silicates, such as dunite and thermomagnesium (TM), represent promising and sustainable nutrient sources for tropical agriculture. These materials, in addition to gradually providing magnesium (Mg) and silicon (Si), contribute to the partial correction of soil acidity, favoring the availability of other essential nutrients and microbiological activity. The application of dunite and TM in greenhouse and field experiments demonstrated significant improvements in nutrition, photosynthesis, and carbohydrate partitioning, resulting in increased productivity in crops such as soybeans and corn. The greater efficiency of finer particles, the interaction with microorganisms, and the influence of factors such as pH, temperature, and humidity reinforce the importance of management strategies that optimize the solubilization and absorption of released nutrients. However, it is important to consider that the slow release of nutrients requires careful planning to ensure that the supply of Mg and Si coincides with critical phases of plant development. Furthermore, the mineralogical variability of natural sources and differences in edaphoclimatic conditions can influence agronomic efficiency, requiring regional characterization studies and specific management recommendations. Thus, magnesium-silicate agrominerals represent a viable alternative to reduce dependence on conventional fertilizers, promote the use of mining byproducts, and stimulate more resilient, economical, and environmentally responsible agricultural systems in highly weathered tropical soils.

5 Declaration of Competing Interest

The authors declare that they have no known competing financial interests or personal relationships that could have appeared to influence the work reported in this paper.

Acknowledgements We are grateful for support from the Center for Carbon Research in Tropical Agriculture (CCARBON) at the University of São Paulo (USP), which is sponsored by the São Paulo Research Foundation (FAPESP) under grant 2021/10573–4. The first author would like to thank the National Council for Scientific and Technological Development (CNPq) for an award for excellence in research.

References

Abd-Allah EF, Hashem A, Alam P, Ahmad P (2019) Silicon Alleviates Nickel-Induced Oxidative Stress by Regulating Antioxidant Defense and Glyoxalase Systems in Mustard Plants. J Plant Growth Regul 38:1260–1273. https://doi.org/10.1007/s00344-019-09931-y

Alcarde JC (2005) Soil acidity correctives: Characteristics and technical interpretations [Corretivos da acidez dos solos: Características e interpretacões técnicas]. São Paulo

Alves EEN (2014) Effect of calcium metasilicate on properties chemicals of a cultivated red-yellow Latosol with rice and beans [Efeito do metassilicato de cálcio nas propriedades químicas de um latossolo vermelho-amarelo cultivado com arroz e feijão]. Dissertation, Federal University of Viçosa

Ameen N, Amjad M, Murtaza B, et al (2019) Biogeochemical behavior of nickel under different abiotic stresses: toxicity and detoxification mechanisms in plants. Environmental Science and Pollution Research 26:10496–10514. https://doi.org/10.1007/s11356-019-04540-4

Assis LG, Dias FM (2007) Estudo da viabilidade da utilização da rocha dunito, proveniente de Catas Altas - MG, como agregado para concreto. In: SEGeT – Simpósio de Excelência em Gestão e Tecnologia

Auler AC, Caires EF, Pires LF et al (2019) Lime effects in a no-tillage system on inceptisols in Southern Brazil. Geoderma Reg 16:e00206. https://doi.org/10.1016/j.geodrs.2019.e00206

Awad A, Koster Van Groos AF, Guggenheim S (2000) Forsteritic olivine: effect of crystallographic direction on dissolution kinetics. Geochim Cosmochim Acta 64:1765–1772. https://doi.org/10.1016/S0016-7037(99)00442-1

Bai S, Zhang M, Tang S, et al (2024) Effects and Impact of Selenium on Human Health, A Review. Molecules 30:50. https://doi.org/10.3390/molecules30010050

Bakhshandeh S, Khormali F, Dordipour E et al (2011) Comparing the weathering of soil and sedimentary palygorskite in the rhizosphere zone. Appl Clay Sci 54:235–241. https://doi.org/10.1016/j.clay.2011.09.007

de Bang TC, Husted S, Laursen KH et al (2021) The molecular–physiological functions of mineral macronutrients and their consequences for deficiency symptoms in plants. New Phytol 229:2446–2469. https://doi.org/10.1111/NPH.17074

Blum A, Lasaga A (1988) Role of surface speciation in the low-temperature dissolution of minerals. Nature 331:431–433. https://doi.org/10.1038/331431a0

Bonneville S, Morgan DJ, Schmalenberger A et al (2011) Tree-mycorrhiza symbiosis accelerate mineral weathering: evidences from nanometer-scale elemental fluxes at the hypha–mineral interface. Geochim Cosmochim Acta 75:6988–7005. https://doi.org/10.1016/j.gca.2011.08.041

Bossolani JW, Moretti LG, Portugal JR et al (2021) Thermomagnesium: a by-product of Ni ore mining as a clean fertilizer source for maize. Agronomy 11:525. https://doi.org/10.3390/agronomy11030525

Bray AW, Oelkers EH, Bonneville S et al (2015) The effect of pH, grain size, and organic ligands on biotite weathering rates. Geochim Cosmochim Acta 164:127–145. https://doi.org/10.1016/j.gca.2015.04.048

Cai L, Xiao H-R, Huang S-M et al (2013) Solubilization of magnesium-bearing silicate minerals and the subsequent formation of glushinskite by aspergillus niger. Geomicrobiol J 30:302–312. https://doi.org/10.1080/01490451.2012.688924

Cakmak I, Yazici AM (2010) Magnesium: a forgotten element in crop production. Better Crops 94:23–25

Cao R, Chen D-F, Gu H-D et al (2022) Mineralogical characteristics of biotite and chlorite in zuluhong polymetallic deposit: implications for petrogenesis and paragenesis mechanisms of the tungsten and copper. Minerals 12:1280. https://doi.org/10.3390/min12101280

Carvalho AMX (2012) Stonemeal and their interactions in soil environment contributions for use in agroecosystems with agroecological management [Rochagem e suas interações no ambiente solo: Contribuições para aplicação em agroecossistemas sob manejo agroecológico]. Federal University of Viçosa

Castro GSA, Crusciol CAC (2015) Effects of surface application of dolomitic limestone and calcium-magnesium silicate on soybean and maize in rotation with green manure in a tropical region. Bragantia 74:311–321. https://doi.org/10.1590/1678-4499.0346

Ceylan Y, Kutman UB, Mengutay M, Cakmak I (2016) Magnesium applications to growth medium and foliage affect the starch distribution, increase the grain size and improve the seed germination in wheat. Plant Soil 406:145–156. https://doi.org/10.1007/s11104-016-2871-8

Correia DMB (2001) Magnesita. In: Balanço Mineral Brasileiro. DNPM, Departamento Nacional de Produção Mineral, Brasília, DF

Crusciol CAC, de Campos M, Momesso L et al (2022) Nickel ore mining waste as a promising source of magnesium and silicon for a smart-agricultural management. Front Environ Sci 10:880613. https://doi.org/10.3389/fenvs.2022.880613

Crusciol CAC, Moretti LG, Bossolani JW et al (2019) Can dunite promote physiological changes, magnesium nutrition and increased corn grain yield? Commun Soil Sci Plant Anal 50:2343–2353. https://doi.org/10.1080/00103624.2019.1659304

Dillenburg S, Laybauer L, Mexias A, Dani N, Barboza EG, Lummertz CN (2000) Significado estratigráfico de minerais glauconíticos da planície coteira do rio grande do sul, região da laguna de Tramandaí. Revista Brasileira de Geociências 30(4):649–654. https://doi.org/10.25249/0375-7536.2000304649654

Farhang F, Oliver TK, Rayson MS et al (2019) Dissolution of heat activated serpentine for CO_2 sequestration: the effect of silica precipitation at different temperature and pH values. J CO_2 Util 30:123–129. https://doi.org/10.1016/j.jcou.2019.01.009

Farhat N, Elkhouni A, Zorrig W et al (2016) Effects of magnesium deficiency on photosynthesis and carbohydrate partitioning. Acta Physiol Plant 38:145. https://doi.org/10.1007/s11738-016-2165-z

Fernandes AJ, Maldaner CH, Sobrinho JMA, et al (2010) Estratigrafia dos Derrames de Basaltos da Formação Serra Geral (Ribeirão Preto - SP) Baseada na Geologia Física, Petrografia e Geoquímica. Geologia USP Série Científica 10:73–99. https://doi.org/10.5327/Z1519-874X2010000200006

Ferreira ERNC, Almeida JA, Mafra AL (2009) Pó de basalto, desenvolvimento e nutrição do feijão comum (Phaseolus vulgaris) e propriedades químicas de um Cambissolo Húmico. Revista de Ciências Agroveterinárias, 8(2):111–121

Gailhanou H, Rogez J, van Miltenburg JC et al (2009) Thermodynamic properties of chlorite CCa-2. Heat capacities, heat contents and entropies. Geochim Cosmochim Acta 73:4738–4749. https://doi.org/10.1016/j.gca.2009.04.040

Galina J, Ribeiro GM, Baretta D, Baretta CRDM (2024) Olivine melilitite powder applied in association with bacterial inoculation impacts soil microbiological attributes. Sci Agric 81:e20230214. https://doi.org/10.1590/1678-992x-2023-0214

Gransee A, Führs H (2013) Magnesium mobility in soils as a challenge for soil and plant analysis, magnesium fertilization and root uptake under adverse growth conditions. Plant Soil 368:5–21. https://doi.org/10.1007/s11104-012-1567-y

Hanisch A., Fonseca JBA., Spagnollo, E. et al (2013). Efeito de pó de basalto no solo e em culturas anuais durante quatro safras, em sistema de plantio direto. Revista Brasileira de Agropecuária Sustentável. 3(2):100–107

Hu Y, Jun Y-S (2012) Biotite dissolution in Brine at varied temperatures and CO_2 pressures: its activation energy and potential CO_2 intercalation. Langmuir 28:14633–14641. https://doi.org/10.1021/la3028995

Hänchen M, Prigiobbe V, Storti G et al (2006) Dissolution kinetics of fosteritic olivine at 90–150 °C including effects of the presence of CO_2. Geochim Cosmochim Acta 70:4403–4416. https://doi.org/10.1016/J.GCA.2006.06.1560

Hébert R, Bideau D, Hekinian R (1983) Ultramafic and mafic rocks from the Garret transform fault near 13°30′S on the East Pacific Rise: igneous petrology. Earth Planet Sci Lett 65:107–125. https://doi.org/10.1016/0012-821X(83)90193-0

Imaizumi CR, Rossi, Fortes L (2018) Potential of dunite as a source of silicon and magnesium for agriculture [Potencialidades do dunito como fonte de silício e magnésio para a agricultura]. Cmag Fertilizantes, Jaboticabal, Brazil

Ishfaq M, Wang Y, Yan M et al (2022) Physiological essence of magnesium in plants and its widespread deficiency in the farming system of China. Front Plant Sci 13:802274. https://doi.org/10.3389/fpls.2022.802274

Jaques AL, Chappell BW, Taylor SR (1983) Geochemistry of cumulus peridotites and gabbros from the Marum ophiolite complex, northern Papua New Guinea. Contrib Miner Petrol 82:154–164. https://doi.org/10.1007/BF01166610

Kaur H, Kaur H, Kaur H, Srivastava S (2023) The beneficial roles of trace and ultratrace elements in plants. Plant Growth Regul 100:219–236. https://doi.org/10.1007/s10725-022-00837-6

Kelemen PB (1990) Reaction between ultramafic rock and fractionating basaltic magma I. Phase relations, the origin of calc-alkaline magma series, and the formation of discordant dunite. J Petrol 31:51–98. https://doi.org/10.1093/petrology/31.1.51

Khaliq A, Ali S, Hameed A, et al (2016) Silicon alleviates nickel toxicity in cotton seedlings through enhancing growth, photosynthesis, and suppressing Ni uptake and oxidative stress. Arch Agron Soil Sci 62:633–647. https://doi.org/10.1080/03650340.2015.1073263

Kierczak J, Pietranik A, Pędziwiatr A (2021) Ultramafic geoecosystems as a natural source of Ni, Cr, and Co to the environment: A review. Science of The Total Environment 755:142620. https://doi.org/10.1016/J.SCITOTENV.2020.142620

Kuwahara Y, Aoki Y (1995) Dissolution process of phlogopite in acid solutions. Clays Clay Miner 43:39–50. https://doi.org/10.1346/CCMN.1995.0430105

Lebedev VA, Parfenov AV, Vashakidze GT et al (2018) Chronology of magmatic activity and petrologic-mineralogical characteristics of lavas of Kazbek Quaternary Volcano, Greater Caucasus. Petrology 26:1–28. https://doi.org/10.1134/S086959111801006X

Lin F-C, Clemency CV (1981) The dissolution kinetics of brucite, antigorite, talc, and phlogopite at room temperature and pressure. Am Miner 66:801–806

Lisboa LAM, de Oliveira Dias GH, Sacco HAA et al (2021) Urochloa brizantha cultivated in aluminum-toxic soil: changes in plant growth and ultrastructure of root and leaf tissues. Trop Grasslands Forrajes Trop 9:23–33. https://doi.org/10.17138/tgft(9)23-33

Liu M, Liu XX, He XL et al (2017) Ethylene and nitric oxide interact to regulate the magnesium deficiency-induced root hair development in Arabidopsis. New Phytol 213:1242–1256. https://doi.org/10.1111/nph.14259

Liu H, Yang X (2020) Solubility of hydroxyl groups in pyroxenes: effect of oxygen fugacity at 0.2–3 GPa and 800–1200 °C. Geochim Cosmochim Acta 286:355–379. https://doi.org/10.1016/j.gca.2020.07.034

Liu Z, Yin X, Xiao N et al (2024) Organic acids released by submerged macrophytes with damaged leaves alter the denitrification microbial community in rhizosphere. Sci Total Environ 946:174059. https://doi.org/10.1016/j.scitotenv.2024.174059

Liu H, Xiao C, Qiu T, et al (2022) Selenium Regulates Antioxidant, Photosynthesis, and Cell Permeability in Plants under Various Abiotic Stresses: A Review. Plants 12:44. https://doi.org/10.3390/plants12010044

Luce RW, Bartlett RW, Parks GA (1972) Dissolution kinetics of magnesium silicates. Geochim Cosmochim Acta 36:35–50. https://doi.org/10.1016/0016-7037(72)90119-6

Macias-Benitez S, Garcia-Martinez AM, Caballero Jimenez P et al (2020) Rhizospheric organic acids as biostimulants: monitoring feedbacks on soil microorganisms and biochemical properties. Front Plant Sci 11:00633. https://doi.org/10.3389/fpls.2020.00633

Maguire ME, Cowan JA (2002) Magnesium chemistry and biochemistry. Biometals 15:203–210. https://doi.org/10.1023/A:1016058229972

Malavolta E (2006) Manual of mineral nutrition of plants. Ceres, São Paulo

Martin RF, Schumann D, Dharmapriya PL (2022) Dolomitic marble in a context of UHT metamorphism: possible signs of melting, Lenadora Quarry, Sri Lanka. J Petrol 63:egac057. https://doi.org/10.1093/petrology/egac057

Mathews J, NG JY, Siew SP, Azizah MJ (2024) Evaluation of ground serpentinite rock as a magnesium source for oil palm nursery seedlings. Plant 100:006. https://doi.org/10.56333/tp.2024.006

de Melo CVCB, Moretti LG, Bossolani JW et al (2022) The influence of particle size on the efficiency of Ni mining slag as a fertilizer. J Soil Sci Plant Nutr 22:4693–4702. https://doi.org/10.1007/s42729-022-00952-3

Menzel MD, Garrido CJ, López Sánchez-Vizcaíno V et al (2019) Subduction metamorphism of serpentinite-hosted carbonates beyond antigorite-serpentinite dehydration (Nevado-Filábride Complex, Spain). J Metamorph Geol 37:681–715. https://doi.org/10.1111/jmg.12481

Miotto A, Tiecher T, Kaminski J et al (2020) Soil acidity and aluminum speciation affected by liming in the conversion of a natural pasture from the Brazilian Campos Biome into no-tillage system for grain production. Arch Agron Soil Sci 66:138–151. https://doi.org/10.1080/03650340.2019.1605164

Moreira A, Malavolta E, Heinrichs R, Tanaka RT (2003) Magnesium influence on manganese and zinc uptake by excised roots of soybean. Pesqui Agropecu Bras 38:95–101

Moretti LG, Bossolani JW, Crusciol CAC et al (2019) Dunite in agriculture: physiological changes, nutritional status and soybean yield. Commun Soil Sci Plant Anal 50:1775–1784. https://doi.org/10.1080/00103624.2019.1635143

Moretti LG, Crusciol CAC, Bossolani JW et al (2021) Dunite solubilization kinetics in silicon-magnesium fertilization. J Plant Nutr 44:2435–2446. https://doi.org/10.1080/01904167.2021.1899214

Nogueira LB, Oliveira VQ, Sampaio GM et al (2017) Elemental and stable isotopes geochemistry of Paleoproterozoic dolomites from Fecho do Funil Formation, Quadrilátero Ferrífero—Brazil. J South Am Earth Sci 79:525–536. https://doi.org/10.1016/J.JSAMES.2017.09.010

Nolla A, da Silva Alves EO, da Silva TG, Bordin AV (2020) Correction of soil acidity and availability of phosphorus and potassium in an oxisol submited to surfaced and incorporated liming. Braz J Anim Environ Res 3:2478–2487. https://doi.org/10.34188/bjaerv3n3-156

Oelkers EH (2001a) General kinetic description of multioxide silicate mineral and glass dissolution. Geochim Cosmochim Acta 65:3703–3719. https://doi.org/10.1016/S0016-7037(01)00710-4

Oelkers EH (2001b) An experimental study of forsterite dissolution rates as a function of temperature and aqueous Mg and Si concentrations. Chem Geol 175:485–494. https://doi.org/10.1016/S0009-2541(00)00352-1

Oelkers EH, Declercq J, Saldi GD et al (2018) Olivine dissolution rates: a critical review. Chem Geol 500:1–19. https://doi.org/10.1016/J.CHEMGEO.2018.10.008

Oelkers EH, Schott J (2001) An experimental study of enstatite dissolution rates as a function of pH, temperature, and aqueous Mg and Si concentration, and the mechanism of pyroxene/pyroxenoid dissolution. Geochim Cosmochim Acta 65:1219–1231. https://doi.org/10.1016/S0016-7037(00)00564-0

Olsen AA, Hausrath EM, Rimstidt JD (2015) Forsterite dissolution rates in Mg-sulfate-rich Mars-analog brines and implications of the aqueous history of Mars. J Geophys Res Planets 120:388–400. https://doi.org/10.1002/2014JE004664

Papenfuß KH, Schlichting E (1979) Bestimmende Faktoren des Mg-Haushaltes von Böden in der Bundesrepublik Deutschland. Magnes Bull 1:12–14

Passos PB, Leandro WM, de Castro JPV et al (2024) Agronomic evaluation study: use of Caiapônia Shale as magnesium supplementation for soybean (glycine max) and maize (zea mays) cultivation. J Sustain Dev 17:94. https://doi.org/10.5539/jsd.v17n5p94

Pavan MA, Bingham FT, Pratt PF (1982) Toxicity of aluminum to coffee in ultisols and oxisols amended with $CaCO_3$, $MgCO_3$, and $CaSO_4 \cdot 2H_2O$. Soil Sci Soc Am J 46:1201–1207. https://doi.org/10.2136/sssaj1982.03615995004600060017x

Saif MS, Shanour AS, Abdelaziz GE et al (2023) Influence of blended powders on properties of ultra-high strength fibre reinforced self compacting concrete subjected to elevated temperatures. Case Stud Constr Mater 18:e01793. https://doi.org/10.1016/j.cscm.2022.e01793

Schachtschabel P, Blume H-P, Thiele-Bruhn S (1976) Lehrbuch der bodenkunde. Enke Stuttgart, Germany

Schroth G, Lehmann J, Barrios E (2002) Soil nutrient availability and acidity. Trees, crops and soil fertility: concepts and research methods. CABI Publishing, UK, pp 93–130

Simandl GJ, Paradis S, Irvine M (2007) Brucite–industrial mineral with a future. Geosci Can 34:57–64

Sperber CM, Wilkinson BH, Peacor DR (1984) Rock composition, dolomite stoichiometry, and rock/water reactions in dolomitic carbonate rocks. J Geol 92:609–622. https://doi.org/10.1086/628901

Sun C, Yao Z, Wang Q et al (2023) Theoretical study on the organic acid promoted dissolution mechanism of forsterite mineral. Appl Surf Sci 614:156063. https://doi.org/10.1016/J.APSUSC.2022.156063

Tinker PB, Nye PH (2000) Solute movement in the rhizosphere. Oxford University Press

Tränkner M, Tavakol E, Jákli B (2018) Functioning of potassium and magnesium in photosynthesis, photosynthate translocation and photoprotection. Physiol Plant 163:414–431. https://doi.org/10.1111/ppl.12747

Van Ranst E, Kips P, Mbogoni J, et al (2020) Halloysite-smectite mixed-layered clay in fluvio-volcanic soils at the southern foot of Mount Kilimanjaro, Tanzania. Geoderma 375:114527. https://doi.org/10.1016/j.geoderma.2020.114527

Wang F, Giammar DE (2013) Forsterite dissolution in saline water at elevated temperature and high CO_2 pressure. Environ Sci Technol 47:168–173. https://doi.org/10.1021/es301231n

Wen T, Yuan J, He X et al (2020) Enrichment of beneficial cucumber rhizosphere microbes mediated by organic acid secretion. Hortic Res 7:154

Agronomic Characterization of Potassium Soil Remineralizers and Silicate Fertilizers in Tropical Conditions

Carlos Alexandre Costa Crusciol, Sirlene Lopes Oliveira, José Roberto Portugal, and Éder de Souza Martins

Abstract This chapter examines the agronomic characterization of potassium soil remineralizers and silicate fertilizers under tropical conditions, highlighting the crucial role of potassium (K) as an essential macronutrient for plant growth. K is involved in various physiological processes, including enzyme activation, photosynthesis, osmotic regulation, and stress response, making its availability vital for crop yield and quality. In tropical soils, potassium exists in multiple forms, including soluble, exchangeable, non-exchangeable, and structural potassium. The dynamics of these forms are significantly influenced by soil properties, weathering processes, and microbial activity. The chapter discusses the dissolution of potassium from silicate minerals, such as K-feldspar, biotite, and glauconite, elucidating the factors that affect potassium availability, including particle size, pH, and the presence of organic acids and microorganisms. Potential processing technologies for enhancing K availability from silicate rocks are evaluated, including physical, chemical, and biological

C. A. C. Crusciol (✉) · S. L. Oliveira · J. R. Portugal
Department of Crop Science, College of Agricultural Sciences, São Paulo State University (UNESP), Botucatu, São Paulo, Brazil
e-mail: carlos.crusciol@unesp.br

S. L. Oliveira
e-mail: sirlene.oliveira@ifmt.edu.br; sirlene.lopes@unesp.br

J. R. Portugal
e-mail: jose.portugal@unesp.br

C. A. C. Crusciol
Center for Carbon Research in Tropical Agriculture (CCARBON), University of São Paulo (USP), Piracicaba, São Paulo, Brazil

S. L. Oliveira
Instituto Federal de Mato Grosso (IFMT), Guarantã do Norte, Mato Grosso, Brasil

É. de Souza Martins
Agrogeological Laboratory, Brazilian Agricultural Research Corporation - Embrapa Cerrados, BR 020, km 18, Planaltina 73310970, Distrito Federal, Brazil
e-mail: eder.martins@embrapa.br

É. de Souza Martins and S. Huff Theodoro (eds.), *Soil Remineralizers and Silicate Fertilizers*, Frontier Studies in Soil Science,
https://doi.org/10.1007/978-3-032-14656-4_10

methods. Integrating high-temperature pretreatments, solubilizing microorganisms, and organic matter can enhance the effectiveness of silicate-derived potassium fertilizers. Field and greenhouse studies are referenced, highlighting the need to expand long-term studies under field conditions to validate the results found in controlled and short-term conditions. Overall, the efficiency of these products is still low, particularly in the medium to long term, when the natural reserves of nutrients in the soil are depleted. However, adopting a sustainable cultivation system can substantially improve the growing environment for profitable crops.

Keywords K silicates · Bioweathering · Good agricultural practices

1 Importance of Potassium for Plant Growth

Potassium (K) is an essential macronutrient with multiple roles in plant development and production. K activates approximately 60 enzymes, including enzymes involved in catabolism, sugar metabolism, and protein synthesis (Cui and Tcherkez 2021). In photosynthesis, K is essential for the biosynthesis and stabilization of chloroplast ultrastructure, as well as for the synthesis of ATP and Rubisco, and affects stomatal conductance, transpiration rate, energy transfer, phosphorylation activity, and the net photosynthetic rate (Imtiaz et al. 2023). K is also involved in osmotic adjustment, cell expansion, pH homeostasis, cell turgor, and regulation of membrane electrical potential (Cui and Tcherkez 2021). In addition, K plays a crucial role in plant responses to abiotic stressors such as drought, salinity, and extreme temperatures, as it contributes to maintaining ionic balance and activating antioxidant enzymes, thereby mitigating damage caused by reactive oxygen species (ROS). Plants with adequate K levels are more resistant to pathogens and herbivores, as K strengthens cell walls and reduces the concentrations of sugars and amino acids that attract pests. On the other hand, a decrease in K levels can lead to an increase in the synthesis of jasmonic acid, which is involved in defense mechanisms (Hasanuzzaman et al. 2018).

The concentration of K in the cytoplasm typically ranges between 80 and 200 mM (Walker et al. 1996; Rodríguez-Navarro 2000). Additionally, plants accumulate significant amounts of this ion in their vacuoles, often exceeding their nutritional needs (Ragel et al. 2019). K is the most abundant cation in plants, accounting for approximately 6–10% of the plant's dry weight (Hafsi et al. 2014; Ragel et al. 2019). K is a mobile element within plants and is translocated from older leaves to growth zones under conditions of potassium (K) deficiency (de Bang et al. 2021). One of the most pronounced symptoms of K deficiency is chlorosis, which may appear as leaf spots or along the leaf margins due to the accumulation of putrescine and eventually progress to necrosis at the leaf tips, edges, and veins (Cui et al. 2020; de Bang et al. 2021). Other common symptoms include leaf bronzing due to polyamine accumulation and leaf curling (Gupta et al. 2013) (Fig. 1). Stems can also be affected, often becoming thin with shortened internodes (Gerardeaux et al. 2010).

Fig. 1 Symptoms of potassium deficiency: **a** margin chlorosis of peanut leaves, **b** necrosis of corn leaves, **c** margin chlorosis of soybean leaves, and **d** bronzing of soybean leaves

In K-deficient plants, soluble carbohydrates and soluble nitrogen compounds accumulate, whereas starch content decreases (Sung et al. 2015; Hu et al. 2018). These metabolic changes occur because regulatory enzymes, such as pyruvate kinase and phosphofructokinase, are highly dependent on K^+ (Storey 1976; Ambasht and Kayastha 2002). Starch synthase, which catalyzes the transfer of glucose to starch molecules, is also highly reliant on monovalent cations, with K^+ being the most efficient (Cui and Tcherkez 2021). K deficiency also reduces the efficiency of translocating assimilates to sink organs, such as fruits and seeds, ultimately reducing crop quality and yield (Yang et al. 2023; Ishikawa et al. 2023).

2 Potassium Dynamics in Tropical Soils

There are two primary forms of K in soil: a liquid fraction (the soil solution) and a solid fraction (bound to minerals and organic particles). These can be further categorized into four distinct pools: (1) soil solution K, (2) exchangeable K, (3) non-exchangeable (fixed) K, and (4) structural (mineral) K (Fig. 2). The most readily available form for plant and microbial uptake is soil solution K, which exists in low concentrations (0.1–2%) but remains in dynamic equilibrium with the exchangeable K fraction (Sparks 2000; Brouder 2015). Exchangeable K represents a larger reserve compared to soil solution K. Exchangeable K is adsorbed onto clay surfaces or organic matter and can be rapidly released into the soil solution via ion exchange. By contrast, non-exchangeable K is immobilized within the interlayers of clay minerals (e.g., illite and vermiculite) and is released only slowly, depending on soil physicochemical conditions (Brouder 2015). K dynamics reflect the balance between these different forms, which tends towards decreasing availability (Rosolem and Steiner 2017). The largest and most stable reservoir of K is found in its structural form, which is encapsulated within the crystal lattices of primary minerals such as feldspar and mica. This fraction becomes available for plant uptake only through prolonged chemical weathering, a process that occurs at a rate too slow to have significant agricultural implications (Brouder 2015).

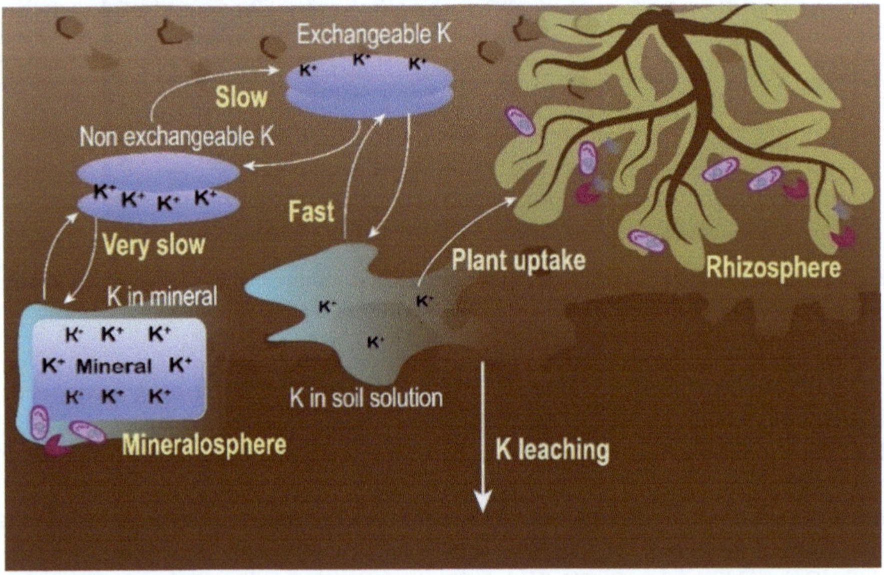

Fig. 2 Interrelationship of K forms in the soil and the effects of soil microorganisms and root exudates

Weathering occurs more rapidly in humid tropical regions. When combined with dry periods, weathering leads to modifications of the mineral structure and the coexistence of 2:1 phyllosilicates with varying degrees of weathering, ranging from micas to smectites, including interstratified minerals where K is bound with different energies. Tropical soils are predominantly composed of low-activity clays such as aluminum hydroxides, which, when associated with high acidity, promote leaching and reduce fertilization efficiency. Exchangeable K is the largest pool of K available for plant absorption (Rosolem and Steiner 2017). When soluble and exchangeable potassium (K) decreases due to plant uptake or leaching, non-exchangeable K can be released from the interlayers of clays and primary minerals, thereby contributing to plant nutrition in the short and medium term (Simonsson et al. 2009). Soils with higher clay content have greater K retention capacity, which reduces leaching losses. Rosolem and Steiner (2017) investigated the effects of soil texture (a sandy clay loam and a clayey loam) and K fertilizer application rates on K dynamics and balance in a soil–plant system under no-tillage. They found that in the sandy clay loam soil, increasing K fertilizer application rates intensified leaching losses below 1 m depth, which accounted for 16–52% of the K added as fertilizer. By contrast, in the clayey soil, application of up to 150 kg K ha^{-1} per year did not result in significant leaching below 1 m. These results demonstrate that the clay content is important in retaining K in the soil profile. Likewise, soils with higher organic matter content have a lower potential for nutrient loss to leaching due to the presence of negatively charged functional groups [such as carboxyl (−COOH), hydroxyl (−OH) and phenols] that attract and retain cations (K^+, Ca^{2+}, Mg^{2+}, NH_4^+) through electrostatic bonds (Lombardi et al. 2006, de Sá Mendonça et al. 2006; Marinho et al. 2014).

The presence of microorganisms can affect the availability of K to plants. Microorganisms are attracted to photoassimilates produced by plants and released into the rhizosphere. The rhizosphere is the region of soil closely associated with plant roots where chemical, biological, and physical interactions occur between the plant, microorganisms, and the soil matrix (Qu et al. 2020). In this region, microorganisms can solubilize forms of K and other nutrients that are not readily absorbable by plants, making them available in the soil solution. Uroz et al. (2015) explored the mineralosphere, demonstrating that rocks and minerals support the development of microbiomes capable of weathering minerals in the rhizosphere. These interactions are complex, dynamic, and challenging to measure under field conditions. However, understanding the mechanisms and factors involved in the mineral dissolution process can facilitate the development of management techniques to increase K bioavailability for crops.

3 Dissolution of Mineral Sources of Potassium

It is estimated that approximately 90% of soil potassium (K) is bound to minerals or in a structural form. The main primary mineral sources of K are K-feldspars (such as orthoclase, microcline, and sanidine) and micas (such as muscovite and biotite), feldspathoids (such as kalsilite, nepheline, and leucite), glauconite, K-zeolites (such as chabazite-K and clinoptilolite-K), illite, and potassium-taranakite. Micas, particularly trioctahedral micas (biotite and phlogopite), have a chemical composition that includes essential nutrients for plant growth, such as K, Mg, Fe, B, Zn, and Mn (Tischendorf et al. 2001). However, nutrient availability varies between different minerals; muscovite, for example, has a dioctahedral structure and is resistant to dissolution (Basak et al. 2017; Kicińska and Pomykała 2023). Biotite and phlogopite, by contrast, have a trioctahedral structure in which the micas exhibit great repulsion, facilitating weathering (Basak et al. 2017). Kalinowski and Schweda (1996) observed preferential release of K from biotite and phlogopite in aqueous solution in a pH range of 1–4. A study of dissolution by sulfuric acid for 36 h reported 67% recovery of K from biotite and 33% recovery from phlogopite (0.85 mL of H_2SO_4 per g of rock); K recovery from biotite increased proportionally with the addition of H_2SO_4 (Pereira and Gomes 2021). The release of K from the intermediate layer is relatively rapid, and within a few days, the movement of K becomes regulated by diffusion (Malmström and Banwart 1997).

Feldspars are a group of primary minerals that are widely found in various types of rocks. Weathering of feldspars forms soil components that are a direct source of nutrients for plants, such as clay minerals. They are also major constituents of the sand and silt fractions in soil, providing important pedological information (Radoslovich 1975). K-feldspars are the most abundant sources of K in granites with low K concentrations (~4.2 wt% K_2O) or ultrapotassic syenites with high K concentrations (~14.3 wt% K_2O) (Ciceri et al. 2017).

From an agronomic point of view, Ciceri et al. (2017) emphasize that the indiscriminate use of the label "K-feldspar" to describe the raw material disregards the numerous variations in elemental compositions, polymorphic structures, and petrographic characteristics. In fact, K silicates present a range of properties that alter their dissolution kinetics. A K-feldspar has approximately 3–4 times more K than a nepheline but is several times more resistant to dissolution. Thus, the dissolution rate of the minerals present in a rock is more important than the overall concentration of K (Swoboda et al. 2022). Table 1 summarizes the dissolution rate constants for the most common primary potassium silicates, as well as their relative rates in comparison to K-feldspar.

Kalsilite ($KAlSiO_4$) contains up to 30% K_2O and exhibits high solubility in water (Li et al. 2019), surpassing that of nepheline. However, it is the rarest potassium silicate, primarily found in synnerite-type intrusive rocks (Panina et al. 2022), which occur in tropical regions such as Northeast Brazil (Candido et al. 2024) and Northwest Africa (Bea et al. 2014). Kalsilite occurs in variable proportions in alkaline ultramafic volcanic rocks, such as kamafugites (Oliveira et al. 2022).

In a review, Yuan et al. (2019) explored the main mechanisms proposed to explain the dissolution of K-feldspars:

Table 1 Potassium mineral dissolution rate constants (25 °C, pH = 0) and relative dissolution rates to K-feldspar (Palandri and Kharaka 2004; Manning and Theodoro 2020)

Mineral sub-group	Mineral	Fórmula Química	% K_2O	Dissolution rate (log mol·m^{-2}·s^{-1})	Relative dissolution rate
K-feldspar	Orthoclase	$KAlSi_3O_8$	14–17	− 10.06	1
Feldspathoid	Nepheline	$(Na,K)AlSiO_4$	6–8	− 2.73	21.400.000
Feldspathoid	Leucite	$KAlSi_2O_6$	17–21	− 6.00	11.500
Mica	Glauconite	$(K,Na)(Fe^{3+},Al,Mg)_2(Si,Al)_4O_{10}(OH)_2$	4–8	− 4.80	182.000
Mica	Biotite	$K(Fe,Mg)_3(AlSi_3O_{10})(OH)_2$	9–11	− 9.84	1.66
Mica	Muscovite	$KAl_2(AlSi_3O_{10})(OH)_2$	9–11	− 11.85	0.016

4 Surface Reaction-Controlled Dissolution Mechanism

This process occurs when the rate of dissolution of a feldspar is determined by chemical reactions that occur directly on its surface (Lagache 1976; Berner and Holdren 1979; Chou and Wollast 1984; Yuan et al. 2019). Surface reaction-controlled feldspar dissolution occurs more frequently in alkaline environments (Chou and Wollast 1984). Initially, dissolution is congruent; that is, all elements (K, Al, and Si) are released in the same proportions as they exist in the original mineral. Over time, as the system approaches chemical equilibrium, dissolution may become incongruent, as Al and silica begin to be consumed in the formation of secondary minerals, such as clays or zeolites. This mechanism is characterized by linear kinetics (constant rate) and the formation of etch pits in areas of the surface with higher energy, such as crystal defects (Berner and Holdren 1979; Alekseyev et al. 1997; Yuan et al. 2019).

5 Dissolution by Preferential Leaching and Diffusion

In acidic environments, feldspar dissolution occurs incongruently, with some elements released more rapidly than others. Ions such as K^+ and Al^{3+} are preferentially leached, leaving a surface layer depleted in these elements and enriched in silica. This residual layer acts as a barrier to the diffusion of other ions, gradually reducing the dissolution rate. Therefore, the kinetics of this process are parabolic (the rate decreases with time). This mechanism explains why, under acidic conditions, feldspars form altered layers rich in silicon, while cations such as aluminum are released more rapidly into solution (Correns and von Engelhard 1938; Correns 1940; Yuan et al. 2019).

5.1 Interfacial Dissolution and Reprecipitation Mechanism

In this mechanism, feldspar dissolves congruently, but the released elements do not completely move away from the surface. Instead, they precipitate almost immediately, forming a porous layer of secondary minerals such as amorphous silica or clays. Unlike in the leaching mechanism, there is no cation-depleted layer; instead, a secondary layer forms over the original mineral. This process can occur under a variety of pH conditions and is particularly relevant in systems where local supersaturation leads to the rapid precipitation of secondary phases (Hellmann et al. 2003; Yuan et al. 2019).

5.2 *Diffusion–Precipitation Controlled Dissolution Mechanism*

This mechanism is an evolution of the leaching–diffusion model, but occurs when the system is closer to chemical equilibrium. The elements released by the dissolution of feldspar diffuse through an altered layer and then precipitate, forming a double structure: an inner sublayer depleted in K and Al (due to leaching) and an outer sublayer enriched in Al or silica (due to precipitation). This double layer acts as a dynamic barrier, progressively reducing the dissolution rate of the material. This mechanism is important in systems where precipitation of secondary minerals occurs close to the feldspar surface, such as in hydrothermal environments or in advanced weathering (Wollast 1967; Helgeson 1971, 1972; Chou and Wollast 1984; Nugent et al. 1998).

5.3 *Factors Affecting the Dissolution of Potassium Minerals*

The dissolution rate of feldspars is influenced by crystal structure, Al/Si ordering, temperature, pH, surface area, acidic solutions (organic acids, rain—due to carbonic acid), chemical affinity, and precipitation of secondary minerals (Yuan et al. 2019). More complex ordering of Al–Si in the structure increases the stability of the structural K located in the interstitial sites of K-feldspar (Ciceri et al. 2017). Furthermore, K-feldspar dissolution is an anisotropic process in which surface orientation and chemical affinity impact the dissolution rate. Thus, different faces of a single crystal may have varying dissolution rates, influenced by the nucleation of corrosion cavities (Pollet-Villard et al. 2016). Surface morphology, including the presence of open pores and roughness, influences the dissolution rate. High porosity increases the contact area between surfaces and the solution, as well as their permeability, allowing for greater water infiltration and thereby increasing the weathering rate (Jeschke and Dreybrodt 2002; Navarre-Sitchler et al. 2015).

- **Particle size**

Particle size is typically inversely related to the dissolution rate: the smaller the particle size, the larger the surface area (Fig. 3), and the greater the exposure of the mineral surface to weathering reactions (Karagöz and Kuşlu 2017; Aracena et al., 2018). For K-feldspar particle sizes of 0.075, 0.09, and 0.11 μm, the dissolved fractions were 85.1%, 79.0%, and 6.0%, respectively, after 120 min in a $CaCl_2$ solution at 75 °C (Ayinla et al. 2018). According to calculations by Rinder and von Hagke (Rinder and von Hagke 2021), smaller basalt particles (<1 μm) have a much larger specific surface area (67.42 m^2/g) and dissolve completely in only 10.7 years, whereas larger particles (<100 μm) have a smaller specific surface area (10.53 m^2/g) and take approximately 1696 years to dissolve completely. However, the process is not linear—approximately 87% of a 100 μm particle dissolves in half the total time.

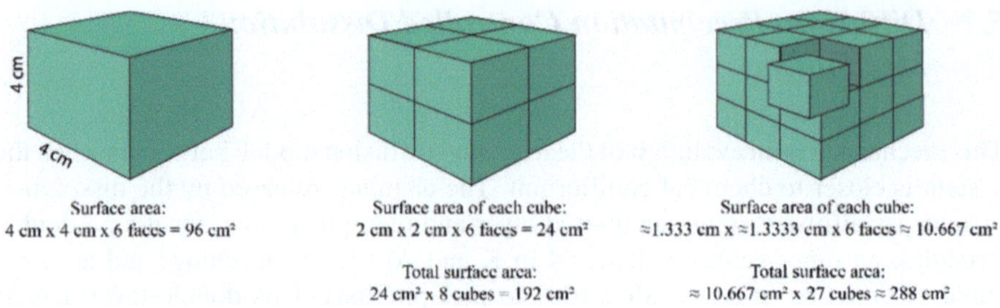

Fig. 3 Specific surface area according to particle size

When considering a 97.5% dissolution, the time frames are 1205 years (<100 μm), 96 years (<10 μm), and 7.6 years (<1 μm). Since the actual distribution includes particles of various sizes, more than 99.9% of smaller grains dissolve within these timeframes.

- **pH range**

Acidity is a crucial factor for increasing the dissolution of K minerals. The primary mechanism involves the protonation of the mineral surface, where H^+ ions adsorb to the oxygen atoms of the Si-O and Al-O bonds, thereby weakening the structure (Fogler et al. 1975; Welch and Ullman 1996; Cagnon et al. 2022). In acidic media, aluminum (Al^{3+}) in AlO_4 tetrahedra is especially reactive and is released into solution together with cations such as Na^+, K^+, or Ca^{2+}. Hydrolysis of the Si-O-Si bonds forms soluble silicic acid (H_4SiO_4) (Ur Rahman et al. 2024). At very low pH (<4), dissolution is accelerated, which may maintain Al^{3+} in solution. However, at moderate pH levels (4–6), aluminum may precipitate as clays (e.g., kaolinite) (De Oliveira et al. 2023).

A study of the action of organic acids in association with high temperatures on the dissolution of K-feldspar revealed that K-feldspar dissolution is driven mainly by organic acids (such as acetic and oxalic) and inorganic acids. At high temperatures, organic acids have greater efficiency due to two main mechanisms: (1) proton exchange, where H^+ ions replace Na^+ and K^+ on the mineral surface, releasing these cations; and (2) complexation, in which carboxylates (such as oxalate) bind to Al and Si, reducing the activation energy and accelerating dissolution (Fig. 4). Dicarboxylic acids (e.g., oxalic) are more effective than monocarboxylic acids (e.g., acetic), as they form stable five-membered ring complexes with metals. This dissolution generates secondary pores in sandstone reservoirs, increasing permeability, but can be partially inhibited by the precipitation of secondary minerals (such as clays and zeolites) (Xiu-ting et al. 2021).

Although most studies focus on the effects of low pH (acidity) due to the ability of H^+ to break crystal structure bonds, basic pH can have a similar effect, with reactions involving hydroxyl ions (OH^-). At pH values close to neutrality, the dissolution rates of silicates tend to be slower and almost independent of pH (Hellmann 1994; Bagheri et al. 2022). Hellmann (1994) showed that the dissolution of albite (a Na-feldspar)

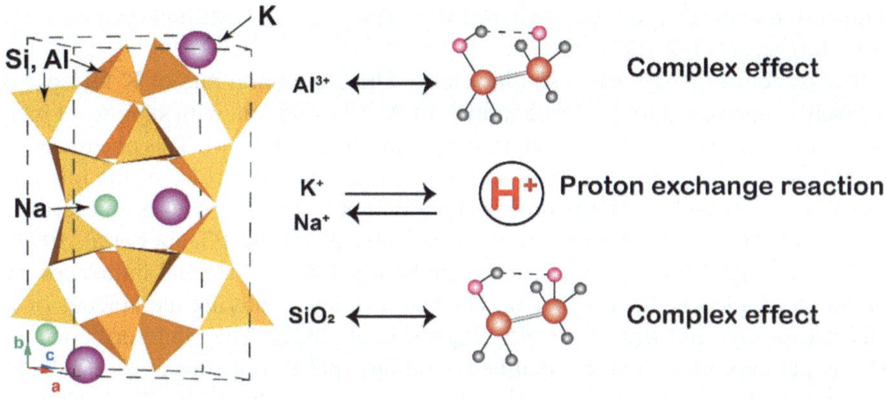

Fig. 4 Dissolution mechanism of a feldspar mediated by organic acids. Adapted from Xiu-ting et al. (2021)

follows a "U"-type curve with minimum dissolution rates at neutral pH (Fig. 5). Furthermore, the magnitude of the pH dependence of the dissolution rate can vary with temperature (Hellmann 1994), mineral composition (Welch and Ullman 1996), and the presence of other solutes (Gautier et al. 1994).

- **Chemical and biological binders**

  (a) **Ions$_2$**

Ions such as Na^+ or Ca^{2+} can contribute to the dissolution of K-feldspar through an ion exchange mechanism. In alkaline medium, these ions also increase the release of K by facilitating the destruction of the mineral structure (Gardner 1983; Stillings and Brantley 1995; Demir et al. 2003; Li et al. 2025). Na^+ is preferentially released in the early stages of dissolution, along with K^+ and Ca^{2+}; however, in Na^+-rich environments, Na^+ can reduce the dissolution rate, especially at acidic pH, due to

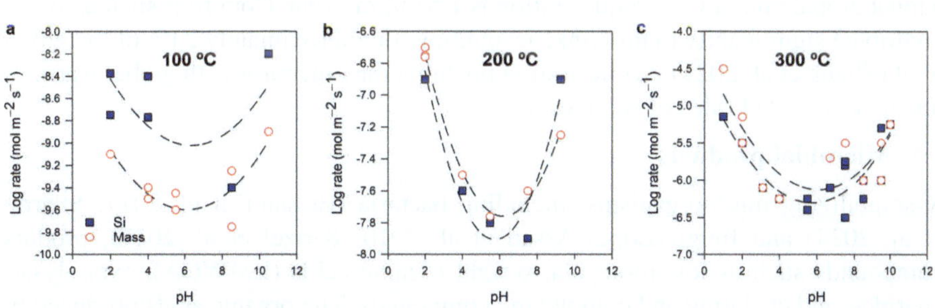

Fig. 5 Dissolution rate of albite at different temperatures as a function of the initial pH in situ. The dissolution rate in acidic and basic regions is higher, as is the rate at higher temperatures. Adapted from Hellmann (1994)

competition with H^+ at surface exchange sites (Gardner 1983; Stillings and Brantley 1995; Bagheri et al. 2022).

The presence of Al^{3+} has a similar effect. The dissolution rate of K-feldspar is inversely proportional to the concentration of Al^{3+} in solution, with significant inhibition observed even at Al^{3+} concentrations as low as 5×10^{-9} M. This phenomenon is attributed to the preferential reabsorption of Al^{3+} on the surface of the mineral, which interferes with the formation of silica-rich precursor complexes and leads to decreases in overall dissolution and, consequently, K release (Gautier et al. 1994). Additionally, Al^{3+} modulates ion exchange between K^+ and H^+ and influences the formation of a leached and silica-enriched surface layer, altering the kinetics of K release (Stillings and Brantley 1995; Bagheri et al. 2022). The inhibitory effect of Al^{3+} is pH-dependent. Under alkaline conditions (pH 9) (Gautier et al. 1994), its presence markedly reduces dissolution. By contrast, in acidic media, Al^{3+} enhances the inhibition of silica dissolution, resulting in distinct patterns of K-feldspar decomposition under different geochemical conditions (Bagheri et al. 2022). The effects of Al^{3+} are particularly significant in tropical soils, where the Al^{3+} concentration is high and can significantly impact the overall dissolution rate of K-feldspars used as a K fertilizer.

(b) **Atmospheric CO_2**

The CO_2 concentration also affects mineral dissolution rates and accelerates the solubilization of K. Carbonic acid (H_2CO_3), which is present in rainwater, accelerates the acid hydrolysis of K-feldspar. This reaction consumes CO_2 by forming bicarbonate, increasing the fixation of inorganic and organic carbon in the soil (Xiao et al. 2016). The dissolution of K-feldspar can be explained by Eq. 1:

$$\begin{aligned} &2\ KAlSi_3O_8(s) + 2H_2CO_3(aq) +_9 H_2O(l) \\ &\rightarrow 2K^+(aq) + 2\ HCO_3^-(aq) + 4\ H_4SiO_4(aq) + Al_2Si_2O_5(OH)_4(s) \end{aligned} \quad (1)$$

In the long term, the HCO_3^- produced by the weathering of silicate may represent a stable sink precipitated as calcite or dolomite in the ocean. The estimated annual global rate of CO_2 sequestration is 1.67 t/km^2/year. Countries such as Brazil contribute significantly to this sink, accounting for approximately 24% of the global total (Zhang et al. 2021). Sequestration potential may increase with global warming due to accelerated rates of weathering.

(c) **Microbial products**

K-solubilizing microorganisms, including bacteria (Etesami et al. 2017; Sharma et al. 2024) and fungi (Lopes-Assad et al. 2010; Kirtzel et al. 2017), produce compounds such as low-molecular-weight organic acids (LMWOA), exopolysaccharides, and chelating and complexing compounds. The organic acids produced by K-solubilizers are acetic acid, citric acid, fumaric acid, glycolic acid, phenylacetic acid, 2-ketogluconic acid, lactic acid, oxalic acid, propionic acid, succinic acid, and tartaric acid, as reviewed by Sharma et al. (2024). Organic acids solubilize minerals through acidolysis and complexolysis reactions in which the release of H^+ weakens the mineral's bonds and carboxylic and hydroxyl groups form soluble complexes with mineral cations, facilitating their removal from the mineral matrix (Lazo et al. 2017).

Rhizosphere microorganisms produce various types of polysaccharides, including exopolysaccharides (EPSs), capsular polysaccharides (CPSs), lipopolysaccharides (LPSs), and β-1,2-glucans. EPSs bind strongly to organic acids, forming complexes that adsorb to mineral surfaces, concentrating these acids around the mineral and reducing the local pH. This acidification favors the solubilization of K from minerals such as feldspar and illite. Mucilaginous polysaccharides and metabolic intermediates secreted by bacteria (such as *Bacillus*, *Clostridium,* and *Thiobacillus*) promote mineral degradation, releasing K^+. In addition, EPSs bind K and SiO_2, helping to balance their concentrations between the soil solution and minerals, which facilitates the continuous release of K^+ and increases its availability to plants. In addition, K-solubilizing microorganisms (KSMs) produce organic ligands (such as glucuronic acid, mannuronic acid, and alginates) that chelate cations (Al^{3+}, Si^{4+}) from the crystal structure of minerals, releasing K^+ into solution. KSMs also secrete molecules (such as EDTA and organic acids) that form complexes with Al^{3+}, Si^{4+}, weakening mineral bonds and facilitating the release of K (Sharma et al. 2024).

6 Agronomic Outcomes of Applying Silicate Rocks as Sources of Potassium Under Tropical Conditions

The results of studies involving the application of potassium source rocks in tropical soils are not unanimous. Although the concept of using ground rocks as a fertilizer source dates back to Hensel (1894), it was only at the end of the twentieth century that the scientific community began to seriously consider the possibility of effectively utilizing them as agricultural inputs. Table 2 compiles recent studies that utilize silicate rocks as a potassium source in tropical soils. Most studies are short-term, which makes it difficult to assess their effectiveness as fertilizers. Additionally, many are conducted in greenhouses, which do not fully represent the complexity of field conditions (Table 2). Generally, KCl, the conventional potassium source, showed better results in agronomic efficiency, which is expected, as it is obtained through an industrial process and is highly soluble.

In contrast, K silicates, particularly K-feldspars, exhibit extremely low relative dissolution rates, even under controlled conditions (Table 1). Despite this, some short-term field studies have shown that crop yields under K-silicates are very close to those of KCl (Mancuso et al. 2014; Soratto et al. 2021a; Crusciol et al. 2022). However, new studies have shown that in the medium to long term, its efficiency can decrease in conventional crop systems. The combination of K-silicate and organic compound may improve crop productivity (Tavares et al. 2018), but no study has measured the actual contribution of the organic compound to the bioavailability of potassium derived from rock. The following subsections detail results from agronomic studies

Table 2 Results of agronomic experiments evaluating the efficiency of K-silicate rocks as potassium fertilizer in tropical soils

Crops	Rock/mineral	Experiment type	Duration	Results	References
Corn	Nepheline syenite, phonolite	Pots	45 days	KCl showed the highest agronomic efficiency index. Nepheline syenite and phonolite were similar to KCl at high K doses	Nogueira et al. (2021)
Soybean, wheat, corn, millet, soybean	Hydrothermalized phonolite, phonolite	Field	16 months	K-silicate rocks had similar agronomic efficiency to KCl	Soratto et al. (2021a)
Rice and common beans	Hydrothermalized phonolite, phonolite	Field	6 months	In rice, KCl provided greater K foliar content, but phonolite reaches a similar result to grain yield. In common beans, phonolite provided similar results to KCl	Soratto et al. (2021b)
Soybean, barley	Siltstone, tephrite, and olivine melilitite	Pots	7 months	Olivine melilitite rock powder, pure or mixed with siltstone, showed the highest agronomic potential for soil remineralization, while tephrite showed little response and siltstone may not meet requirements. All treatments with rock powder were inferior to the conventional treatment (KCl)	Cunha and Almeida (2021)
Soybean, corn, common bean, upland rice	Phonolite, hydrothermalized phonolite	Field	15 months	Phonolite increased grain yield for all crops, while hydrothermalized phonolite was efficient for soybean, corn, and common bean	Crusciol et al. (2022)

(continued)

Table 2 (continued)

Crops	Rock/mineral	Experiment type	Duration	Results	References
Coffee	Phonolite	Field	2 years	Phonolite powder showed similar efficiency to KCl at the recommended rate (150 kg ha^{-1} K_2O)	Mancuso et al. (2014)
Alfalfa, centrosema crops	Alkaline ultramafic rock	Pots	7 months	Silicate rock increased K uptake but did not overcome the biomass yield promoted by KCl	Moreira et al. (2023)
Corn	Phonolite and verdete	Pots	4 months	Thermal pretreatment at 300-600°C of phonolite and *verdete* rocks, and, especially when mixed with coffee husks, was efficient to increase biomass and macronutrient content in corn plants	Pessoa et al. (2023)
Sweet potato	Hydrothermalized phonolite	Field	5 months	Castor meal + ground hydrothermalized phonolite combined with synthetic fertilizers improved sweet potato nutrition, yield, and quality	Parecido et al. (2024)
Urochloa decumbens	Phonolite	Field	84 days	Phonolite did not change leaf K content or shoot dry mass. Its efficiency increased when enriched with organic compost, though there's no evidence the compost enhanced its solubility	Tavares et al. (2018)

(continued)

Table 2 (continued)

Crops	Rock/mineral	Experiment type	Duration	Results	References
Urochloa ruziziensis, U. decumbens, U. humidicola, Andropogon gayanus	Phonolite	Pots	70 days	Phonolite powder increased exchangeable, non-exchangeable, structural, and total K in soil, enhancing K uptake and dry matter production in *Urochloa species*, especially *U. decumbens* compared to the control treatment (no phonolite fertilizer). In this study, phonolite was not compared with a conventional K source	da Rocha Neto et al. (2024)
Ornamental sunflower	Glauconite	Pots	35 days	Glauconite fertilizers (especially organomineral and fine-textured) improved agronomic and postharvest traits but were less efficient than KCl for growth and flower size	Torqueti et al. (2016)
Corn and common beans	Glauconitic siltstone	Field	1 year	Beans required 4 times more K_2O from rock than KCl for superior yields, while corn needed only double the dose	Brasil et al. (2020)
Corn, *Panicum maximum* cv. Mombaça, eucalyptus	Verdete rock	Pots	90–105 days	Calcined/acidified verdete rock effectively released K, with agronomic results similar to KCl for corn and grass (82% recovery). Lower for eucalyptus (~52–60%) but still competitive	Santos et al. (2016)

(continued)

Table 2 (continued)

Crops	Rock/mineral	Experiment type	Duration	Results	References
Rice	Breccia, biotite schist, ultramafic, chapada by-product, mining by-product, phlogopite	Pots	120 days	Highest rice grain yield was achieved with Ultramafic rock at 360 kg K_2O ha^{-1}. Biotite Schist and Phlogopite showed the lowest yields regardless of K dose. Conventional source not tested	Silva et al. (2014)
(1) Soybean/off-season; (2) Soybean/Urochloa brizantha; (3) Soybean/pasture intercropped with niger, forage radish, buckwheat, cowpea (18 months); (4) Soybean/pasture intercropped with niger, forage turnip, buckwheat, cowpea (20 months); (5) Soybean/pasture intercropped with niger, turnip, buckwheat, cowpea with grazing animals	Mica schist	Field	3 years	The remineralizing rock powder did not significantly modify total K reserves in the short term but initiated incipient mineralogical transformations. Diverse cropping systems enhanced K cycling, reducing readily available K (due to higher uptake by productive crops) but resulting in positive K balance and significantly increased soybean/cotton yields in sandy-loam Cerrado soils even with rock powder application	Nogueira et al. (2024)

(continued)

Table 2 (continued)

Crops	Rock/mineral	Experiment type	Duration	Results	References
Corn, soybean	Basalt	Pots	47 days	The release of K from basalt in the clayey soil was limited, but this K was assimilated by the plants and contributed to the increase in dry mass. The release of K from the rock was very low in the clay loam soil, and soybean did not reach leaf K concentrations classified as adequate	Luchese et al. (2023)
Urochloa brizantha	Basalt	Pots	240 days	Basalt increased K accumulation in *U. brizantha* and raised shoot dry mass by 6% in a Typic Quartzipsamment soil and by 27% in an Oxisol	Rodrigues et al. (2024)

that evaluated K-silicates as alternative potassium fertilizer sources. However, new studies have shown that in the medium to long term, its efficiency can decrease in conventional crop systems. The combination of K-silicate and organic compound may improve crop productivity (Tavares et al. 2018), but no study has measured the actual contribution of the organic compound to the bioavailability of potassium derived from rock. The following subsections detail results from agronomic studies that evaluated K-silicates as alternative potassium fertilizer sources.

(a) **Phonolite**

A series of studies have examined the agronomic efficiency of applying phonolite rock, which is mainly composed of feldspars and feldspathoids (Table 3). Phonolite is an alkaline extrusive igneous rock with an aphanitic or porphyritic texture that typically exhibits shades of green to nearly black. It occurs in various forms, including lava flows, dikes, and stocks. It closely resembles other syenite rocks chemically and mineralogically and primarily differs in texture. Another potassic rock formed through natural hydrothermal processes, referred to here as *hydrothermalized phonolite*. These alkaline rocks are fine-grained, K-rich, and primarily composed of microcline ($KAlSi_3O_8$) and illite ($K_{0.7}Al_2(Si,Al)_4O_{10}(OH)_2$). They typically exhibit light beige to pale gray hues, reflecting their felsic mineral composition and formation conditions. The rocks examined in these studies were collected in Poços de Caldas, Minas Gerais State, Brazil.

The hydrothermal process on phonolite concentrated K_2O and caused the loss of Na_2O, which may be beneficial for plant growth (Table 4). However, this also caused the loss of nepheline, a mineral whose dissolution rate is several orders of magnitude greater than that of feldspars, decreasing the solubility of the phonolite rock (Table 3).

Table 3 Mineral composition of phonolite and hydrothermalized phonolite rocks

Phonolite		Hydrothermalized phonolite	
Mineral composition (wt%)		Mineral composition (wt%)	
Analcime	16.6%	Anorthoclase	15.0%
Augite	9.6%	Augite	1.6%
Gibbsite	6.8%	Gibbsite	0.3%
Microcline	11.0%	Illite	5.2%
Nepheline	10.1%	Microcline	10.9%
Orthoclase	20.0%	Muscovite	2.8%
Phlogopite	2.1%	Nepheline	7.0%
Sanidine	23.8%	Orthoclase	22.0%
		Fuchsite	17.0%
		Sanidine	18.0%

Table 4 Chemical composition of potassic rocks

Element	Phonolite	Hydrothermalized phonolite
P_2O_5 (wt%)	0.04	0.11
K_2O (wt%)	8.0	12.12
CaO (wt%)	1.63	0.61
MgO (wt%)	0.29	0.21
Fe_2O_3 (wt%)	3.51	4.29
Zn (mg kg^{-1})	148	147
Mn (mg/kg^{-1})	230	170
Al_2O_3 (wt%)	19.45	19.4
Na_2O (wt%)	6.41	0.32

Particle size: 100% passing through the 50 mesh sieve (297 μm); 99% passing through the 100 mesh sieve (149 μm); 87% passing through the 200 mesh sieve (74 μm); 58% passing through the 635 mesh sieve (20 μm)

Four field experiments were conducted in two agricultural years in Botucatu, SP, Brazil, in an Oxisol. The K sources [KCl, K2 (phonolite), and K3 (hydrothermalized phonolite)] were ground and applied superficially (Crusciol et al. 2022). The K doses corresponded to 0, 0.5, 1.0, and 2.0 times the recommended dose for soybean, corn, common bean, and upland rice. Soybean foliar K content reached ideal values (17–25 g kg^{-1}) at a K_2O dose of 21.7 kg ha^{-1}, regardless of the source used. Regarding crop productivity, the data revealed that silicate rocks can replace KCl without compromising agricultural yield. For soybeans, applying K2 and K3 provided yields equivalent to those provided by applying KCl, with K3 standing out at intermediate doses. For corn, all sources promoted similar increases in grain production, with a quadratic response to dose. For upland rice, K3 showed comparable performance to KCl at the recommended dose, whereas K2 was not suitable for this crop. Fascinating results were observed for common bean, for which the K3 source increased productivity by up to 6.9% compared with KCl. The agronomic efficiency analysis highlighted the superiority of the K3 source for common bean, with an agronomic efficiency index 4.5 times higher than that of KCl and 2.5 times higher than that of K2. Specifically, each kg of K_2O applied via K3 generated an increase in bean productivity of 9.4 kg ha^{-1}, versus 4.8 kg ha^{-1} for K2 and 3.5 kg ha^{-1} for KCl. In addition to their nutritional benefits, the silicate rocks increased foliar silicon content, especially in common bean and upland rice, potentially conferring greater plant resistance to biotic and abiotic stresses (Crusciol et al. 2022).

In two other experiments, the same treatments were applied to soybean–wheat–corn succession or corn–millet–soybean succession (Soratto et al. 2021a, b). Regardless of the source, the application of K increased the productivity of all crops. The agronomic efficiencies of the alternative sources, K1 (hydrothermalized phonolite) and K2 (phonolite), were similar to that of KCl. Furthermore, the alternative sources had greater residual effects approximately one year after their application. In another study in which phonolite rock was applied to Arabica coffee via the soil at 0.5, 1, and 2 times the amount recommended for the crop (75, 150, and 300 kg ha^{-1} K_2O), the results showed that the effect of phonolite application was similar to that of KCl at the recommended rate for the crop (150 kg ha^{-1} K_2O) (Mancuso et al. 2014).

The combination of castor meal and hydrothermalized phonolite exhibited synergistic effects on sweet potato production in tropical soils, demonstrating comparable marketable root yields to those achieved with synthetic sources of nitrogen and potassium, while also improving soil quality indices, according to a study conducted by Parecido et al. (2024). This research highlighted that integrating organic sources with low-solubility minerals can yield synergistic benefits, similar to findings in a greenhouse study by Busato et al. (2022), which showed enhanced growth and nutrient accumulation in corn tissues when humic substances were combined with biotite syenite.

Toledo et al. (2011) showed that compared with applying KCl, the physiological quality of soybean seeds was superior when alternative K sources were applied, which resulted in the production of heavier seeds with lower seed coat permeability. In addition, the application of phonolite and hydrothermalized phonolite was more beneficial than KCl application for soybean nodulation, especially under early application, 30 days before planting, which may have favored the solubilization of K and Si until the moment of effective crop growth (Barbosa et al. 2025).

For peanut (unpublished data), considering two growing seasons in separate areas, the early application of phonolite and hydrothermalized phonolite, 30 days before planting (Fig. 6), in a sandy Oxisol, also increased leaf K content and pod yield but remained lower than the application of KCl.

Applying phonolite did not increase the dry matter yield of *Urochloa brizantha* cv. Paiaguás compared with applying KCl, although dry matter yield responded

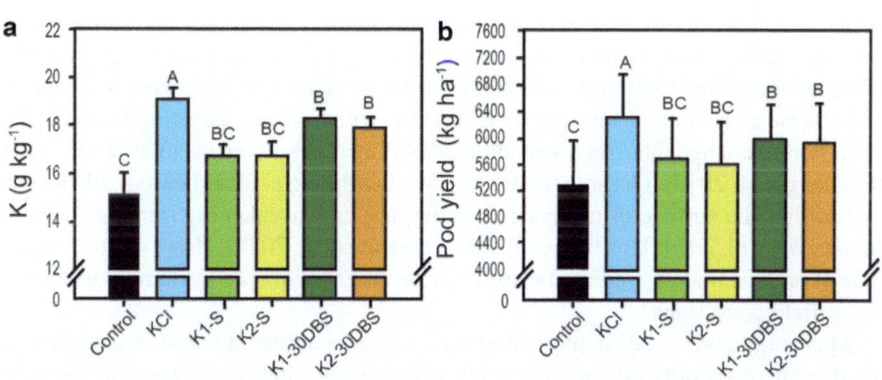

Fig. 6 Potassium (K) content in peanut leaves (A) and pod yield (B) as a function of K sources: no K application (control), potassium chloride (KCl), phonolite applied on the day of sowing (K1-S), hydrothermalized phonolite applied on the day of sowing (K2-S), phonolite applied 30 days before sowing (K1-DBS), and hydrothermalized phonolite applied 30 days before sowing (K2-DBS). Bars with different uppercase letters are significantly different according to the LSD test ($p \leq 0.05$) (Crusciol and Alves, unpublished)

to increasing doses. Moreover, applying a mixture of KCl (30%) and phonolite (70%) increased fertilizer efficiency compared to applying phonolite alone (Cruz et al. 2021). When applied alone (*in natura*), phonolite powder did not significantly increase dry matter production of *Urochloa decumbens*. However, it increased the availability of K and Si in the soil compared to the control. However, applying phonolite powder in association with organic compost dramatically enhanced the release of these nutrients (68% for K and 486% for Si in relation to the pure compost) (Tavares et al. 2018). However, the study did not provide evidence that bioweathering occurred and attributed the enhanced release of K and Si to a possible synergistic effect between the inputs.

In an unpublished study, an experiment was carried out in Botucatu, São Paulo, Brazil, for five years (2019–2024) to evaluate the effect of hydrothermalized phonolite powder on the yield of corn–soybean succession. The effects of a corn–ruzigrass (*Urochloa ruziziensis*) intercropping system and the addition of organic compost on the availability of K for the crops were also studied. The soil was a clayey Oxisol with an initial K content in the 0–20 cm layer of 65 mg dm^{-3}. The treatments consisted of (I) control; (II) KCl (80 kg ha^{-1} K_2O); (III) phonolite (80 kg ha^{-1} K_2O); (IV) phonolite + ruzigrass; (V) phonolite + organic compost (1500 kg ha^{-1}); and (VI) phonolite + ruzigrass + organic compost. Fertilizers were applied before each harvest. After five years of cultivation, leaf K content in the phonolite treatment was equivalent to that in the control treatment. Nutrient levels were inadequate for soybean and corn crops (Fig. 7), revealing severe symptoms of nutritional deficiency (Fig. 1). This study serves as a counterpoint to previously short-term studies (Soratto et al. 2021a; Crusciol et al. 2022), highlighting the need for long-term evaluation to determine the capacity of these rocks to supply potassium after the natural soil reserves are depleted. The corn-ruzigrass intercropping system and the organic compost associated with phonolite significantly improved the nutrition and yield of corn and soybeans. Despite this, it was not possible to establish whether the rock was bioweathered under these cultivation systems. Some studies have reported that ruzigrass mediates high nutrient cycling capacity due to its extensive root system and release of root exudates such as LMWOA, which favors the solubilization of K from non-exchangeable fractions and makes it available to crops (Volf et al. 2018; Rosolem et al. 2019). Organic matter can form carbonic acid and carboxylic groups that acidify the surface of minerals and complex ions, weakening mineral structure (Dontsova et al. 2014; Hoffland et al. 2020; Sharma et al. 2024), However, the magnitude of organic matter-mediated dissolution is unknown in open systems, particularly under field conditions.

Microorganisms play a promising role in accelerating mineral dissolution. A short-term (8-month) greenhouse study investigated changes in the soil bacterial community (SBC) structure in response to different silicate rock powders, including phonolite and three types of basalt and granite, compared with KCl treatment and a control (Reis et al. 2024). The results showed that rock type shaped the bacterial communities. The response was fastest when KCl was applied, but over time, the SBC structure in the KCl treatment approached that of the control. After 8 months, the SBCs were grouped according to the SiO_2 content of the rocks: acid

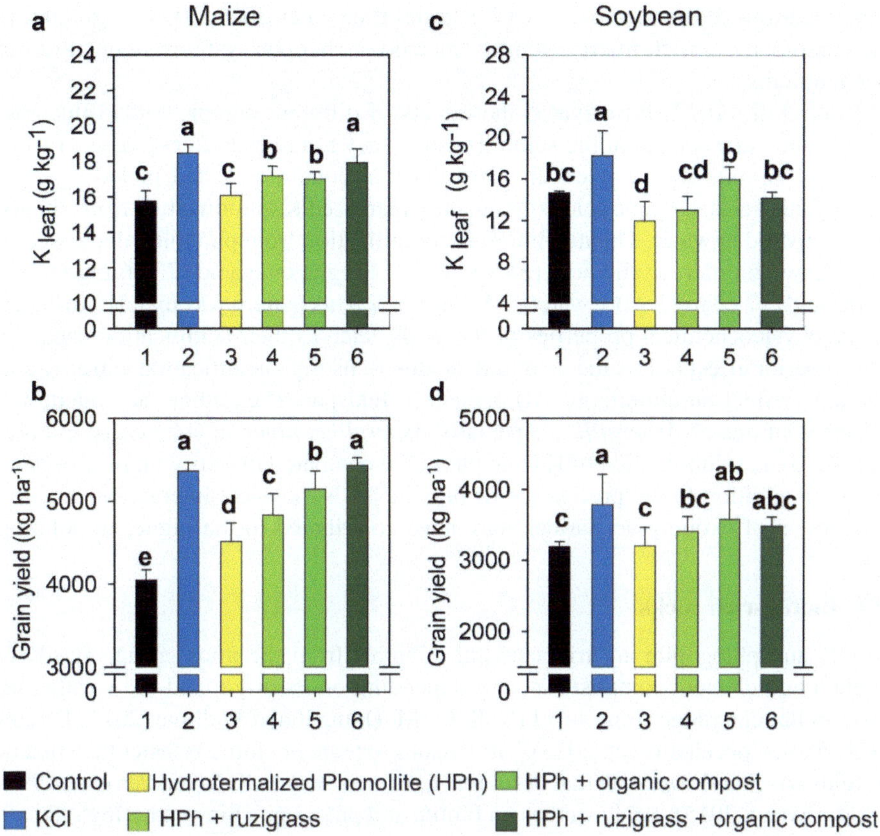

Fig. 7 Potassium (K) content in leaves (**a, c**) and grain yield of soybean and corn (**b, d**) in succession as a function of the application of K sources and production system after 5 years of cultivation: (1) control, (2) KCl, (3) hydrothermalized phonolite (HPh) applied in conventional planting system, (4) phonolite applied in corn–ruzigrass intercropping system, (5) phonolite and organic compost applied in conventional planting system, and (6) phonolite and organic compost applied in corn–ruzigrass intercropping system. Bars with different letters are significantly different according to the LSD test ($p \leq 0.05$) (Crusciol and Oliveira unpublished*)

rock (granite), intermediate rock (phonolite), and basic rocks (basalts). Phonolite tended to group with treatments that shared mineralogical characteristics, such as basalt, which was attributed to the presence of zeolites. Although phonolite application did not cause drastic changes in bacterial diversity, it influenced the activity of some soil enzymes, such as arylsulfatase and acid phosphatase. Studies of changes in the bacterial community in response to chemical and mineralogical composition are useful for isolating and selecting K-solubilizing microorganisms (Sharma et al. 2024). Furthermore, diazotrophic bacteria of the genera *Burkholderia*, *Herbaspirillum*, and *Azospirillum* are considered efficient solubilizers of K and P from phonolite, but in vitro experiments have shown reduced effectiveness of these genera under soil conditions (Florentino et al. 2017; Paredes Filho et al. 2020, 2023), highlighting the complexity of rock-microorganism interactions in open systems. interactions in open systems.

Lodi et al. (2022) investigated the effects of different organic acids (gluconic, oxalic, and citric) on phonolite solubilization. They found that after 40 days of incubation, K solubilization was less than 30% under all conditions tested, including water and organic acids. Organic acids significantly increased K solubilization from phonolite compared to water. The maximum K solubilization from phonolite decreased in the following order: oxalic acid (34.86 mg L^{-1}) > gluconic acid (23.06 mg L^{-1}) > citric acid (17.63 mg L^{-1}) > water (7.51 mg L^{-1}). This pattern was mainly attributed to the physicochemical properties of the acids, such as their acidification capacity. The superior effects of oxalic acid may be due to its high acidification capacity and chelating effect on aluminum (Al) present in feldspar. The authors also identified different strains of *Aspergillus*, especially *Aspergillus niger* in submerged culture, improved the solubilization of K from phonolite compared to tests with pure organic acids. In addition to the production of organic acids, physical interactions and the production of exopolysaccharides may have contributed to the higher dissolution rate.

(b) **Biotite-rich rocks**

Biotite and phlogopite are trioctahedral minerals from the mica group. The K_2O content and element composition vary depending on its origin and the weathering process but can range from 9 to 11% K_2O (McDougall and Wellman 2011; Pereira et al. 2019; Conceição et al. 2023). The dissolution rate of biotite is faster than that of K-feldspars, but slower than that of soluble fertilizers such as KCl (Bakken et al. 1996, 2000; Li et al. 2015). Basic studies on biotite and phlogopite bioweathering indicate that the biological action of the rhizosphere is fundamental to accelerating transformation processes of these micas in hydrobiotite, an interstratification of biotite and vermiculite (Brindley et al. 1983; Hinsinger et al. 1991, 1992, 1993; Krahl et al. 2022).

The application of biotite syenite in clayey Oxisol in the Brazilian Cerrado increased available P and K and net permanent charges in the soil, in addition to increasing K and N uptake and the biomass of corn plants (Santos et al. 2021). The relative efficiency was 128% higher than the control treatment, which contained all nutrients except K, highlighting significant agronomic potential as an alternative fertilizer. Furthermore, Busato et al. (2022) found that the combination of biotite syenite and humic substance enhanced nutrient accumulation in corn tissues, outperforming the effects of isolated biotite syenite.

However, better results were obtained when biotite was applied in combination with nutrient-solubilizing microorganisms. Co-inoculation with *Paenibacillus polymyxa* + *Ralstonia solanacearum* increased K solubility by 177.77% compared to non-inoculated biotite (da Silva et al. 2017). Despite its potential to solubilize nutrients, inoculation of *Ralstonia solanacearum* should be limited to controlled conditions due to its phytopathogenic potential. The application of biotite + *Acidithiobacillus* increased soil K and Mg content due to acidification by the bacteria, which produces H_2SO_4 in the presence of elemental sulfur. Consequently, liming is recommended before inoculation (Stamford et al. 2009). The bacterial genera *Pseudomonas*, *Bacillus*, *Arthrobacter* and *Agrobacterium* and fungal genera *Penicillium* and *Neurospora* have also been described as potential biotite solubilizers (Hopf et al. 2009; Keshavarz Zarjani et al. 2013; Ahmed and Holmström 2015; Sindhu et al. 2016; Saha et al. 2016; Sarikhani et al. 2018; Basak et al. 2022; Sheng et al. 2024).

(c) **Glauconite**

Glauconite rocks are of sedimentary origin, including glauconite siltstone and sandstone, often referred to as greensand. These rocks are rich in K, containing 6–11% K_2O and are composed of glauconite, K-feldspar, illite, kaolinite, and quartz (Dasi et al. 2024). Glauconite is classified as an interlayer-deficient dioctahedral, characterized by $(K + Na) < 0.85$, with $(Mg + Fe) > Al$ and $Fe^{3+} > Al$ in the octahedral sites, essentially placing it within the celadonite subgroup of minerals. Among the micas, glauconite exhibits the highest solubility (Table 1). However, it is dioctahedral, with octahedral sites, along with its interlayer deficiency.

The occupancy range for interlayer cation sites is defined as 0.6–0.85, while in the octahedral layer, (Mg + Fe) exceeds Al. Glauconite may either represent a distinct subgroup similar to the Celadonite subgroup or merely signify varieties within that subgroup (Rieder et al. 1998). The structural composition of glauconite as defined by (Velde 2014):

$K_{(x+y)}(Fe^{3+}, Al, Mg, Fe^{2+})(OH)_{\sim 2}(Si_{4-x}Al_x)O_{10}$ with $Fe^{3+} \gg Al$ and $x = 0.2$–0.6; y (= sum of the dioctahedral cations) = 0.4–0.6 (with $Mg > Fe^{2+}$).

Glauconite is an authigenic mineral formed in shallow, reducing marine environments through reverse weathering processes under warm climatic conditions. During this process, dissolved silicates react with cations released from continental weathering, leading to the formation of authigenic phyllosilicates (Baldermann et al. 2025). The authors indicate that the formation of glauconite has increased over geological time, accounting for 30% of authigenic minerals in the Precambrian, e.g., Jingeryu formation, North China (Zhang et al. 2023a), 33% in the Paleozoic, e.g., Bambuí Group, Brazil, (Lima et al. 2023), 51% in the Mesozoic, e.g., Western Desert, Egypt, (Baioumy et al. 2020), and 63% in the Cenozoic, e.g., Barmer basin, India, (Roy Choudhury et al. 2022).

Despite having a relatively high total K_2O content ranging from 6–11%, glauconite rocks typically contain less than 40% glauconite, particularly in deposits from the Precambrian and Paleozoic eras (Baldermann et al. 2025). In these instances, other potassium minerals, such as K-feldspar and illite, also contribute to the overall potassium content, as they are slow-release sources of potassium. This accounts for the low solubility of potassium extracted with 2% citric acid, as observed in the glauconite rocks of the Bambuí Group in Brazil, which contain less than 1/3 extractable of the total K_2O using this extractant (Schimicoscki et al. 2025). T Similarly, *verdete*—a glauconitic rock from Brazil with less than 40% glauconite (Moreira et al. 2021)—contains around 10% of total K_2O. Given that glauconite accounts for up to 40% of the rock, this translates to a contribution of approximately 2.4% K_2O, based on glauconite containing 6% K_2O. Consequently, this leaves 7.6% K_2O attributed to other potassium minerals, primarily in the form of K-feldspar and illite. In other words, over 75% of the total potassium is present as slow or very slow-release minerals. This characteristic helps to explain the low potassium extractions observed when using low molecular weight organic acids, like 2% citric acid, up to 2.4% of total K (Duarte et al. 2022). up to 2.4% of total K (Duarte et al. 2022).

In a study with corn, *verdete* mixed with dolomitic limestone had a lower dry matter production than the control treatment with KCl. Overall, *verdete* resulted in a dry matter production of about 27% lower compared to phonolite (Pessoa et al. 2023).

7 Technologies for Processing Potassium Silicate Agrominerals

Potassic silicate agrominerals with K_2O contents exceeding 8%, such as phonolites, syenites, and glauconite rocks, are primarily composed of potassium minerals that are more resistant to dissolution, particularly K-feldspars, muscovite, and illite. In these rocks with higher potassium content, minerals with greater dissolution potential, such as feldspathoids including kalsilite, nepheline, and leucite, as well as biotite and phlogopite, typically occur in proportions of less than 30%. Even glauconite rocks typically contain glauconite proportions below 40%, resulting in less than 25% of the total potassium being available for release into the soil and for plant nutrition. Thus, a key approach to studying the agronomic behavior of potassic rocks is to evaluate their reactivity based on the mineralogy and granulometry of the fertilizer. Relying exclusively on the total potassium content for managing these sources is not an effective strategy.

Rocks with a higher proportion of reactive potassium minerals typically contain less than 8% total K_2O. This is illustrated by mica schists and gneisses, which are used as sources for soil remineralizers and exhibit K_2O levels ranging from 3 to 5%. Notably, nearly half or more of the potassium in these rocks is present as biotite, a mineral known for its high potassium availability.

Mineral beneficiation processes can be categorized into three main methods: physical, chemical, and biological. The most crucial physical process is comminution, which entails reducing particle size through crushing and grinding. To enhance the dissolution potential of K-feldspar, the particle size must be reduced to less than 150 μm through grinding.

Mechanochemical activation processes can enhance the chemical reactions between potassium silicate rocks and salt additives, thereby increasing potassium availability (Zhang et al. 2023b). However, despite recent advancements, these mechanochemical methods still incur high production costs and are not yet economically feasible for developing technologies to produce potassium fertilizers from silicate sources (Wang et al. 2023).

Thermal mineral transformation processes can significantly enhance the availability of potassium in silicate rocks. For instance, using $CaCl_2$ flux combined with natural potassium silicates demonstrates that heating after 80 min to 900 °C enables complete dissolution of potassium from muscovite and glauconite, while achieving 70% dissolution from K-feldspar and 60% from sericite (Tanvar and Dhawan 2022). Li et al. (2019) demonstrated that sintering potassium silicate with 33% limestone at 1220 °C results in optimal energy consumption while achieving high solubility. Pessoa et al. (2023) conducted tests on mixtures of limestone and *verdete* in dry thermal processes, demonstrating that potassium availability for corn increases starting at temperatures of 300 °C.

Alkaline hydrothermal processes also have significant potential for producing highly available potassium fertilizers from potassium silicate agrominerals (Ciceri et al. 2017). These thermal processes generate high-solubility and reactive silicates, such as kalsilite, wollastonite, and an amorphous silicate phase, which contains potassium, silica, and calcium, making them available to toints (Ciceri et al. 2019).

Chemical processes for extracting and producing fertilizers from potassium silicate agrominerals are efficient but often face challenges related to the economic viability of high-cost procedures. However, these processes can become feasible when they generate valuable by-products. For example, extracting glauconite rocks with sulfuric acid not only facilitates the production of potassium sulfate fertilizer but also yields by-products such as aluminum, iron, and magnesium sulfate (Schimicoscki et al. 2020).

Potassium extraction from silicate rocks using organic acids shows limited efficiency, especially for more resistant minerals such as K-feldspar and muscovite. While oxalic acid offers greater extraction potential than citric acid, the overall extraction remains below 10% of the total potassium (Duarte et al. 2022). Consequently, this method does not represent a viable approach for the mineral beneficiation of low-solubility potassium silicate minerals.

Biological processes for dissolving potassium silicate rocks show promise, yet a viable route has not been fully developed. For example, the microbiological approach has proven efficient in solubilizing potassium from *verdete*, with *Burkholderia* sp. and *Bacillus* sp. facilitating the extraction of 71.3 and 53.6% of potassium, respectively, in Aleksandrov medium (Schueler et al. 2021). Similarly, Badr et al. (2006) demonstrated that certain bacterial strains can enhance the dissolution of potassium silicates. However, they also noted that the practical application of these strains is limited by competition with existing soil microbial communities. The study suggests that these solubilizing microorganisms are widespread in soils, and effective management practices should be implemented to optimize their performance in facilitating potassium release.

The development of bioinputs, consisting of efficient and multifunctional microbial communities, especially in the form of biofertilizers or organic compost, can significantly enhance the efficiency of potassium release from silicate minerals (Sindhu et al. 2016). Composting processes that incorporate potassium silicate sources not only enhance nutrient availability but also introduce beneficial microbial communities and stable organic matter into agricultural soils (Basak et al. 2017).

8 Conclusions

Silicate rocks have been studied as alternative sources of potassium for agriculture, aiming to provide a more sustainable and economically viable solution in light of the dependence on conventional fertilizers. However, most available studies are short-term, which limits our understanding of the long-term agronomic effects of these sources, especially under real field conditions. The main limitation observed is the low solubilization rate of potassium-bearing minerals, which may restrict K availability during critical stages of crop development. Strategies such as thermal pretreatment, combination with organic matter or solubilizing microorganisms, and diversification of production systems have been investigated to enhance the effectiveness of these sources. Additionally, effective management practices can play a crucial role in enhancing the efficiency of these alternative sources. The adoption of conservation systems, such as no-till farming and crop rotation with species that promote greater root biomass and exudation of organic compounds, may facilitate the gradual release of nutrients. Although some studies have reported agronomic performance comparable to KCl under certain conditions, the results still show variability, strongly influenced by edaphoclimatic and management factors. In addition, the long-term use of these alternative potassium sources may generate distinct environmental outcomes, such as enhanced carbon sequestration through silicate weathering and the potential accumulation of functional stable minerals, which should be weighed when recommending their adoption. Therefore, recommendation to use these material solely as a potassium source should be made with caution and supported by robust technical evidence, preferably derived from long-term studies.

Acknowledgements We are grateful for support from the Center for Carbon Research in Tropical Agriculture (CCARBON) at the University of São Paulo (USP), which is sponsored by the São Paulo Research Foundation (FAPESP) under grant 2021/10573–4. The first author would like to thank the Nationalb Council for Scientific and Technological Development (CNPq) for an award for excellence in research.

Author's Contribution Carlos Alexandre Costa Crusciol conceived and supervised the study, contributed to the interpretation of the agronomic findings, and critically reviewed the manuscript.

Sirlene Lopes Oliveira was the main contributor to the experimental work and data interpretation, including field trials derived from her doctoral research, and made significant contributions to the drafting of the manuscript.

José Roberto Portugal contributed to the literature review, assisted in data synthesis, and helped with editing and formatting the manuscript.

Éder de Souza Martins provided specialized knowledge in soil mineralogy and weathering processes, contributed to the discussion on silicate dissolution mechanisms, potential technological processing of potassium silicates, and reviewed the manuscript for scientific and intellectual content.

Declaration of Competing Interest The authors declare that they have no known competing financial interests or personal relationships that could have appeared to influence the work reported in this paper.

References

Ahmed E, Holmström SJM (2015) Microbe–mineral interactions: the impact of surface attachment on mineral weathering and element selectivity by microorganisms. Chem Geol 403:13–23. https://doi.org/10.1016/j.chemgeo.2015.03.009

Alekseyev VA, Medvedeva LS, Prisyagina NI et al (1997) Change in the dissolution rates of alkali feldspars as a result of secondary mineral precipitation and approach to equilibrium. Geochim Cosmochim Acta 61:1125–1142. https://doi.org/10.1016/S0016-7037(96)00405-X

Ambasht PK, Kayastha AM (2002) Plant Pyruvate Kinase. Biol Plant 45:1–10. https://doi.org/10.1023/A:1015173724712

Aracena A, Sanino A, Jerez O (2018) Dissolution kinetics of molybdite in KOH media at different temperatures. Trans Nonferrous Metals Soc China 28:177–185. https://doi.org/10.1016/S1003-6326(18)64651-5

Ayinla IK, Baba AA, Raji MA et al (2018) Quantitative assessment of potassium from k-feldspar ore by roast-leaching process. J Chem Soc Niger 43(4)

Badr MA, Shafei AM, Sharaf El-Deen SH (2006) The dissolution of K and P-bearing minerals by silicate dissolving bacteria and their effect on sorghum growth

Bagheri M, Lothenbach B, Shakoorioskooie M, Scrivener K (2022) Effect of different ions on dissolution rates of silica and feldspars at high pH. Cem Concr Res 152:106644. https://doi.org/10.1016/J.CEMCONRES.2021.106644

Baioumy H, Farouk S, Al-Kahtany K (2020) Paleogeographic, paleoclimatic and sea-level implications of glauconite deposits in Egypt: a review. J Afr Earth Sc 171:103944. https://doi.org/10.1016/j.jafrearsci.2020.103944

Bakken AK, Gautneb H, Myhr K (1996) The potential of crushed rocks and mine tailings as slow-releasing K fertilizers assessed by intensive cropping with Italian ryegrass in different soil types. Nutr Cycl Agroecosyst 47:41–48. https://doi.org/10.1007/BF01985717

Bakken AK, Gautneb H, Sveistrup T, Myhr K (2000) Crushed rocks and mine tailings applied as K fertilizers on grassland. Nutr Cycl Agroecosyst 56:53–57. https://doi.org/10.1023/A:1009709914578

Baldermann A, Banerjee S, Löhr SC et al (2025) Exploring reverse silicate weathering across geological time: a review. Clay Miner 60(1):1–27. https://doi.org/10.1180/clm.2024.27

Barbosa MC, Fernandes GC, Lima BH et al (2025) The effects of potassium dose, timing, and source in Soybean Crops in Brazilian Savannah Oxisol. Sustainability 17:934. https://doi.org/10.3390/su17030934

Basak BB, Maity A, Ray P et al (2022) Potassium supply in agriculture through biological potassium fertilizer: a promising and sustainable option for developing countries. Arch Agron Soil Sci 68:101–114. https://doi.org/10.1080/03650340.2020.1821191

Basak BB, Sarkar B, Biswas DR et al (2017) Bio-intervention of naturally occurring silicate minerals for alternative source of potassium, pp 115–145

Bea F, Montero P, Haissen F et al (2014) Kalsilite-bearing plutonic rocks: the deep-seated Archean Awsard massif of the Reguibat Rise, South Morocco, West African Craton. Earth Sci Rev 138:1–24. https://doi.org/10.1016/j.earscirev.2014.08.003

Berner RA, Holdren GR (1979) Mechanism of feldspar weathering—II. Observations of feldspars from soils. Geochim Cosmochim Acta 43:1173–1186. https://doi.org/10.1016/0016-7037(79)90110-8

Brasil EPF, Leandro WM, Teixeira WG et al (2020) Chemical extractors to assess potassium availability in glauconitic siltstone. J Agric Sci 12:166. https://doi.org/10.5539/jas.v12n9p166

Brindley GW, Zalba PE, Bethke CM (1983) Hydrobiotite, a regular 1: 1 interstratification of biotite and vermiculite layers. Am Miner 68:420–425

Brouder S (2015) Potassium cycling. Soil management: building a stable base for agriculture. Soil Science Society of America, Madison, WI, USA, pp 79–102

Busato JG, dos Santos LF, Monteiro de Paula A et al (2022) Can co-application of silicate rock powder and humic-like acids increase nutrient uptake and plant growth in weathered tropical

soil? Acta Agric Scand B Soil Plant Sci 72:761–774. https://doi.org/10.1080/09064710.2022.2078222
Cagnon B, Daval D, Cabié M et al (2022) A comparative study of the dissolution mechanisms of amorphous and crystalline feldspars at acidic pH conditions. Npj Mater Degrad 6:34. https://doi.org/10.1038/s41529-022-00240-6
Candido JSS, Silveira Braga FC, Melo LV (2024) Ultrapotassic Alkaline Magmatism in the Borborema Province: the Brejinho Alkaline Pluton. Pi/Ce, Brazil
Chou L, Wollast R (1984) Study of the weathering of albite at room temperature and pressure with a fluidized bed reactor. Geochim Cosmochim Acta 48:2205–2217. https://doi.org/10.1016/0016-7037(84)90217-5
Ciceri D, Close TC, Barker AV, Allanore A (2019) Fertilizing properties of potassium feldspar altered hydrothermally. Commun Soil Sci Plant Anal 50:482–491. https://doi.org/10.1080/00103624.2019.1566922
Ciceri D, de Oliveira M, Stokes RM et al (2017) Characterization of potassium agrominerals: correlations between petrographic features, comminution and leaching of ultrapotassic syenites. Miner Eng 102:42–57. https://doi.org/10.1016/J.MINENG.2016.11.016
Conceição H, Sotero A, Angélica R, Paz S (2023) Biotita-xisto da suíte intrusiva tracuateua do nordeste do pará e seu potencial agromineral. Boletim do Museu de Geociências da Amazônia 10:1–18. https://doi.org/10.31419/ISSN.2594-942X.v102023i1agromineraisa6HMC
Correns CW (1940) Die chemische Verwitterung der Silikate. Naturwissenschaften 28:369–376
Correns CW, von Engelhardt W (1938) Neue Untersuchungen über die Verwitterung des Kalifeldspates. Naturwissenschaften 26:137–138
Crusciol CAC, Soratto RP, Gilabel AP et al (2022) Broadcast application of ground silicate rocks as potassium sources for grain crops. Pesqui Agropecu Bras 57:02443. https://doi.org/10.1590/s1678-3921.pab2022.v57.02443
Cui J, Pottosin I, Lamade E, Tcherkez G (2020) What is the role of putrescine accumulated under potassium deficiency? Plant Cell Environ 43:1331–1347. https://doi.org/10.1111/pce.13740
Cui J, Tcherkez G (2021) Potassium dependency of enzymes in plant primary metabolism. Plant Physiol Biochem 166:522–530. https://doi.org/10.1016/j.plaphy.2021.06.017
Cunha GODM, Almeida JAD (2021) Agronomic potential of four rock powders, pure or mixed, as soil remineralizers. Res Soc Dev 10:e169101724828. https://doi.org/10.33448/rsd-v10i17.24828
da Cruz CAC, Orioli Júnior V, Bernardes JVS et al (2021) Efficiency of phonolite as a potassium source for Paiaguás palisadegrass. Pesqui Agropecu Trop 51:67857. https://doi.org/10.1590/1983-40632021v5167857
da Rocha Neto AR, de Mello PR, da Cruz MCP (2024) The impact ground phonolite rock's potassium solubilization in tropical soil depends on the cultivated forage species. Plants 13:199. https://doi.org/10.3390/plants13020199
da Silva VN, de Souza Fernandes da Silva LE, da Silva AJN et al (2017) Solubility curve of rock powder inoculated with microorganisms in the production of biofertilizers. Agric Nat Resour 51:142–147. https://doi.org/10.1016/J.ANRES.2017.01.001
Dasi E, Rudmin M, Banerjee S (2024) Glauconite applications in agriculture: a review of recent advances. Appl Clay Sci 253:107368. https://doi.org/10.1016/j.clay.2024.107368
de Bang TC, Husted S, Laursen KH et al (2021) The molecular–physiological functions of mineral macronutrients and their consequences for deficiency symptoms in plants. New Phytol 229:2446–2469. https://doi.org/10.1111/NPH.17074
de Lima MF, Caxito FA, Large R et al (2023) Trace elements in sedimentary pyrite track redox and nutrient fluctuations in the Ediacaran/Cambrian Bambuí Group, Brazil. Chem Geol 635:121625. https://doi.org/10.1016/j.chemgeo.2023.121625
De Oliveira MW, Bhatt R, Nikpay A et al (2023) Reactivity of two acidity correctives in a medium texture soil. Revista Caribeña de Ciencias Sociales 12:264–276. https://doi.org/10.55905/rcssv12n1-015

de Sá ME, Rowell DL, Guarçoni Martins A, Paiva da Silva A (2006) Effect of pH on the development of acidic sites in clayey and sandy loam Oxisol from the Cerrado Region, Brazil. Geoderma 132:131–142. https://doi.org/10.1016/j.geoderma.2005.05.001

Demir C, Bentli I, Gülgönül I, Çelik MS (2003) Effects of bivalent salts on the flotation separation of Na-feldspar from K-feldspar. Miner Eng 16:551–554. https://doi.org/10.1016/S0892-6875(03)00078-5

Dontsova K, Zaharescu D, Henderson W et al (2014) Impact of organic carbon on weathering and chemical denudation of granular basalt. Geochim Cosmochim Acta 139:508–526. https://doi.org/10.1016/j.gca.2014.05.010

dos Santos LF, Sodré FF, Martins ÉS et al (2021) Effects of biotite syenite on the nutrient levels and electrical charges in a Brazilian Savanna Ferralsol. Pesqui Agropecu Trop 51:e66691. https://doi.org/10.1590/1983-40632021v5166691

Duarte LM, Xavier LV, Rossati KF et al (2022) Potassium extraction from the silicate rock Verdete using organic acids. Sci Agric 79:e20200164. https://doi.org/10.1590/1678-992x-2020-0164

Etesami H, Emami S, Alikhani HA (2017) Potassium solubilizing bacteria (KSB): mechanisms, promotion of plant growth, and future prospects—a review. J Soil Sci Plant Nutr 17:897–911. https://doi.org/10.4067/S0718-95162017000400005

Florentino LA, Rezende AV, Miranda CCB et al (2017) Potassium solubilization in phonolite rock by diazotrophic bacteria. Comunicata Scientiae 8:17. https://doi.org/10.14295/cs.v8i1.1292

Fogler HS, Lund K, McCune CC (1975) Acidization III—the kinetics of the dissolution of sodium and potassium feldspar in HF/HCl acid mixtures. Chem Eng Sci 30:1325–1332. https://doi.org/10.1016/0009-2509(75)85061-5

Gardner LR (1983) Mechanics and kinetics of incongruent feldspar dissolution. Geology 11:418. https://doi.org/10.1130/0091-7613(1983)11%3c418:MAKOIF%3e2.0.CO;2

Gautier JM, Oelkers EH, Schott J (1994) Experimental study of K-feldspar dissolution rates as a function of chemical affinity at 150°C and pH 9. Geochim Cosmochim Acta 58:4549–4560. https://doi.org/10.1016/0016-7037(94)90190-2

Gerardeaux E, Jordan-Meille L, Constantin J et al (2010) Changes in plant morphology and dry matter partitioning caused by potassium deficiency in Gossypium hirsutum (L.). Environ Exp Bot 67:451–459. https://doi.org/10.1016/J.ENVEXPBOT.2009.09.008

Gupta K, Dey A, Gupta B (2013) Plant polyamines in abiotic stress responses. Acta Physiol Plant 35:2015–2036. https://doi.org/10.1007/s11738-013-1239-4

Hafsi C, Debez A, Abdelly C (2014) Potassium deficiency in plants: effects and signaling cascades. Acta Physiol Plant 36:1055–1070. https://doi.org/10.1007/s11738-014-1491-2

Hasanuzzaman M, Bhuyan M, Nahar K et al (2018) Potassium: a vital regulator of plant responses and tolerance to abiotic stresses. Agronomy 8:31. https://doi.org/10.3390/agronomy8030031

Helgeson HC (1971) Kinetics of mass transfer among silicates and aqueous solutions. Geochim Cosmochim Acta 35:421–469. https://doi.org/10.1016/0016-7037(71)90043-3

Helgeson HC (1972) Kinetics of mass transfer among silicates and aqueous solutions: correction and clarification. Geochim Cosmochim Acta 36:1067–1070. https://doi.org/10.1016/0016-7037(72)90022-1

Hellmann R (1994) The albite-water system: part I. The kinetics of dissolution as a function of pH at 100, 200 and 300°C. Geochim Cosmochim Acta 58:595–611. https://doi.org/10.1016/0016-7037(94)90491-X

Hellmann R, Penisson J-M, Hervig RL et al (2003) An EFTEM/HRTEM high-resolution study of the near surface of labradorite feldspar altered at acid pH: evidence for interfacial dissolution-reprecipitation. Phys Chem Miner 30:192–197. https://doi.org/10.1007/s00269-003-0308-4

Hensel J (1894) Bread from stones: a new rational system of land fertilization and physical regeneration. Philadelphia, PA, A.J. Tafel, translated from the German, 140 p. https://tile.loc.gov/storage-services/public/gdcmassbookdig/breadfromstones00hens/breadfromstones00hens.pdf

Hinsinger P, Elsass F, Jaillard B, Robert M (1993) Root-induced irreversible transformation of a trioctahedral mica in the rhizosphere of rape. J Soil Sci 44:535–545. https://doi.org/10.1111/j.1365-2389.1993.tb00475.x

Hinsinger PH, Jaillard B, Dufey JE (1992) Rapid weathering of a trioctahedral mica by the roots of ryegrass. Soil Sci Soc Am J 56:977–982. https://doi.org/10.2136/sssaj1992.03615995005600030049x

Hinsinger PH, Dufey JE, Jaillard B (1991) Biological weathering of micas in the rhizosphere as related to potassium absorption by plant roots. Dev Agric Manag for Ecol 24:98–105. https://doi.org/10.1016/B978-0-444-89104-4.50018-9

Hoffland E, Kuyper TW, Comans RNJ, Creamer RE (2020) Eco-functionality of organic matter in soils. Plant Soil 455:1–22. https://doi.org/10.1007/s11104-020-04651-9

Hopf J, Langenhorst F, Pollok K et al (2009) Influence of microorganisms on biotite dissolution: an experimental approach. Geochemistry 69:45–56. https://doi.org/10.1016/j.chemer.2008.11.001

Hu W, Loka DA, Fitzsimons TR et al (2018) Potassium deficiency limits reproductive success by altering carbohydrate and protein balances in cotton (Gossypium hirsutum L.). Environ Exp Bot 145:87–94. https://doi.org/10.1016/j.envexpbot.2017.10.024

Imtiaz H, Mir AR, Corpas FJ, Hayat S (2023) Impact of potassium starvation on the uptake, transportation, photosynthesis, and abiotic stress tolerance. Plant Growth Regul 99:429–448. https://doi.org/10.1007/s10725-022-00925-7

Ishikawa J, Okamura M, Arai-Sanoh Y et al (2023) Close relationship between potassium accumulation in internodes and translocation of non-structural carbohydrate to grains in rice cultivars with different yields. Field Crops Res 300:109007. https://doi.org/10.1016/J.FCR.2023.109007

Jeschke AA, Dreybrodt W (2002) Dissolution rates of minerals and their relation to surface morphology. Geochim Cosmochim Acta 66:3055–3062. https://doi.org/10.1016/S0016-7037(02)00893-1

Kalinowski BE, Schweda P (1996) Kinetics of muscovite, phlogopite, and biotite dissolution and alteration at pH 1–4, room temperature. Geochim Cosmochim Acta 60:367–385. https://doi.org/10.1016/0016-7037(95)00411-4

Karagöz Ö, Kuşlu S (2017) Dissolution kinetics of colemanite in potassium dihydrogen phosphate solution (KH_2PO_4). Int J Hydrogen Energy 42:23250–23259. https://doi.org/10.1016/j.ijhydene.2017.04.051

Keshavarz Zarjani J, Aliasgharzad N, Oustan S et al (2013) Isolation and characterization of potassium solubilizing bacteria in some Iranian soils. Arch Agron Soil Sci 59:1713–1723. https://doi.org/10.1080/03650340.2012.756977

Kicińska A, Pomykała R (2023) Incongruent dissolution of silicates and its impact on the environment: an example of a talc mine. Sci Rep 13:22519. https://doi.org/10.1038/s41598-023-50143-y

Kirtzel J, Siegel D, Krause K, Kothe E (2017) Stone-Eating Fungi: mechanisms in bioweathering and the potential role of laccases in black slate degradation with the basidiomycete Schizophyllum commune. Adv Appl Microbiol 99:83–101. https://doi.org/10.1016/bs.aambs.2017.01.002

Krahl LL, Marchi G, Paz SPA et al (2022) Increase in cation exchange capacity by the action of maize rhizosphere on Mg or Fe biotite-rich rocks. Pesqui Agropecu Trop 52:e72376. https://doi.org/10.1590/1983-40632022v5272376

Lagache M (1976) New data on the kinetics of the dissolution of alkali feldspars at 200°C in CO_2 charged water. Geochim Cosmochim Acta 40:157–161. https://doi.org/10.1016/0016-7037(76)90173-3

Lazo DE, Dyer LG, Alorro RD (2017) Silicate, phosphate and carbonate mineral dissolution behaviour in the presence of organic acids: a review. Miner Eng 100:115–123. https://doi.org/10.1016/J.MINENG.2016.10.013

Li G, Cao Y, Liu S et al (2025) Efficient leaching of potassium from K-feldspar in KOH-$Ca(OH)_2$ solution via one-step hydrothermal process. Sep Purif Technol 360:130999. https://doi.org/10.1016/j.seppur.2024.130999

Li T, Wang H, Wang J et al (2015) Exploring the potential of phyllosilicate minerals as potassium fertilizers using sodium tetraphenylboron and intensive cropping with perennial ryegrass. Sci Rep 5:9249. https://doi.org/10.1038/srep09249

Li X, Yu W, Zhao B (2019) Effect of sintering temperature of potassium feldspar-limestone/dolomite on composition and microstructure of silicate fertilizers. Int J Ceram Eng Sci 1:185–193. https://doi.org/10.1002/ces2.10022

Lodi LA, Klaic R, Bortoletto-Santos R et al (2022) Unveiling the solubilization of potassium mineral rocks in organic acids for application as K-fertilizer. Appl Biochem Biotechnol 194:2431–2447. https://doi.org/10.1007/s12010-022-03826-7

Lombardi KC, Mangrich AS, Wypych F et al (2006) Sequestered carbon on clay mineral probed by electron paramagnetic resonance and X-ray photoelectron spectroscopy. J Colloid Interface Sci 295:135–140. https://doi.org/10.1016/j.jcis.2005.08.015

Lopes-Assad ML, Avansini SH, Rosa MM et al (2010) The solubilization of potassium-bearing rock powder by Aspergillus niger in small-scale batch fermentations. Can J Microbiol 56:598–605. https://doi.org/10.1139/W10-044

Luchese AV, de Castro G, Leite IJ, da Silva Giaretta AP et al (2023) Use of quarry waste basalt rock powder as a soil remineralizer to grow soybean and maize. Heliyon 9:e14050. https://doi.org/10.1016/j.heliyon.2023.e14050

Malmström M, Banwart S (1997) Biotite dissolution at 25°C: the pH dependence of dissolution rate and stoichiometry. Geochim Cosmochim Acta 61:2779–2799. https://doi.org/10.1016/S0016-7037(97)00093-8

Mancuso MAC, Soratto RP, Crusciol CAC, Castro GSA (2014) Effect of potassium sources and rates on arabica coffee yield, nutrition, and macronutrient export. Rev Bras Cienc Solo 38:1448–1456. https://doi.org/10.1590/S0100-06832014000500010

Manning DAC, Theodoro SH (2020) Enabling food security through use of local rocks and minerals. Extr Ind Soc 7:480–487. https://doi.org/10.1016/J.EXIS.2018.11.002

Marinho EB, de Oliveira AL, Zandonadi DB et al (2014) Organic matter pools and nutrient cycling in different coffee production systems in the Brazilian Cerrado. Agrofor Syst 88:767–778. https://doi.org/10.1007/s10457-014-9723-4

McDougall I, Wellman P (2011) Calibration of GA1550 biotite standard for K/Ar and 40Ar/39Ar dating. Chem Geol 280:19–25. https://doi.org/10.1016/j.chemgeo.2010.10.001

Moreira A, Moraes LAC, Silva SR, Delfim J (2023) Silicate rock with different granulometry as a potassium source for alfalfa and centrosema crops. Pesqui Agropecu Bras 58:02806. https://doi.org/10.1590/s1678-3921.pab2023.v58.02806

Moreira DS, Uhlein A, Uhlein GJ et al (2021) Ediacaran/Early Cambrian Serra da Saudade Formation, Bambuí Group: the sedimentary record of a foreland basin in Southeastern Brazil. Brazilian Journal of Geology 51:e20210029. https://doi.org/10.1590/2317-4889202120210029

Navarre-Sitchler A, Brantley SL, Rother G (2015) How porosity increases during incipient weathering of crystalline silicate rocks. Rev Mineral Geochem 80:331–354. https://doi.org/10.2138/rmg.2015.80.10

Nogueira BKA, Silva LS, Gasques LR et al (2024) Spatial variability of potassium and agricultural productivity in sandy loam soil with rock dust under functional diversity in the Brazilian Cerrado. J Soil Sci Plant Nutr 24:3441–3458. https://doi.org/10.1007/s42729-024-01766-1

Nogueira TAR, Miranda BG, Jalal A et al (2021) Nepheline syenite and phonolite as alternative potassium sources for maize. Agronomy 11:1385. https://doi.org/10.3390/agronomy11071385

Nugent MA, Brantley SL, Pantano CG, Maurice PA (1998) The influence of natural mineral coatings on feldspar weathering. Nature 395:588–591. https://doi.org/10.1038/26951

Oliveira ÍL, Brod JA, Junqueira-Brod TC et al (2022) The IUGS nomenclature on kalsilite-bearing volcanic rocks: a critical appraisal and recommendations. J Petrol 63:egac026. https://doi.org/10.1093/petrology/egac026

Palandri JL, Kharaka YK (2004) A compilation of rate parameters of water-mineral interaction kinetics for application to geochemical modeling. US Geological Survey

Panina LI, Rokosova EY, Ryabukha MA (2022) Ultrapotassic aluminosilicate melts: specifics of formation by the example of synnyrites from the Synnyr massif. Russ Geol Geophys 63:153–164. https://doi.org/10.2113/RGG20204251

Parecido RJ, Soratto RP, Fernandes AM et al (2024) Castor meal and ground hydrothermalized phonolite optimize sweet potato nutrition, yield, and quality. Horticulturae 10:775. https://doi.org/10.3390/horticulturae10080775

Paredes Filho MV, Mesquita AC, Mantovani JR et al (2020) Bacterial strains decrease solubility of potassium in the soil. Biosci J 36:1577–1582. https://doi.org/10.14393/BJ-v36n5a2020-42326

Paredes Filho MV, da Silva AB, Florentino LA (2023) Phonolite associated with organic compound and potassium solubilizing bacteria in tomato cultivation. Revista Ceres 70:133–141. https://doi.org/10.1590/0034-737x202370010015

Pereira AC, Gomes MRDS (2021) Biotita e flogopita como fonte alternativa de potássio para fertilizante: extração hidrometalúrgica. Geosci Geociências 40:781–786. https://doi.org/10.5016/geociencias.v40i03.13764

Pereira AC, dos Santos Gomes MR, Rocha SDF (2019) Rocha contendo biotita como fonte alternativa de potássio para fertilizante após processamento térmico com aditivos. Holos Environ 19:502. https://doi.org/10.14295/holos.v19i4.12323

Pessoa RS, Zonta E, Pozza AAA et al (2023) Agricultural use of phonolite and verdete rocks as a source of k for corn crop. Revista Ceres 70:e70510. https://doi.org/10.1590/0034-737x202370050010

Pollet-Villard M, Daval D, Ackerer P et al (2016) Does crystallographic anisotropy prevent the conventional treatment of aqueous mineral reactivity? A case study based on K-feldspar dissolution kinetics. Geochim Cosmochim Acta 190:294–308. https://doi.org/10.1016/j.gca.2016.07.007

Qu Q, Zhang Z, Peijnenburg WJGM et al (2020) Rhizosphere microbiome assembly and its impact on plant growth. J Agric Food Chem 68:5024–5038. https://doi.org/10.1021/acs.jafc.0c00073

Radoslovich EW (1975) Feldspar minerals. Soil components. Springer, Berlin and Heidelberg, pp 433–448

Ragel P, Raddatz N, Leidi EO et al (2019) Regulation of K^+ nutrition in plants. Front Plant Sci 10:281. https://doi.org/10.3389/fpls.2019.00281

Reis BR, Vasconcelos ALS, Silva AMM et al (2024) Changes in soil bacterial community structure in a short-term trial with different silicate rock powders. Chemical and Biological Technologies in Agriculture 11:61. https://doi.org/10.1186/s40538-024-00586-w

Rieder M, Cavazzini G, D'yakonov YS et al (1998) Nomenclature of the Micas. Clays Clay Miner 46:586–595. https://doi.org/10.1346/CCMN.1998.0460513

Rinder T, von Hagke C (2021) The influence of particle size on the potential of enhanced basalt weathering for carbon dioxide removal—insights from a regional assessment. J Clean Prod 315:128178. https://doi.org/10.1016/J.JCLEPRO.2021.128178

Rodrigues M, Bortolini PC, Neto CK et al (2024) Unlocking higher yields in Urochloa brizantha: the role of basalt powder in enhancing soil nutrient availability. Discover Soil 1:4. https://doi.org/10.1007/s44378-024-00006-3

Rodríguez-Navarro A (2000) Potassium transport in fungi and plants. Biochim Biophys Acta (BBA) Rev Biomembr 1469:1–30. https://doi.org/10.1016/S0304-4157(99)00013-1

Rosolem CA, Neto LO, Costa VE, da Silva GC (2019) Ruzigrass root persistence and soybean root growth. Plant Soil 442:333–341. https://doi.org/10.1007/s11104-019-04198-4

Rosolem CA, Steiner F (2017) Effects of soil texture and rates of K input on potassium balance in tropical soil. Eur J Soil Sci 68:658–666. https://doi.org/10.1111/ejss.12460

Roy Choudhury T, Khanolkar S, Banerjee S (2022) Glauconite authigenesis during the warm climatic events of Paleogene: case studies from shallow marine sections of Western India. Glob Planet Change 214:103857. https://doi.org/10.1016/j.gloplacha.2022.103857

Saha M, Maurya BR, Meena VS et al (2016) Identification and characterization of potassium solubilizing bacteria (KSB) from Indo-Gangetic Plains of India. Biocatal Agric Biotechnol 7:202–209. https://doi.org/10.1016/j.bcab.2016.06.007

Santos WO, Mattiello EM, Vergutz L, Costa RF (2016) Production and evaluation of potassium fertilizers from silicate rock. J Plant Nutr Soil Sci 179:547–556. https://doi.org/10.1002/jpln.201500484

Sarikhani MR, Oustan S, Ebrahimi M, Aliasgharzad N (2018) Isolation and identification of potassium-releasing bacteria in soil and assessment of their ability to release potassium for plants. Eur J Soil Sci 69:1078–1086. https://doi.org/10.1111/ejss.12708

Schimicoscki RS, Oliveira KD, Ávila-Neto CN (2020) Potassium recovery from a Brazilian glauconitic siltstone via reaction with sulfuric acid in hydrothermal conditions. Hydrometallurgy 191:105251. https://doi.org/10.1016/j.hydromet.2020.105251

Schimicoscki RS, Oliveira KD, Mendes GDO, Ávila Neto CND (2025) Production of potassium fertilizer from Verdete: a review. J Eng Exact Sci 11:21569. https://doi.org/10.18540/jcecvl11iss1pp21569

Schueler TA, Dourado ML, Videira SS, da Cunha CD, Rizzo AC (2021) Biosolubilization of verdete: an alternative potassium source for agriculture fertilizer. Biocatal Agric Biotechnol 34:102031. https://doi.org/10.1016/j.bcab.2021.102031

Sharma R, Sindhu SS, Glick BR (2024) Potassium solubilizing microorganisms as potential biofertilizer: a sustainable climate-resilient approach to improve soil fertility and crop production in agriculture. J Plant Growth Regul 43:2503–2535. https://doi.org/10.1007/s00344-024-11297-9

Sheng Q, Gao S, Dong W et al (2024) The quorum-sensing TraI/TraR system contributes to the release of Fe and Al from biotite by agrobacterium pusense SX41. Chem Geol 670:122419. https://doi.org/10.1016/j.chemgeo.2024.122419

Silva DRG, Spehar CR, Marchi G et al (2014) Yield, nutrient uptake and potassium use efficiency in rice fertilized with crushed rocks. Afr J Agric Res 9:455–464. https://doi.org/10.5897/AJAR2013.7638

Simonsson M, Hillier S, Öborn I (2009) Changes in clay minerals and potassium fixation capacity as a result of release and fixation of potassium in long-term field experiments. Geoderma 151:109–120. https://doi.org/10.1016/J.GEODERMA.2009.03.018

Sindhu SS, Parmar P, Phour M, Sehrawat A (2016) Potassium-solubilizing microorganisms (KSMs) and its effect on plant growth improvement. In: Potassium solubilizing microorganisms for sustainable agriculture. Springer India, New Delhi, pp 171–185

Soratto RP, Crusciol CAC, de Campos M et al (2021a) Efficiency and residual effect of alternative potassium sources in grain crops. Pesqui Agropecu Bras 56:e02686. https://doi.org/10.1590/s1678-3921.pab2021.v56.02686

Soratto RP, Crusciol CAC, de Campos M et al (2021b) Silicate rocks as an alternative potassium fertilizer for upland rice and common bean crops. Pesqui Agropecu Bras 56:e01411. https://doi.org/10.1590/s1678-3921.pab2021.v56.01411

Sparks DL (2000) 1.4 bioavailability of soil potassium. In Handbook of soil. CRC Press, New York

Stamford NP, Moura PM, Lira Júnior MA et al (2009) Chemical attributes of an Argisoil of the Vale do São Francisco after melon growth with phosphate and potash rocks biofertilizers. Hortic Bras 27:447–452. https://doi.org/10.1590/S0102-05362009000400008

Stillings LL, Brantley SL (1995) Feldspar dissolution at 25°C and pH 3: reaction stoichiometry and the effect of cations. Geochim Cosmochim Acta 59:1483–1496. https://doi.org/10.1016/0016-7037(95)00057-7

Storey KB (1976) Purification and properties of adductor muscle phosphofructokinase from the oyster, Crassostrea virginica. Eur J Biochem 70:331–337. https://doi.org/10.1111/j.1432-1033.1976.tb11022.x

Sung J, Lee S, Lee Y et al (2015) Metabolomic profiling from leaves and roots of tomato (Solanum lycopersicum L.) plants grown under nitrogen, phosphorus or potassium-deficient condition. Plant Sci 241:55–64. https://doi.org/10.1016/j.plantsci.2015.09.027

Swoboda P, Döring TF, Hamer M (2022) Remineralizing soils? The agricultural usage of silicate rock powders: a review. Sci Total Environ 807:150976. https://doi.org/10.1016/j.scitotenv.2021.150976

Tanvar H, Dhawan N (2022) Kinetic and thermodynamic study of potassium recovery from silicate rocks. Miner Process Extr Metall 131:1–13. https://doi.org/10.1080/25726641.2019.1699360

Tavares LDF, de Carvalho AMX, Camargo LGB et al (2018) Nutrients release from powder phonolite mediated by bioweathering actions. Int J Recycl Org Waste Agric 7:89–98. https://doi.org/10.1007/s40093-018-0194-x
Tischendorf G, Förster H-J, Gottesmann B (2001) Minor- and trace-element composition of trioctahedral micas: a review. Mineral Mag 65:249–276. https://doi.org/10.1180/002646101550244
Toledo MZ, Castro GSA, Crusciol CAC et al (2011) Physiological quality of soybean and wheat seeds produced with alternative potassium sources. Revista Brasileira De Sementes 33:363–371. https://doi.org/10.1590/S0101-31222011000200019
Torqueti STDS, Boldrin KVF, Nascimento ÂMP et al (2016) Alternative potassium source for the cultivation of ornamental sunflower. Ciência e Agrotecnologia 40:257–264. https://doi.org/10.1590/1413-70542016403036115
Ur Rahman S, Han J-C, Ahmad M et al (2024) Aluminum phytotoxicity in acidic environments: a comprehensive review of plant tolerance and adaptation strategies. Ecotoxicol Environ Saf 269:115791. https://doi.org/10.1016/j.ecoenv.2023.115791
Uroz S, Kelly LC, Turpault MP et al (2015) The mineralosphere concept: mineralogical control of the distribution and function of mineral-associated bacterial communities. Trends Microbiol 23:751–762. https://doi.org/10.1016/J.TIM.2015.10.004
Velde B (2014) Green clay minerals. In: Treatise on geochemistry. Elsevier, pp 351–364
Volf MR, Guimarães TM, Scudeletti D et al (2018) Potassium dynamics in ruzigrass rhizosphere. Rev Bras Cienc Solo 42:201703070. https://doi.org/10.1590/18069657rbcs20170370
Walker DJ, Leigh RA, Miller AJ (1996) Potassium homeostasis in vacuolate plant cells. Proc Natl Acad Sci 93:10510–10514. https://doi.org/10.1073/pnas.93.19.10510
Wang Q, Shan C, Zhang P et al (2023) The combination of nanotechnology and potassium: applications in agriculture. Environ Sci Pollut Res 31:1890–1906. https://doi.org/10.1007/s11356-023-31207-y
Welch SA, Ullman WJ (1996) Feldspar dissolution in acidic and organic solutions: compositional and pH dependence of dissolution rate. Geochim Cosmochim Acta 60:2939–2948. https://doi.org/10.1016/0016-7037(96)00134-2
Wollast R (1967) Kinetics of the alteration of K-feldspar in buffered solutions at low temperature. Geochim Cosmochim Acta 31:635–648. https://doi.org/10.1016/0016-7037(67)90040-3
Xiao L, Sun Q, Yuan H et al (2016) A feasible way to increase carbon sequestration by adding dolomite and K-feldspar to soil. Cogent Geosci 2:1205324. https://doi.org/10.1080/23312041.2016.1205324
Xiu-ting S, Mei-rong L, Jun-tao X et al (2021) The complex effect of organic acids on the dissolution of feldspar at high temperature. Environ Earth Sci 80:244. https://doi.org/10.1007/s12665-021-09537-2
Yang H, Peng L, Chen L et al (2023) Efficient potassium (K) recycling and root carbon (C) metabolism improve K use efficiency in pear rootstock genotypes. Plant Physiol Biochem 196:43–54. https://doi.org/10.1016/J.PLAPHY.2023.01.024
Yuan G, Cao Y, Schulz HM et al (2019) A review of feldspar alteration and its geological significance in sedimentary basins: from shallow aquifers to deep hydrocarbon reservoirs. Earth Sci Rev 191:114–140. https://doi.org/10.1016/J.EARSCIREV.2019.02.004
Zhang S, Bai X, Zhao C et al (2021) Global CO_2 consumption by silicate rock chemical weathering: its past and future. Earths Futur 9:e2020EF001938. https://doi.org/10.1029/2020EF001938
Zhang Q, Tian H, Radwan AE et al (2023a) Mineralogical characteristics and genesis mechanism of glauconite in the meso-neoproterozoic and cambrian strata in north China. Mar Pet Geol 153:106280. https://doi.org/10.1016/j.marpetgeo.2023.106280
Zhang X, He Z, Jia W et al (2023b) Mechanism of potassium release from feldspar by mechanical activation in presence of additives at ordinary temperatures. Materials 17:144. https://doi.org/10.3390/ma17010144

The Potential of Remineralizers and Bioinputs as a Strategy for Sovereignty in Tropical Agriculture

Caroline Siqueira Gomide, Fernanda de Paula Medeiros, Andreia Matheus, Viviane Sousa Oliveira, Luiza Coimbra Oliveira, Luiz Henrique Gomes de Moura, Claudete Gindri Ramos, and Suzi Huff Theodoro

Abstract The Brazilian territory, primarily located in the tropical region, is one of the largest exporters of food and agricultural commodities. However, it depends on the import of inputs to facilitate its production. Regional/locally available sources, such as remineralizers (REM) and bioinputs, can reduce this dependence and become a sustainable strategy for the country and other tropical regions with a similar production profile. These inputs are multi-nutrient sources that guarantee productivity and increase agricultural resilience. REM are products derived from certain types of crushed silicate rocks that offer a wide range of macro and micronutrients. Bioinputs are products or processes of biological origin that impact the growth, development, and response mechanisms of animals and plants. This article aims to demonstrate the high potential of the combined or individual use of these inputs in maintaining Brazil's leading role in the production of food and/or commodities. We utilized data from researchers from various groups, including our own, to demonstrate the significant potential of these inputs, particularly for countries that rely heavily on them. We identified the benefits and challenges that tropical countries, particularly Brazil, must overcome so that these sources can be used optimally to contribute to food and/or commodity production. We conclude that utilizing regionally available sources interacts with the soil in a more comprehensive and complex manner, thereby favoring the

C. S. Gomide · F. de Paula Medeiros · A. Matheus · S. H. Theodoro (✉)
Postgraduate Program in Environment and Rural Development (PPG-MADER), FUP-UnB, University of Brasília-Área Universitária, Planaltina, DF, Brazil
e-mail: suzitheodoro@unb.br

C. S. Gomide · F. de Paula Medeiros · A. Matheus · V. S. Oliveira · L. C. Oliveira · S. H. Theodoro
Stonemeal Technology Research Group, CNPq-National Council for Scientific and Technological Development, Brasilia, Brazil

L. H. G. de Moura
Agrarian Studies Center, NEAGRI/CEAM-UnB, Brasilia, Brazil

C. G. Ramos
Department of Civil and Environmental, Universidad de La Costa, Barranquilla, Atlántico, Colombia

É. de Souza Martins and S. Huff Theodoro (eds.), *Soil Remineralizers and Silicate Fertilizers*, Frontier Studies in Soil Science,
https://doi.org/10.1007/978-3-032-14656-4_11

production of more nutritious foods, enhancing productive capacity, and preserving natural resources.

Keywords Crushed rocks · Microorganisms · Fertility · Food security

1 Introduction

Agriculture is an anthropogenic activity with a complex relationship to climate change (Rosenzweig et al. 2020). Agricultural production is responsible for 29% of global greenhouse gas emissions, with the manufacture and application of fertilizers being one of the primary sources of emissions (Vermeulen et al. 2012; Alencar et al. 2023).

The predominant agricultural development model and the effects of climate change have had direct impacts on food sovereignty at both the national and global levels, affecting food systems. According to the Intergovernmental Panel on Climate Change, global greenhouse gas (GHG) emissions continue to rise (IPCC 2023), leading to a high number of people in situations of food and nutritional vulnerability (Rosenzweig et al. 2020). To limit warming to 1.5 °C (a limit established in COP 21—Paris Agreement), economic activities must be rethought to reduce emissions by 2025. However, the recurrence of extreme weather events (droughts, cyclones, and heat waves) shows that this objective is not being achieved (IPCC 2023).

Brazil is one of the largest agricultural producers, but it is highly dependent on inputs (Brazil, Dec. 10, 1991/2022; Globalfert 2023). Added to this dependence on inputs are the limits on increasing the productivity of the main crops in Brazilian agribusiness, as identified by recent research institutes (IBGE 2017; Brazil 2022). Although the soybean-maize binomial has been an extremely efficient system for decades, maintaining it over successive harvests has had significant impacts due to the absence of crop rotation (Contini et al. 2019). Particularly in soybean management, the expected productivity for the next ten years is expected to remain at around 3.57 tons per hectare (Brazil 2021).

In Brazil, combating deforestation, promoting (re)forestation, improving land-use practices, promoting sustainable herd management, and incorporating new technologies that reduce greenhouse gas emissions are fundamental mechanisms for mitigating and adapting to climate change (Zalles et al. 2018). These actions are important because they affect soil quality and agricultural production (Gomes et al. 2020). According to Agegnehu and Amede (2017), the productivity of crops has been increasingly compromised by soil loss that exceeds the natural rates of its formation. Cultivated soil can deteriorate over years of agricultural use due to intensive production and the export of nutrients with crops, as well as the loss of nutrients through leaching and surface runoff. A periodic supply of organic matter and nutrients (macro and micronutrients) is necessary for good production, as well as soil protection and crop rotation (Primavesi 2002).

The efficiency of nutrients applied via high-solubility fertilizers has decreased over time, and in the last 40 years, the effect of nitrogen sources has decreased by two-thirds, despite a sevenfold increase in consumption per hectare (Angus 2016). Compounds derived from the excessive use of this mineral source to enhance short-cycle crops (for example, soybeans and maize), contribute significantly to the accumulation of GHG (Menegat et al. 2022).

It is important to note that soil fertility is a systemic component (Anghinoni and Vezzani 2021) and is the result of different interactions or needs (mainly plants, microorganisms, mesofauna, and macrofauna) that form a soil food web. In this approach, the establishment of conservationist and integrated management systems, utilizing cover crops, green manure, bioinputs, bio-defensive agents, and remineralizers, has contributed to reestablishing soil health and maintaining and/or boosting productivity (Theodoro et al. 2022; Steffen et al. 2015).

Some of the more promising alternatives for reducing the import of fertilizers, for enhancing soil life, for the biological fixation of nitrogen (Zhang et al. 2020), for the solubilization of phosphorus, through better use of nutrients present in the soil are organic waste cycle (Tavares et al. 2018) and stonemeal technology (Leonardos et al. 2000; Theodoro and Leonardos 2015; Medeiros et al. 2024).

Among the various dimensions of the global agricultural model, this article focuses on fertility and the management of soil. We will present the potential and limitations of agricultural production in Brazil, highlighting remineralizers and bioinputs (used in combination or isolation) as important inputs to maintain the country's productive potential and leadership.

2 Agricultural Production: Success or Crisis?

The implementation of the green revolution and the acceleration of urbanization are transforming the relationship between human beings and Nature (Head et al. 2022; Theodoro and Leonardos, 2021). Several authors, including Steffen et al. (2015), mention the impacts generated by a production model that is increasingly dependent on the intensive use of inputs, machines, implements, and technology applied to cultivation. Although initially proposed to increase food production and labor assets over the years, it has become increasingly geared toward generating/accumulating profit (Frank and Schlenker 2016) and excluding a large contingent of rural families (FAO 2018).

The excessive use of natural resources is changing the dynamics between the soil, plants, and biodiversity that govern the changes to the pedosphere. There is ample evidence of changes already underway, including climate change, species extinction, ocean acidification, and warming (IPCC 2023), as well as alterations in biogeochemical cycles. Regarding biogeochemical cycles, Steffen et al. (2015) suggest that nitrogen (N) and phosphorus (P), which are incorporated through excessive use of soluble fertilizers, already pose a high level of risk to planetary boundaries. Authors such as Nesme et al. (2018) report that most biogeochemical cycles are controlled by microbial activity, and the methane emissions generated from agricultural activities (250 Tg per year) are significant factors in the growing global climate crisis.

The synthetic N fertilizer supply chain alone was responsible for estimated emissions of 1.13 Gt CO_2, representing 10.6% of agricultural emissions and 2.1% of global GHG emissions in 2018 (Menegat et al. 2022). Agri-food systems are responsible for one-third of global anthropogenic GHG emissions (Crippa et al. 2021). This same process is verified in the biogeochemical cycle of P, which is accentuated by the use of phosphate fertilizers, responsible for 70–80% of the phosphorus flow (Ng et al. 2023). At this point, Brazil is a significant contributor, having added more than 132 million tonnes of phosphate fertilizers over the past 60 years (FAO 2023).

At the same time, the utilitarian view of land use, especially forests, generated new use and exchange values on ecological processes and new possibilities for exploring income from land or nature (Bartra 2014). While it is true that valuable initiatives have been implemented to reduce the impact of the agricultural sector, these initiatives give the false impression that they are viable solutions to the environmental and climate crisis. These initiatives, known as green capitalism, involve the commodification and marketization of nature and the natural world. The most demonstrative example of these kinds of solutions is quantifiable carbon capture mechanisms or methodologies, where credits generated or avoided are negotiable on a voluntary or regulated basis, thus converting into a new currency (Moreno et al. 2016), where countries or companies with higher emission rates buy credits from those who have reduced emission levels, thus acquiring the right to pollute.

Particularly in Brazil, the high degree of availability of land, water, and natural resources has resulted in the expansion of agriculture, with an exaggerated image of its agricultural potential. This, in conjunction with the movement towards the internationalization of agriculture, intensified the control of transnational agribusiness companies in the country. The purchase of land by national and/or foreign groups has resulted in high rates of land concentration. This dominance (achieved through globalized market dynamics) leads to further social, economic, political, and environmental crises as well as the marginalization of farmers and increased poverty and hunger in rural areas, which, ironically, is the place where food production occurs (FAO 2022).

Brazil has become both a protagonist and a hostage to the production model; a protagonist because it has constantly achieved record harvests, but to reach these levels it has become hostage and dependent on the import of highly soluble fertilizers (average of 85% in the last 10 years), particularly potassium and nitrogen (Manning and Theodoro 2018; Martins et al. 2024; Brazil 2021). An overview of this dependency is shown in Figs. 1 and 2.

Figure 1 presents data related to the use, import, and export of high-solubility fertilizers from the top five agricultural countries in 2021 (China, India, the United States, Brazil, and Russia). China exports nitrogen and phosphorus (and imports only a small amount for storage purposes to make less use of its reserves), and imports only potassium (52.7% of demand). India imports 25.7% of the nitrogen it consumes, 32% of the phosphorus, and 83.1% of the potassium. The United States imports 39.2% of its nitrogen, 40.4% of phosphorus, and more than 100% of its potassium demand

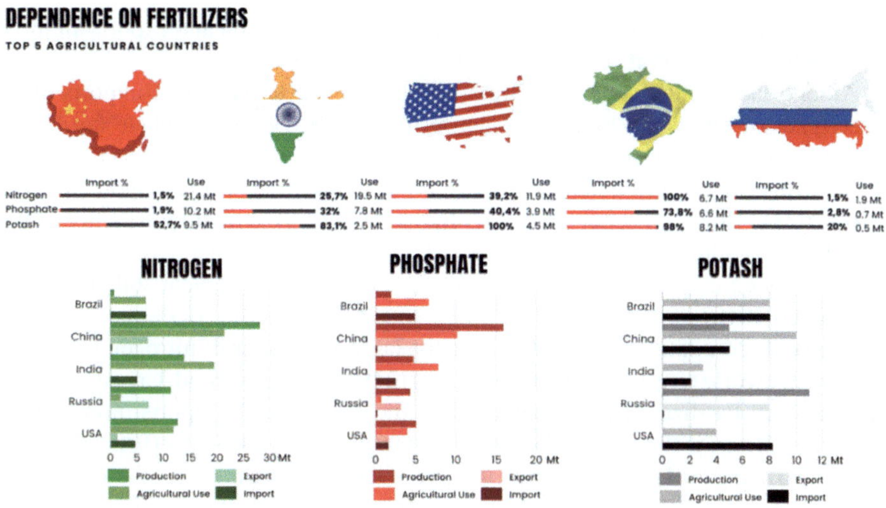

Fig. 1 Dependence on high-solubility fertilizers in 2022. *Source* Authors based on data from the FAOSTAT, Rome (2024). Inputs: Fertilizer by Nutrient

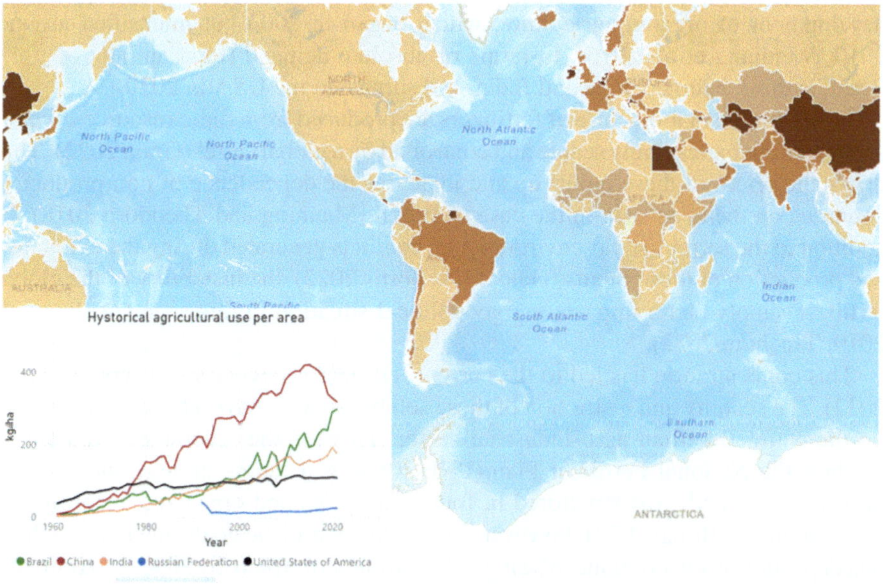

Fig. 2 Global use of soluble fertilizers per area of cropland (kg ha^{-1}) in 2021, with a historical record of the top 5 agricultural countries. *Source* Authors based on data from the FAOSTAT/Rome (2023). Inputs: Fertilizer by Nutrient

(for storage). Russia produces enough for its own consumption and exports all three types of fertilizers, yet it still imports 20% of its potassium.

Among the top five countries, Brazil is the only one to have high imports of all three types of inputs: 100% of nitrogen fertilizers, 73.8% of phosphate fertilizers, and 96% of potassium fertilizers. Since 2006, Brazil has been the largest importer of fertilizers (in terms of quantity), representing 17.6% of all global imports in 2021 (FAO 2023). This high dependence highlights the serious, structural fragility of the Brazilian agricultural model (Manning and Theodoro 2018; Theodoro et al. 2022), which is based on information about nutritional poverty in tropical soils (Cardoso and Fávero 2018).

Global import data show that Brazil is the largest importer of nitrogen fertilizers (14% of the world total), phosphate fertilizers (21%), and the second-largest importer of potassium fertilizers (20%), behind only the USA, which imports more than it consumes. Brazil is the world's largest importer of soluble fertilizers (FAO 2023). In addition to being the largest importer, it is one of the countries that uses these products in greater quantities (Fig. 2). When compared to the top agricultural countries, Brazil is the second-largest country in terms of quantity per hectare of cultivation, with only China surpassing it. As shown in Fig. 2, China has been decreasing its consumption per area in recent years, which is likely a reflection of its policies to support the development of organic agriculture, which began in 2003 but intensified around 2010 (Wenliang et al. 2016). In terms of absolute demand (total amount of use), Brazil is third, behind China and India, and surpassed the USA in 2020.

Inputs for the manufacture of fertilizers are produced from finite resources, which are becoming increasingly scarce and controlled by a global cartel (Grain 2022). This drives the prices of these inputs up and increases the dependence of non-producing countries on them, as seen in the case of Brazil (Manning and Theodoro 2018). In addition to the serious socio-environmental conflicts generated during the extraction and production of these inputs (Gale 2019; Fortis 2022), the instability of the global fertilizer supply chain impacts the geopolitical situation in this segment (Li et al. 2019; Tanchum 2022).

This concentration has led to the current fertilizer access crisis (Theodoro et al. 2022). The scarcity and restriction of these soluble sources have a transversal impact on the entire Brazilian agro-industrial sector. This led the federal government to produce the National Fertilizer Plan (PNF-2050) as a future strategy for the next 28 years. The PNF was structured in four-year cycles and organized according to five guidelines (Brazil 2022); however, despite the optimistic goals, the PNF predicts that dependence will continue to be high. According to the plan, by 2030 Brazil should produce 1.9 Mt of N, 4.2 Mt of different formulations of P_2O_5, and 2 Mt of K_2O. These numbers are modest and far below the sector's demand, representing approximately 28.3, 63.6, and 24.4% of national consumption (using 2021 as a reference) (Brazil 2022). These estimates suggest that the country does not foresee any alternatives or mechanisms that could reduce its dependence on the international fertilizer trade. In addition to Brazil being a large consumer of fertilizers, it also buys and uses the most number of pesticides and thus has become one of the main growth possibilities and market sources for large agrochemical companies (Bombardi 2017). Data from the Food and Agriculture Organization of the United Nations (FAO 2023) shows that Brazil used more pesticides than the United States and China combined each year (from 2012 to 2019). In 2020 and 2021, the United States used more pesticides, getting close to Brazil.

Just as serious is the country's insistence on using a production model that weakens its sovereignty and is linked to large estates that specialize in producing reduced commodities (Theodoro et al. 2022). Maintaining these arrangements weakens the agricultural sector and hinders the opportunities provided by production diversification. It appears that the country has not learned from the major crises (coffee, rubber, and sugar cane) it experiences (Galeano 1971; Castro 1984) regarding the centralization of production chains that meet global demands. Soy, corn, and sugarcane make up around 73% of all imported inputs and represent around 75% of the total agribusiness. Although there are warnings regarding this fragility linked to a profile centered on very few products, the change in these determinants is not evident in short- and medium-term scenarios (Brazil 2022).

In this sense, adopting technological approaches that utilize local/regional sources, such as soil remineralizers (REM) and bio-inputs, could help address the fertilizer supply crisis and serve as mechanisms to advance the ecological transition. This opportunity is one of the new strategies to maintain the leading role of Brazilian agriculture (Goulet 2021; Croplife 2022; FAO 2022; Olmo et al. 2022; Medeiros et al. 2024). Both REM and bioinputs are referred to as multi-nutrient inputs as they ensure a diverse supply of nutrients for the soil. This is particularly important in tropical regions where soils are subjected to more severe weathering processes (Formoso 2006). Remineralizers have the advantage of guaranteeing greater autonomy for the sector and the country due to the immense diversity of rocks found throughout the national territory. This guarantees significantly lower supply and prices (Martins et al. 2024; Theodoro et al. 2022).

Based on these perspectives, the individual or combined use of remineralizers and bio-inputs can help guarantee the fertility and health of crops, thus becoming a differentiator for tropical agriculture and reaffirming its leading role in agricultural production. In this regard, in this paper we will discuss the importance of these two types of inputs, highlighting their direct effects (pH, CTC, soil fertility, production) and indirect effects (facilitating CO_2 sequestration), a mechanism defined as enhanced rock weathering, as suggested by several authors including Manning (2018), Beerling et al. (2020, 2024), Guo et al. (2023) and Kantola et al. (2023).

3 Remineralizers

The use of rock dust (in Brazil these are referred to as remineralizers, or REM), according to the assumptions of stonemeal technology, has been quickly disseminated and incorporated among farmers. For many authors, especially Leonardos et al. (2000), Kronberg et al. (1987), Theodoro (2000), Hinsinger and Gilkes (1997), and

Carvalho et al. (2018), the stonemeal technology strengthens and favors the remineralization and rejuvenation of soils (Theodoro and Leonardos 2006) by adding macro and micronutrients needed by plants (Carvalho 2012), which are not available in soluble fertilizers (Theodoro et al. 2022). Remineralizers have lower solubility than synthetic inputs but offer a greater diversity of nutrients for longer periods (Van Straaten 2002; 2022; Medeiros et al. 2025); and are thus more resilient (Camara et al. 2023; Luchese et al. 2023; Swoboda et al. 2022; Cunha and Almeida 2021; Souza et al. 2017). As plants absorb only what is necessary for their growth/development, the remaining nutrients remain in the system, in the form of secondary minerals, becoming a nutrient bank available for a longer period (Theodoro and Leonardos 2006). In the medium and long term, they can exceed productivity and reduce losses due to leaching and adsorption in relation to the use of soluble fertilizers (Ramos et al. 2022).

Furthermore, the use of these inputs generates the possibility of using mining by-products/co-products (Martins et al. 2023), in addition to reservoir sediments (Theodoro et al. 2013), which help reduce costs (of materials already explored) and mitigate recurring environmental liabilities in the mineral sector (Theodoro et al. 2021). These benefits reduce the prices of REM to some extent, making them cheaper than soluble fertilizers (Manning and Theodoro 2018). This is a significant possibility, especially at a time when fertilizer supplies on the international market are being constrained, as is evident at the beginning of the war in Eastern Europe and, more recently, in the Middle East.

The association between bioactivators (Alovisi et al. 2023) and certain microorganisms that dissolve chemical compounds, efficient microorganisms, or the addition of organic matter increases the availability of nutrients from minerals (Assunção et al. 2024; Curtis et al. 2023; Fang et al. 2023; Silva et al. 2023, 2017; Azeem et al. 2022; Sousa et al. 2021; Basak et al. 2018; 2021; Li et al. 2021; Ribeiro et al. 2020) and can add nutrients in the composting process, as is the case with vermicompost (Souza et al. 2013; Medeiros et al. 2024). These techniques favor the formation of aggregates in the soil and contribute to its structuring (Primavesi 2002; Churchman et al. 2020), allowing the circulation of organic acids, the formation of phenols that contribute to respiration, phosphorus absorption, and vegetable health.

Weathering and the efficient use of remineralizers depend on the interaction of several factors, such as soil management, plant species, rock mineral, dissolution rate, particle size and applied quantity (Swoboda et al. 2022; Ramos et al. 2022). The dissolution rate of silicate minerals can be increased by the combined effects of organic acids (Tavares et al. 2018) and CO_2 produced in the soil or removed from the atmosphere (Beerling et al. 2018, 2024). According to Basak et al. (2018), organic molecules produced by microorganisms (including bacteria and mycorrhizal fungi) are capable of dissolving mineral compounds, originating a type of biofilm which tends to create a congenital microenvironment that facilitates the dissolution of rock-forming minerals (Zhang et al. 2020). Plant roots can also accelerate the wear of minerals through the prevalence of microbial activity points, production of organic acids, and a low pH environment (Basak et al. 2021). This process is accelerated when rocks are ground into different grain sizes or when primary minerals are already undergoing alteration and forming clay minerals (Manning 2018).

Research by Kelland et al. (2020) has shown that the minerals that form basalt rocks, especially diopside and olivine, had their surfaces colonized by mycorrhizal fungi, which accelerated the physical and chemical dissolution rates of the grains when compared to the smooth and angular shapes of the original material. Similar results were verified by Arcand and Schneider (2006) for phosphate rocks (apatite crystals) associated with microorganism activities. Harley and Gilkes (2000) suggest that the rate of dissolution will occur, preferentially, in the "defects" of the mineral surfaces, at different stages of change in the crystalline structure.

Currently, one of the biggest challenges in using REM is related to expanding the supply of primary macronutrients, especially potassium (K) and phosphorus (P), which are the nutrients most demanded by agricultural crops. Although there is a significant quantity and variety of rocks that are possible sources of these nutrients, the mechanisms by which K and P compounds are released from the crystalline structure of minerals (van Straaten 2022) still need to be better understood, particularly the acceleration or inhibition of chemical reactions. Studies (Manning 2010; Mbissik et al. 2021; Santos et al. 2021) have shown significant results in the release of K from minerals such as leucite, biotite, muscovite, and K-feldspars. Research by Setiawati and Mutmainnah (2016) highlights that the solubilization of K can vary according to the nature and composition of minerals, their crystalline structure, and the degree of alteration and genesis. However, the presence of organic acids, secreted by the roots of plants and microorganisms, favors increased potassium solubility (Carvalho 2012; Medeiros et al. 2025).

With these results in mind, the commercialization and use of REM in Brazil have grown significantly, especially over the last three years, due to the problems of scarcity and the prices of soluble fertilizers. The fact that Brazil is a leader in research on the use of crushed rocks in agriculture (REM) and has already established a legal framework (Brazil 2013) regulating its use has led to marked growth (Theodoro et al. 2022). The production and supply of REM have doubled year after year. Around three million tons of remineralizers were sold in 2022, and around 10% of Brazil's agricultural area already uses this input in the production process (Martins et al. 2024), a result of the increased number of registrations with the Ministry of Agriculture, Livestock, and Supply. Over a span of four years (2019 to 2023) registrations increased by 400%, starting at 17 registered products and increasing up to 68 registered products (Martins et al. 2024). There are currently 79 products registered (2025, September). The capacity for this number to grow even further is very possible, considering the country's mineral profile and the increased use, which shows that the PNF's projections (18 million tons of remineralizers per year by 2050) could be very quickly shortened. Figure 3 shows the locations of REMs registered with the Ministry of Agriculture, Livestock, and Supply.

The registered products are rock dust from mica schist, chalcoxist, basalt, kamafugite, amphibolite, serpentinite, granite, olivine melilitite, tephrite, phonolith, syenite, dunite, pyroxenite, gabbro, diabase, dacite, biotite gneisses, and others (Oliveira et al. 2024; Araujo et al. 2023; Nogueira et al. 2021). Despite the significant

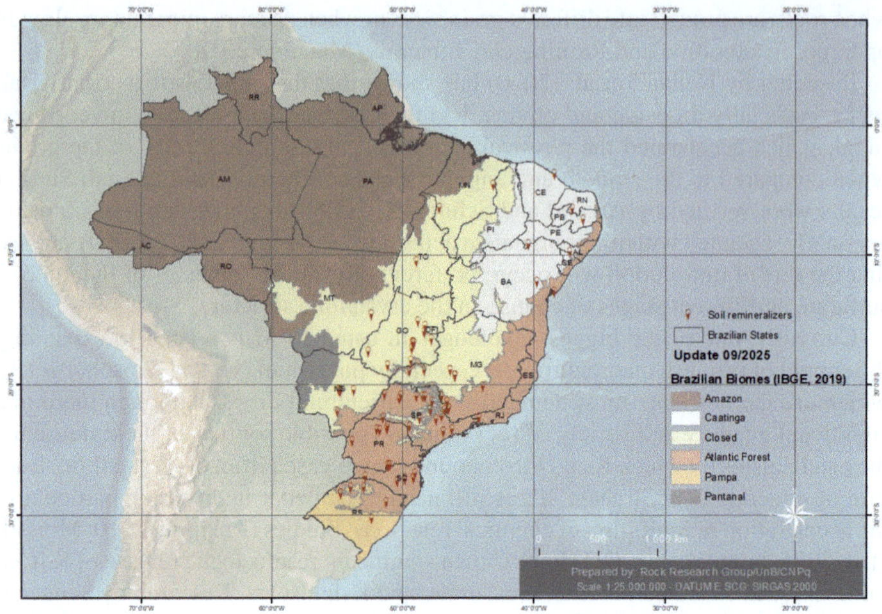

Fig. 3 Remineralizers registered with the Ministry of Agriculture, Livestock and Supply. *Source* Authors based on data from the Ministry of Agriculture, Livestock and Supply (MAPA)

number of rocks suitable for this, it is important to highlight that the lithochemistry of some rocks can present very significant levels of potentially toxic elements (PTEs), or other harmful elements. For example, steatite (soapstone) may contain nickel and/or chromium levels above 1500 mg kg^{-1} (Carvalho 2012), which would render it unable to meet standards and, consequently, unsuitable for use in agricultural soils. As demonstrated by Souza et al. (2013) and Medeiros et al. (2025), even though these elements may not be available to plants, it is necessary to understand that not all rocks are suitable for this purpose. Problems of this nature (although not yet scientifically resolved) require further studies to address the immobilization and chelation of metals in the soil. This can be achieved through associations with bio-inputs, composting, and organic management techniques, such as vermicomposting (Medeiros et al. 2024).

There are studies that apply and prove the effectiveness of rock dust in various crops, such as soybeans, corn, beans, wheat, coffee, tomatoes, quinoa, oats, strawberries, sugar cane, palm, etc. (Medeiros et al. 2024; Almeida Júnior et al. 2022; Burbano et al. 2022; Ramos et al. 2021; Chiomento et al. 2021; Sousa et al. 2021). In these cases, the use of remineralizers favors the elevation of pH and can, in some cases, replace liming (Gillman et al. 2001; Boniao et al. 2002), reduce exchangeable aluminum and increase water retention in the soil (Theodoro et al. 2022), conserve nutrients and preserve water quality (Leonardos et al. 2000; Van Straaten 2006), and increase soil CEC (Momesso et al. 2022). They can reduce evaporation loss and conserve soil moisture (Groenevelt et al. 1989; Jiménez et al. 2017).

Some of the indirect effects are that they have great potential to capture CO_2, thus increasing the potential for a more sustainable form of agriculture (Liu and Zhao 1999; Renforth et al. 2015; Guo et al. 2023; Kantola et al. 2023; Santos et al. 2023; Beerling et al. 2024). According to Beerling et al. (2020), the fluids resulting from this wear tend to react with basic cations (Ca^{2+} and Mg^{2+}), forming new mineral compounds which trap CO_2 in the form of carbonate. Defined as enhanced rock weathering, by Beerling et al. (2020) and Kelemen et al. (2020), this mechanism is an important way in which to enable the capture and storage of atmospheric CO_2 and can help regulate the global climate (Fan et al. 2014). In addition to capturing CO_2 and remineralizing the soil, rock dust enhances the development of plants which need to capture carbon from the atmosphere to grow (Soares et al. 2022). Enhanced rock weathering is also enhanced and controlled by biological processes and the presence of organic matter (Wood et al. 2023; Vicca et al. 2022).

These possibilities make this an important technological approach in maintaining the leading role of tropical agriculture. The association with bioinputs helps understand the dissolution kinetics of rocks and how the biogeochemical cycles of the main macro and micronutrients affect the soil, soil biology, and plants.

4 Bioinputs

Bioinputs represent a new technological approach that helps the agricultural sector adapt to its diverse demands by offering innovative solutions at various scales, serving as an alternative to the intensive use of synthetic fertilizers and pesticides (Vidal et al. 2020). In this sense, bioinputs can be included as one of the technologies to shape a new paradigm of agriculture, one based on agroecology (Vidal and Dias 2023) and family farming (Cabanillas et al. 2017). In turn, it can also add layers of sustainability to agribusiness (Neves et al. 2023), ensuring economic and environmental gains unrelated to the changing dominant agricultural production model. The soil microbiome plays a crucial role in agroecosystems, as demonstrated in several previous studies.

The modern use of bioinputs in Brazil was established between the 1980s and 1990s (Bortoloti and Sampaio 2022). In the field of biological control, there are consolidated techniques such as insertion within conventional technological packages, emergency use in the Helicoverpa crisis (EMBRAPA 2013), the use of *Rhizobium* (Santos et al. 2019) and *Azospirillum* (Santos et al. 2021), and with nematodes (Araujo et al. 2023). When it comes to organic fertilizers, there are numerous possible techniques. Vermicomposting is widely used throughout the world, and several studies have pointed to its great benefits on the soil and plants (Hussain and Abbasi 2018). Currently, the use of bioinputs shows consistent growth, as can be seen in Fig. 4.

The Bioinput Law (15,070/2024) considers bioinputs as a product, process or technology of plant, animal or microbial origin, intended for use in the production, storage

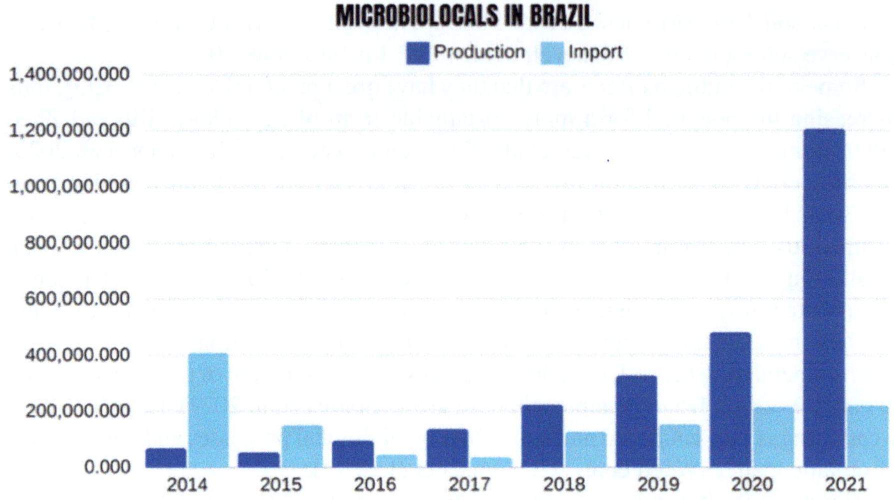

Fig. 4 Production and import of microbiologicals in Brazil. *Source* Authors based on data from the IBAMA (2023)

and processing of agricultural products, in aquatic or planted forestry production systems and positively interfere with the growth, development and response mechanism of animals, plants, microorganisms, and derived substances, also interacting with physical–chemical and biological products and processes (Brazil 2024).

This is a broad category of inputs of natural origin, which involves diverse possibilities and solutions, using products related to plant nutrition and soil health such as inoculants, biofertilizers, organic substances and extracts, stimulants such as phytochemicals/bioactives, biological control such as macro biological control (insects, mites and nematodes), and micro biological control (bacteria, fungi and viruses) (Brazil 2024). Given their wide diversity, bioinputs can be utilized in various agricultural management strategies (König and Sá 2022), primarily in systems employing Integrated Pest Management (IPM), organic, and agroecological approaches, as well as in livestock farming (Oliveira et al. 2023; Vidal and Dias 2023).

Sustainable agriculture is based on the understanding of the complex interactions of ecology. It considers the mechanisms that nature uses to maintain the ecological cycle in perfect harmony in each biome. It is a technique that connects rock–soil–plant–food–health by rebalancing the biological, chemical, and physical factors, thus reestablishing quality of life (Primavesi 2018). Soil is a living organism derived from interactions between rocks and the environment, resulting from weathering processes influenced by soil formation factors, including climate, organisms, relief, source material/rock, and formation time (Primavesi 2002). In short, the soil is a system made up of the following subsystems: lithosphere, atmosphere, hydrosphere, and biosphere.

In their studies, Ingham et al. (1985) also demonstrated that the complex biomass, in the form of organisms inhabiting tropical soils, is the reason these soils are productive, especially in the top 15 cm of soil. This is a fundamental part of the tripod (chemical, physical, and biological) that makes up the balance of synergy between ecological processes that have been neglected by the "Green Agriculture" model used for decades in Brazil.

Bioinputs have a symbiotic relationship with the structure of plants, particularly their roots, and produce organic compounds that facilitate greater nutrient absorption, pest control, and promote healthy plant growth/development. They restore soil life and balance out the physical, chemical, and biological properties. They are indicated to increase agricultural productivity and quality, improve plant metabolism (photosynthetic capacity), activate/accelerate the decomposition of organic matter, activate root growth, increase germination, flowering, and fruiting, increase resistance to diseases (pesticide and herbicide), and are used for silage, which accelerates the composting process.

Studies conducted by Li et al. (2021) have verified that the soil food web is more complex in organic management practices, with an increase in biomass at all trophic levels. It is especially beneficial for biota at higher trophic levels, accelerating maturation processes of soil microorganisms, which was probably driven by the elimination of synthetic inputs. According to Faria et al. (2021), the microbiota is formed through a series of events that establish stable and diversified consortia based on cooperative interactions or competition. The greater complexity in the rhizosphere can contribute to plant protection. Based on this understanding of the importance of soil biocenosis, a series of techniques has been promoted with the aim of accelerating this process.

The global biopesticides market in 2022 was approximately $6.5 billion, while biofertilizers were valued at $2 billion and biostimulants at $3 billion (Mafra, 2023). The CropLife Brazil association, with several partner companies that work with bioinputs, including world leaders such as BASF, Bayer, Syngenta, and Corteva, points out Brazil's bio-defensive products (or biological products). The biological control part alone generated R$3.3 billion for the 2021/2022 harvest, a growth of 219% compared to the 2019/2020 harvest, while the average rate of adoption of biological control rose from 17% (2019/2020) to 28% (2021/2022) of the 85.7 million hectares of crops (Borsari et al. 2022). On the other hand, there is a significant proliferation of *on-farm* units, also called biofactories. These units multiply isolated microorganisms, including those that exhibit single mechanisms or possess multiple mechanisms. In general, they are controlled by numerous agribusiness companies that have started incorporating bio-input production into their production structure (Almeira 2020).

Brazil has credit and development mechanisms to promote new technological approaches for bioinputs, such as the 2023/2024 Harvest Plan, BNDES, EMBRAPII, and FINEP. These efforts are in addition to those of other research institutions and national programs that share the goal of increasing the use of these inputs and promoting more sustainable Brazilian agriculture and livestock (Brazil 2024).

In 2020, EMBRAPA had 632 researchers involved in 73 different projects related to bio-consumption. These projects were distributed among 40 company units and a bank of 24,000 strains of different microorganisms (Croplife 2020).

The legal provision for using this type of input had already been outlined in the National Bioinput Program (PNB) and in Law No. 15,070/2024, which also established a Strategic Council (Brazil 2024). The main objective of this Law is to expand and increase the use of bioinputs to promote sustainable development in different agroecosystems. Additionally, it has other strategic objectives, including encouraging the construction of biofactories for producing bio-inputs, providing incentives to articulate credit, and developing mechanisms aimed at production, stimulating the generation and dissemination of knowledge, and creating new products and processes (Brazil 2024).

This is a strategic option for Brazil for a few reasons: (i) dependence, previously reported about synthetic fertilizers and agrochemicals; (ii) increased productivity; and (iii) increased societal demand for healthy and diversified foods that are free from pesticides and synthetic fertilizers (Goulet 2021; Rosenzweig et al. 2020). The production and use of bio-inputs then become a key element toward building the foundations for national and food sovereignty, which involves guaranteeing the right for each country to build its production and food strategies to meet the needs of their respective societies. (Rosset 2015).

Although the use of bioinputs represents a small percentage of the market (Oliveira et al. 2023), the expectation is that, at a global level, this market will continue to show a high growth rate and reach values over US$ 10 billion for pesticides and US$ 3 billion for biostimulants (Croplife 2020). Inputs aimed at biological control are expected to increase to approximately $11.1 billion by 2025 (Vidal and Dias 2023).

5 Building a New Soil Fertilization Paradigm

Considering that the integration of different technologies, or regionally available inputs, helps to enrich the soil with different inorganic and organic materials, as well as facilitate the availability of phosphorus, calcium, potassium, and other macro and micronutrients in the soil, thus improving the conditions and health of crops (Cardoso and Fávero 2018), Brazil created a chain of emerging inputs (which include remineralizers and bio-inputs) under the National Fertilizer Program (Brazil 2021).

The positive correlation and complementarity between these inputs ensure reciprocal action among soil, plant, and microbial life, as conceptualized by Primavesi (2002). Anghinoni and Vezzani (2021) report that biological activity in the soil ecosystem enhances the bioavailability of minerals, recirculates them within the system, and promotes the development of plants with improved health and productivity, resulting in dynamic, self-organized ecosystems that house large amounts of nutrients and diverse forms of life.

Experiences of the integrated use of bio-inputs conducted throughout the country have been seen as positive, favoring the cycling of nutrients in production systems, insertion of functional native species, management of spontaneous herbs, energy integration between different subsystems, collection and multiplication of efficiency, and production of grouts and compounds (Marchetti et al. 2023). Likewise, studies related to remineralizers point to several economic (Manning and Theodoro 2018), productive (Ramos et al. 2022) and environmental (Soares et al. 2022) benefits. Integrated soil management with the "Soil Food Web" method, organic compounds, bio-inputs, remineralizes, and bio-defensive products are inputs that are available in all regions and countries. This technological method can ensure compatible soil fertility and agricultural productivity in different agroecosystems. Thus, geodiversity added to biodiversity can facilitate a change or transition to a more sustainable, less dependent agricultural model with more significant potential to capture and store CO_2 in the soil (Manning 2018).

Implementing a new form of land use that yields economic, productive, and environmental benefits brings a fresh perspective to Brazil. Although these options are encouraging and proliferating in the country, it is essential to understand that these inputs have different dynamics in quantity (available content) and agroecosystem management. Unlike soluble sources (with a restricted supply of nutrients), which force plants to respond specifically, remineralizers and bioinputs are characterized as multi-nutrient sources that interact in the soil in a more complete and complex way, favoring the production of more nutritious food (Theodoro 2021). Improving productive capacity and preserving resources can reestablish the connection between human beings and the soil, which is derived from rocks and is essential for making life possible. We have an opportunity to change the standards!

6 Concluding Remarks

Just as important as guaranteeing agricultural production, the use of remineralizers and bioinputs provides a broad spectrum of economic benefits, including a more excellent supply of materials available locally or regionally at lower costs. Additionally, these practices ensure productive and socio-environmental benefits, such as the elevation of pH and CTC, reduction of exchangeable aluminum, increased water and moisture retention in the soil, and increased nutrient diversity, making it a local action with a global benefit. Furthermore, these inputs, either individually or in combination, minimize nutrient loss due to erosion, increase productivity, and protect against herbivores and insectivorous pathogens, resulting in improved plant health. It also reduces the use and cost of synthetic inputs and pesticides, and its vast diversity of nutrients tends to provide increased food and nutritional security. More recent research has suggested that these inputs can reduce GHG emissions and help capture CO_2, thereby becoming an essential tool for global agriculture, as they increase farmers' autonomy, especially those transitioning to agroecological production. This dual technological option benefits Brazilian agriculture and enables the reduction of greenhouse gases. More than just mitigating the adverse effects on the planet's climate, Brazil is developing a new paradigm in food and commodity production.

Building the success established in Brazil, remineralizers and bioinputs are emerging as a powerful and sustainable strategy for agriculture, Kenya, Cameroon, Angola and Malawi are harnessing these technologies to rehabilitate degraded soils and boost crop resilience against climatic stress, thereby mitigating their dependency on costly, imported synthetic fertilizers. India and Philippines are championing their use to advance food security and address soil degradation in pivotal agricultural sectors. Argentina and Colombia are leveraging mineral resources to empower smallholder farmers. The widespread adoption across the Global South represents a collective commitment to a new agricultural model, contributing to global climate change mitigation through reduction of GEE emissions and enhanced carbon sequestration and fostering greater autonomy and economic viability for farmers.

Acknowledgements We want to thank the University of Brasilia (UnB) for its institutional support, the Grantham Foudation (UnB/Finatec - 7358), National Council for Scientific and Technological Development (CNPq 409283/2022-3) for funding research developed by the Stonemeal (Rochagem) research group (UnB/CNPq) and the Santa Catarina Research and Innovation Foundation (FAPESC), Edital 43/2024.

Credit Authorship Contribution Statement Caroline Siqueira Gomide: Writing—original draft, review and editing, Fernanda de Paula Medeiros: Data curation. Andreia Matheus: Writing—original draft, Formal analysis. Viviane Sousa de Oliveira: Data curation. Luiza Coimbra Oliveira: Data curation. Luiz Henrique Gomes de Moura: Data curation and Writing—original draft. Claudete Gindri Ramos: Data curation. Suzi Huff Theodoro: Writing—review & editing, Resources, Conceptualization.

Declaration of Competing Interest The authors declare that they have no known competing financial interests or personal relationships that could have appeared to influence the work reported in this paper.

References

Agegnehu G, Amede T (2017) Integrated soil fertility and plant nutrient management in tropical agro-ecosystems: a review. Pedosphere 27(4):662–680. https://doi.org/10.1016/S1002-0160(17)60382-5

Alencar A, Zimbres B, Souza E et al (2023) Estimativa de emissões de Gases de Efeito Estufa dos sistemas alimentares no Brasil. SEEG, p 89. https://www.oc.eco.br/wp-content/uploads/2023/10/SEEG-Sistemas-Alimentares.pdf. Accessed 24 May 2024

Almeida Jr JJ, Smiljanic KBA, Matos FS, Pereira RM, Netto AML, Perozini AC, Oliveira EJC, Machado Jr AJM, Menezes JFS, Barbosa UR, Miranda BC, Silva VJA, Vilela VC (2022) Consortium between k-forte soil remineralizer and reactive natural phosphate as a source of natural fertilizer for the Cerrado region. Conjecturas. https://doi.org/10.53660/CONJ-1282-X53

Almeira JEM (2020) Biofábricas para produção de micopesticidas no Brasil: oportunidades de negócio e inovações. Curitiba, Almeira. Braz J Anim Environ Res 3(3):2544–2557. https://doi.org/10.34188/bjaerv3n3-162

Alovisi AA, Alovisi AMT, Taques MM, Medeiros ES, Souza LCF, Cassol CJ (2023) Chemical properties of soils submitted to the application of a bioactivator and basalt and serpentinite powders. Braz J Agric Environ Eng 27(10):811–819. https://doi.org/10.1590/1807-1929/agriambi

Anghinoni I, Vezzani FM (2021) Systemic soil fertility as product of system self-organization resulting from management. Revista Brasileira de Ciência do Solo 45. https://doi.org/10.36783/18069657rbcs20210090

Angus I (2016) Facing the Anthropocene—fossil capitalism and the crisis of the earth system. Monthly Review Press, New York

Araujo FG, Teixeira SJC, Souza JC, Arieira CRD (2023) Cover crops and biocontrol agents in the management of nematodes in soybean crop. Rev Caatinga 36(2):243–250. https://doi.org/10.1590/1983-21252023v36n201rc

Arcand MM, Schneider KD (2006) Plant- and microbial-based mechanisms to improve the agronomic effectiveness of phosphate rock: a review. An Acad Bras Ciênc 78(4). https://doi.org/10.1590/S0001-37652006000400013

Assunção RDV, Cassol PC, Granados SB, Mafra AL, Bittencourt RFPM (2024) Microbial-inoculated remineralizers as source of potassium and other nutrients. Braz J Agric Environ Eng 28(2). https://doi.org/10.1590/1807-1929/agriambi

Azeem M, Raza S, Li G, Smith P, Zhu Y (2022) Soil inorganic carbon sequestration through alkalinity regeneration using biologically induced weathering of rock powder and biochar. Soil Ecol Lett 4:293–306. https://doi.org/10.1007/s42832-022-0136-4

Bartra A (2014) El hombre de hierro: límites sociales y naturales del capital en la perspectiva de la gran crisis. ITACA/UAM, Ciudad de Mexico

Basak BB, Sarkar B, Sanderson P, Naidu R (2018) Waste mineral powder supplies plant available potassium: evaluation of chemical and biological interventions. J Geochem Explor 186:114–120. https://doi.org/10.1016/j.gexplo.2017.11.023

Basak BB, Sarkar B, Naidu R (2021) Environmentally safe release of plant available potassium and micronutrients from organically amended rock mineral powder. Environ Geochem Health 43:3273–3286. https://doi.org/10.1007/s10653-020-00677-1

Beerling DJ, Leake JR, Long SP, Scoles JD, Ton J, Nelson PN et al (2018) Farming with crops and rocks to address global climate, food and soil security. Nat Plants 4:138–147. https://doi.org/10.1038/s41477-018-0108-y

Beerling DJ, Kantzas EP, Lomas MR et al (2020) Potential for large-scale CO_2 removal via enhanced rock weathering with croplands. Nature 583:242–248. https://doi.org/10.1038/s41586-020-2448-9

Beerling DJ, Epihov DZ, Kantola IB, Marters MD et al (2024) Enhanced weathering in the US Corn Belt delivers carbon removal with agronomic benefits. PNAS 121(9). https://doi.org/10.1073/pnas.2319436121

Bombardi LM (2017) Geografia do uso de agrotóxicos no Brasil e conexões com a União Europeia. FFLCH–USP, São Paulo

Boniao RD, Shamshuddin J, Van Ranst E, Zauyah S, Omar SRS (2002) Changes in chemical properties and growth of corn in volcanic soils treated with peat, ground basalt pyroclastics, and calcium silicate. Commun Soil Sci Plant Anal 33(7–8):1219–1233. https://doi.org/10.1081/CSS-120003883

Borsari A, Cantos C, Dias R (2022) Sumário executivo Biodefensivos: mercado brasileiro - safra 2021/2022. https://croplifebrasil.org/publicacoes/mercado-de-biodefensivos-safra-2021-22/. Accessed 24 May 2024

Bortoloti G, Sampaio RM (2022) Demandas tecnológicas: os bioinsumos para controle biológico no Brasil. Cadernos De Ciência & Tecnologia Brasília 39(1):1–16. https://doi.org/10.35977/0104-1096.cct2022.v39.26927

Brazil (2013) Lei 12.890 de 10 de dezembro de 2013 - Altera a Lei no 6.894, de 16 de dezembro de 1980. Available: https://www.planalto.gov.br/ccivil_03/_ato2011-2014/2013/lei/l12890.htm. Accessed 24 May 2024

Brazil (2021) Plano Nacional de Fertilizantes 2050 – Uma estratégia para os Fertilizantes no Brazil. Secretaria Especial de Assuntos Estratégicos. Available: https://www.gov.br/agricultura/ptbr/assuntos/insumos-agropecuarios/insumos-agricolas/fertilizantes/plano-nacional-de-fertilizantes. Accessed 24 May 2024

Brazil (2022) Decreto nº 10.991 de 11 de março de 2022. Institui o Plano Nacional de Fertilizantes 2022–2050. Available: https://www.planalto.gov.br/ccivil_03/_ato2019-2022/2022/decreto/d10991.htm. Accessed 24 May 2024

Brazil (2024) Lei 15070 de 20 Dez de 2024. Institui a Bio-inputs. Available: https://www.in.gov.br/en/web/dou/-/lei-n-15.070-de-23-de-dezembro-de-2024-603825827. Accessed 02 Feb 2025

Burbano DFM, Theodoro SH, Carvalho AXM, Ramos CG (2022) Crushed volcanic rock as soil remineralize: a strategy to overcome the global fertilizer crisis. Nat Resour Res 31:2197–2210. https://doi.org/10.1007/s11053-022-10107-x

Cabanillas C, Tablada M, Ferreyra L, Pérez A, Sucani G (2017) Sustainable management strategies focused on native bio-inputs in *Amaranthus cruentus* L. in agro-ecological farms in transition. J Clean Prod 141(1):343–350. https://doi.org/10.1016/j.jclepro.2016.06.065

Camara GR, Santos ES, Gualandi AJZ, Silveira AJZ, Neves MA, Chiodi Filho C, Castro NF (2023) Agronomic potential of using ornamental stones waste as a source of agricultural fertilization. Global Stone Congress.

Cardoso IM, Fávero C (2018) Ressignificar nossas percepções sobre o solo: atitude essencial para manejar agroecossistemas sustentáveis. In: Cardoso MC, Fávero C (eds) Solos e Agroecologia, 4 edn. Embrapa/Brasília, pp 33–60

Carvalho AMX, Cardoso IM, Souza MEP, Theodoro SH (2018) Rochagem: o que se sabe sobre essa técnica? In: Cardoso I, Fávero C. Solos e Agroecologia, vol 4, 1st edn. Embrapa, Brasília, pp 103–128. ISBN: 978-85-7035-774-8

Carvalho AMX (2012) Rochagem e suas interações no ambiente solo: contribuições para aplicação em agroecossistemas sob manejo agroecológico. Ph.D Thesis, Universidade Federal de Viçosa, p 166. https://locus.ufv.br//handle/123456789/1631

Castro J (1984) Geografia da Fome: o dilema brasileiro. Edições Antares, Rio de Janeiro

Chiomento JLT, Korchagin J, Fornari M, Brugnera LP, Trentin TS, Dornelles AG, Zanin E, Borges LM (2021) Agronomic performance of strawberry cultivated in substrate with rock powder. Braz J Dev 7(2):15646–15657. https://doi.org/10.34117/bjdv7n2-267

Churchman GJ, Singh M, Schapel A, Sarkar B, Bolan N (2020) Clay minerals as the key to the sequestration of carbon in soils. Clays Clay Miner 68:135–143. https://doi.org/10.1007/s42860-020-00071-z

Contini E, Mota MM, Marra R et al. (2019) Série desafios do agronegócio brasileiro: Milho – Caracterização e Desafios Tecnológicos. Empresa Brasileira de Pesquisa Agropecuária, p 45. https://ainfo.cnptia.embrapa.br/digital/bitstream/item/195075/1/Milho-caracterizacao.pdf. Accessed 24 May 2024

Crippa M, Solazzo E, Guizzardi D, Monforti-Ferrario F, Tubiello FN, Leip A (2021) Food systems are responsible for a third of global anthropogenic GHG emissions. Nat Food 2:198–209. https://doi.org/10.1038/s43016-021-00225-9

Croplife (2020) New biological products approved in Brazil. Available: https://croplifebrasil.org/noticias/novosprodutos-biologicos-aprovados-no-brasil/ (accessed May 24, 2025). Translated with DeepL.com (free version)

Croplife (2022) Bioinsumos, cruciais para a autonomia do Brasil na nutrição de plantas. Available: https://croplifebrasil.org/noticias/bioinsumos-cruciais-para-a-autonomia-do-brasil-na-nutricao-de-plantas/. Accessed 24 May 2024

Cunha GO, Almeida JA (2021) Agronomic potential of four rock powders, pure or mixed, as soil remineralizers. Res Soc Dev 10(17). https://doi.org/10.33448/rsd-v10i17.24828

Curtis J, Luchese A, Missio R (2023) Application of soil remineralizer to poultry liter as an efficient and sustainable alternative for fertilizing maize crop. J Plant Nutr 423–438. https://doi.org/10.1080/01904167.2022.2071728

Embrapa (2013) Ações emergenciais propostas pela Embrapa para o manejo integrado de *Helicoverpa spp* em áreas agrícolas. Available: https://www.embrapa.br/documents/10180/1602515/A%C3%A7%C3%B5es+emergenciais+propostas+pela+Embrapa+-+Documento+oficial/3a569ce1-c132-4bfa-8314-bc993ce8b920

Fan BL, Zhao ZQ, Tao FX, Liu BJ, Tao ZH, Gao S, Zhang H (2014) Characteristics of carbonate, evaporite and silicate weathering in Huanghe River basin: a comparison among the upstream, midstream and downstream. J Asian Earth Sci 9617–9626. https://doi.org/10.1016/j.jseaes.2014.09.005

Fang Q, Lu A, Hong H et al (2023) Mineral weathering is linked to microbial priming in the critical zone. Nat Commun 14:345. https://doi.org/10.1038/s41467-022-35671-x

FAO (2018) Panorama de la pobreza rural en América Latina y el Caribe. Food and Agriculture Organization of the United Nations, Santiago. Available: http://www.fao.org/3/CA2275ES/ca2275es.pdf. Accessed 24 May 2024

FAO (2022) Biofertilizers and other technologies available to address the fertilizer crisis in the Latin American and Caribbean region. Food and Agriculture Organization of the United Nations Webinar. Available: https://www.fao.org/americas/eventos/ver/en/c/1513177/. Accessed 24 May 2024

FAO (2023) Statistical database. Food and Agriculture Organization of the United Nations, Rome. Available: https://www.fao.org/faostat/en/#data. Accessed 24 May 2024

Faria MR, Costa LSAS, Chiaramonte JB, Bettiol W, Mendes R (2021) The rhizosphere microbiome: functions, dynamics, and role in plant protection. Trop Plant Pathol 46:13–25. https://doi.org/10.1007/s40858-020-00390-5

Formoso MLL (2006) Some topics on geochemistry of weathering: a review. Acad Bras Ciênc 78(4):809–820 [Online]. https://doi.org/10.1590/S0001-37652006000400014

Fortis B (2022) Environmentalists target mountains of fertilizer waste. Available: https://undark.org/2022/04/13/environmentalists-target-mountains-of-fertilizer-waste/. Accessed 24 May 2024

Frank EG, Schlenker W (2016) Balancing economic and ecological goals. Science 353(6300):651–652. https://doi.org/10.1126/science.aaf9697

Gale SJ (2019) Lies and misdemeanours: Nauru, phosphate and global geopolitics. Extract Ind Soc 6(3):737–746

Galeano E (1971) As veias abertas da América Latina. Paz e Terra, São Paulo

Gillman GP, Burkett DC, Coventry RJ (2001) A laboratory study of application of basalt dust to highly weathered soils: effect on soil cation chemistry. Aust J Soil Res 39(4):799–811. https://doi.org/10.1071/SR00073

Globalfert (2023) Outlook Globalfert 2023. Globalfert report, p 181

Gomes L, Miranda HS, Silvério DV, Bustamante MMC (2020) Effects and behaviour of experimental fires in grasslands, savannas, and forests of the Brazilian Cerrado. For Ecol Manage 458:117804. https://doi.org/10.1016/j.foreco.2019.117804

Goulet F (2021) Biological inputs and agricultural policies in South America: between disruptive innovation and continuity. Perspective 53:1–4. https://doi.org/10.19182/perspective/36383

Grain (2022) A fertilizer cartel holds the global food system hostage. Available: https://grain.org/en/article/6882-a-fertiliser-cartel-holds-the-global-food-system-hostage

Groenevelt PH, Van Straaten P, Rasiah V, Simpson J (1989) Modifications in evaporation parameters by rock mulches. Soil Technol 2:279–285. https://doi.org/10.1016/0933-3630(89)90012-3

Guo F, Sun H, Yang J, Zhang L, Mu Y, Wang Y, Wu F (2023) Improving food security and farmland carbon sequestration in China through enhanced rock weathering: field evidence and potential assessment in different humid regions. Sci Total Environ. https://doi.org/10.1016/j.scitotenv.2023.166118

Harley AD, Gilkes RJ (2000) Factors influencing the release of plant nutrient elements from silicate rock powders: a geochemical overview. Nutr Cycl Agroecosyst 56:11–36

Head MJ, Steffen W, Fagerlind D, Waters CN et al (2022) The great acceleration is real and provides a quantitative basis for the proposed Anthropocene series/epoch. Episodes 45(4):359–376. https://doi.org/10.18814/epiiugs/2021/021031

Hinsinger P, Gilkes RJ (1997) Dissolution of phosphate rock in the rhizosphere of five plant species grown in an acid, P-fixing mineral substrate. Geoderma 75(3–4):231–249. https://doi.org/10.1016/S0016-7061(96)00094-8

Hussain N, Abbasi SA (2018) Efficacy of the vermicomposts of different organic wastes as "clean" fertilizers: state-of-the-art. Sustainability 10(4). https://doi.org/10.3390/su10041205

IBAMA – Instituto Brasileiro do Meio Ambiente e dos Recursos Naturais Renováveis (2023) Histórico de comercialização de semioquímicos e microbiológicos 2014–2022. Available: https://www.gov.br/ibama/pt-br/assuntos/quimicos-e-biologicos/agrotoxicos/arquivos/qualidadeambiental/relatorios/2023/. Accessed 24 May 2024

IBGE—Instituto Brasileiro de Geografia e Estatística (2017) Censo Agropecuário 2017 -resultados definitivos. Rio de Janeiro. Available: https://sidra.ibge.gov.br/pesquisa/censo-agropecuario/censo-agropecuario-2017. Accessed 24 May 2024

IPCC (2023) Synthesis report of the IPCC Sixth Assessment Report (AR6). Available: https://www.ipcc.ch/report/sixth-assessment-report-cycle/. Accessed 24 May 2024

Jiménez MN, Pinto JR, Ripoll MA, Sánchez-Miranda A, Navarro FB (2017) Impact of straw and rock-fragment mulches on soil moisture and early growth of holm oaks in a semiarid area. CATENA 152:198–206. https://doi.org/10.1016/j.catena.2017.01.021

Kantola IB, Blanc-Betes E, Masters MD et al (2023) Improved net carbon budgets in the US Midwest through direct measured impacts of enhanced weathering. Glob Change Biol. https://doi.org/10.1111/gcb.16903

Kelemen PB, McQueen N, Wilcox J, Renforth P, Dipple G, Vankeuren AP (2020) Engineered carbon mineralization in ultramafic rocks for CO_2 removal from air: review and new insights. Chem Geol 550:119628. https://doi.org/10.1016/j.chemgeo.2020.119628

Kelland ME, Wade PW, Lewis AL et al (2020) Increased yield and CO_2 sequestration potential with the C_4 cereal *Sorghum bicolor* cultivated in basaltic rock dust-amended agricultural soil. Glob Change Biol. https://doi.org/10.1111/gcb.15089

König CC, Sá CD (2022) Bioinsumos: conceitos, potencial e desafios no Brasil. Available: https://www.insper.edu.br/noticias/bioinsumos-conceitos-potencial-e-desafios-no-brasil/

Leonardos OH, Theodoro SH, Assad ML (2000) Remineralization for sustainable agriculture: a tropical perspective from a Brazilian viewpoint. Nutr Cycl Agroecosyst 56:3–9. https://doi.org/10.1023/A:1009855409700

Li B, Bicknell KB, Renwick A (2019) Peak phosphorus, demand trends and implications for the sustainable management of phosphorus in China. Resour Conserv Recycl 146:316–328. https://doi.org/10.1016/j.resconrec.2019.03.033

Li J, Mavrodi DV, Dong Y (2021) Effect of rock dust-amended compost on the soil properties, soil microbial activity, and fruit production in an apple orchard from the Jiangsu province of China. Arch Agron Soil Sci 67(10):1313–1326. https://doi.org/10.1080/03650340.2020.1795136

Liu Z, Zhao J (1999) Contribution of carbonate rock weathering to the atmospheric CO_2 sink. Environ Geol 39(9):1053–1058. https://doi.org/10.1007/s002549900072

Luchese AV, Leite IJGC, Giaretta APS, Alves ML, Pivetta LA, Missio RF (2023) Use of quarry waste basalt rock powder as a soil remineralizer to grow soybean and maize. Heliyon 9(3). https://doi.org/10.1016/j.heliyon.2023.e14050

Mafra E (2023) O mercado de bioinsumos vai para onde no Brasil? Available: https://forbes.com.br/forbesagro/2023/01/o-mercado-de-bioinsumos-vai-para-onde-no-brasil/. Accessed 24 May 2024

Manning DAC (2010) Mineral sources of potassium for plant nutrition. A review. Agron Sustain Dev 30:281–294. https://doi.org/10.1051/agro/2009023

Manning DAC (2018) Innovation in resourcing geological materials as crop nutrients. Nat Resour Res 27(2):217–227. https://doi.org/10.1007/s11053-017-9347-2

Manning DAC, Theodoro SH (2018) Enabling food security through use of local rocks and minerals. Extract Ind Soc 7(2):480–487. https://doi.org/10.1016/j.exis.2018.11.002

Martins ÉS, Theodoro SH, Bernardez FFG, Luchese AV, Bergmann M, Siqueira DS, Teixeira A, Azevedo A, Curtis JCD (2024) Produção Brasileira de Remineralizadores e Fertilizantes Naturais: 2019 a 2023. Revista Novo Solo. Available: https://abrefen.org.br/2024/11/08/artigos-producao-brasileira-de-remineralizadores-e-fertilizantes-naturais-2019-a-2023/. Accessed 24 Dez 2025

Mbissik A, Elghali A, Ouabid M, Raji O, Bodinier J-L, El Messbahi H (2021) Alkali-hydrothermal treatment of k-rich igneous rocks for their direct use as potassic fertilizers. Minerals 11:140. https://doi.org/10.3390/min11020140

Medeiros FP, Carvalho AMX, Ramos CG, Dotto GL, Cardoso IM, Theodoro SH (2024) Rock powder enhances soil nutrition and coffee quality in agroforestry systems. Sustainability 16:354. https://doi.org/10.3390/su16010354

Medeiros FP, Theodoro SH, Carvalho AMX, Oliveira VS et al (2025) The combination of crushed rock and organic matter enhances the capture of inorganic carbon in tropical soils, Journal of South American Earth Sciences, 152. https://doi.org/10.1016/j.jsames.2024.105254

Menegat S, Ledo A, Tirado R (2022) Greenhouse gas emissions from global production and use of nitrogen synthetic fertilizers in agriculture. Sci Rep 12:14490. https://doi.org/10.1038/s41598-022-18773-w

Momesso L, Crusciol CAC, Nascimento CAC, Soratto RP, Canisares PL, Moretti LG, Rosolem, CA, Trivelin PCO, Kuramae EE, Cantarella H (2022) Feasibility of early fertilization of maize with 15N application to preceding cover crop. Eur J Agron 135. https://doi.org/10.1016/j.eja.2022.126485

Moreno C, Chassé DS, Fuhr L (2016) A métrica do carbono: abstrações globais e epistemicídio ecológico. Fundação Heinrich Boll, Rio de Janeiro

Nesme T, Metson GS, Bennett EM (2018) Global phosphorus flows through agricultural trade. Glob Environ Change 50:133–141. https://doi.org/10.1016/j.gloenvcha.2018.04.004

Neves M, Casagrande BP, Cambaúva V, Teixeira GO, Toledo PJF (2023) Agriculture 6.0: a new proposal for the future of agribusiness. Revista de Gestão Social e Ambiental 17(9):1–16. https://doi.org/10.24857/rgsa.v17n9-021

Ng SJ, Li B, He Z, Han J-C, Munir MT, Wu X, Huang Y (2023) Global phosphorus cycling: the impact of international commercial trading and the path towards sustainable phosphorus management. Resour Policy 85. https://doi.org/10.1016/j.resourpol.2023.103781

Nogueira TAR, Miranda BG, Jalal A, Lessa LGF, Teixeira Fº MCM et al (2021) Nepheline syenite and phonolite as alternative potassium sources for maize. Agronomy 11(7):1385. https://doi.org/10.3390/agronomy11071385

Oliveira VC, Melo LDFA, Melo Jr JLA, Massahud RTR, Grugiki MA (2023) Bioinputs and organic production in Brazil: a study based on the Embrapa's Bioinsumos application. Pesq Agropec Trop Goiania 53. https://doi.org/10.1590/1983-40632023v5376326

Oliveira VS, Oliveira LC, Theodoro SH, Medeiros FP, Almeida RMP, Viana MB, Gomide CS (2024) Panorama dos remineralizadores de solo registrados no Brasil. Cong. Bras. Geologia. BH/MG Available: https://51cbg.com.br/site/cbg2024/

Olmo R, Wetzels SU, Armanhi JSL, Arruda P, Berg G, Cernava T, Wagner M (2022) Microbiome research as an effective driver of success stories in agri-food systems–a selection of case studies. Front Microbiol 13:834622. https://doi.org/10.3389/fmicb.2022.834622

Primavesi AM (2018) A biocenose do solo na produção vegetal e Deficiências minerais em culturas, nutrição e produção vegetal. Expressão Popular, São Paulo

Primavesi AM (2002) Manejo ecológico do solo: a agricultura em regiões tropicais. Ed. Nobel. São Paulo, p 550. ISBN 85-213-0004-2

Ramos CG, Dalmora AC, Kautzmann RM, Hower J, Dotto GL, Oliveira LFS (2021) Sustainable release of macronutrients to black oat and maize crops from organically-altered dacite rock powder. Nat Resour Res 30:1941–1953. https://doi.org/10.1007/s11053-021-09862-0

Ramos CG, Hower JC, Blanc E, Oliveira MLS, Theodoro SH (2022) Possibilities of using silicate rock powder: an overview. Geosci Front 13(1). https://doi.org/10.1016/j.gsf.2021.101185

Renforth P, Von Strandmann PAE, Henderson GM (2015) The dissolution of olivine added to soil: implications for enhanced weathering. Appl Geochem 61:109–118. https://doi.org/10.1016/j.apgeochem.2015.05.016

Ribeiro IDA, Volpiano CG, Vargas LK, Granada CE, Lisboa BB, Passaglia LMP (2020) Use of mineral weathering bacteria to enhance nutrient availability in crops: a review. Front Plant Sci. https://doi.org/10.3389/fpls.2020.590774

Rosenzweig C, Mbow C, Barioni LG (2020) Climate change responses benefit from a global food system approach. Nat Food 1:94–97. https://doi.org/10.1038/s43016-020-0031-z

Rosset PM (2015) Social organization and process in bringing agroecology to scale. In: Proceedings of FAO international symposium. Agroecology for Food Security and Nutrition, Rome, 18–19 Sept. Available: http://www.fao.org/3/a-i4729e.pdf. Accessed 24 May 2024

Santos MS, Nogueira MA, Hungria M (2019) Microbial inoculants: reviewing the past, discussing the present and previewing an outstanding future for the use of beneficial bacteria in agriculture. AMB Expr 9. https://doi.org/10.1186/s13568-019-0932-0

Santos MS, Nogueira MA, Hungria M (2021) Outstanding impact of Azospirillum brasilense strains Ab-V5 and Ab-V6 on the Brazilian agriculture: lessons that farmers are receptive to adopt new microbial inoculants. Rev Bras Cienc Solo. https://doi.org/10.36783/18069657rbcs20200128

Santos RM, Araujo F, Jariwala H, Khalidy R, Haque F, Chiang YW (2023) Pathways, roundabouts, roadblocks, and shortcuts to safe and sustainable deployment of enhanced rock weathering in agriculture. Front Earth Sci 11. https://doi.org/10.3389/feart.2023.1215930

Setiawati TC, Mutmainnah L (2016) Solubilization of potassium containing mineral by microorganisms from sugarcane rhizosphere. Agric Agric Sci Procedia 9:108–117. https://doi.org/10.1016/j.aaspro.2016.02.134

Silva VN, Silva LESF, Silva AJN, Stamford NP, Macedo GR (2017) Solubility curve of rock powder inoculated with microorganisms in the production of biofertilizers. Agric Nat Resour 51(3):142–147. https://doi.org/10.1016/j.anres.2017.01.001

Silva A, Souza A, de Araujo L, Frezarin E, de Souza G, da Silveira C, Rigobelo E (2023) Trichoderma harzianum and Bacillus subtilis in association with rock powder for the initial development of maize plants. Agronomy 13(3):872. https://doi.org/10.3390/agronomy13030872

Soares GJ, Theodoro SH, Carvalho AXM, Burbano DFM, Ramos CG (2022) Remineralizadores de suelos y sistemas agroforestales: una opción para la captura de CO_2. Rev Agroecol 15(1). https://doi.org/10.59187/revistaagroecologia.v15i1

Sousa WS, Campos TS, Souza AGV, Cintra PHN, Faria LO, Santos TEB (2021) Performance of lettuce submitted to the rock dust remineralizer and doses of efficient microorganisms. Rev Agric Neotrop 8(2). https://doi.org/10.32404/rean.v8i2.5526

Souza MEP, Carvalho AMX, Deliberali DC, Jucksch I, Brown GG, Mendonça ES, Cardoso IM (2013) Vermicomposting with rock powder increases plant growth. Appl Soil Ecol 69:56–60. https://doi.org/10.1016/j.apsoil.2013.01.016

Souza FNS, Oliveira CG, Martins ÉS, Alves JM (2017) Efeitos condicionador e nutricional de um remineralizador de solos obtido de resíduos de mineração. AgriEnviron Sci. https://doi.org/10.36725/agries.v3i1.204

Steffen W et al (2015) Planetary boundaries: guiding human development on a changing planet. Science 347:1259855. https://doi.org/10.1126/science.1259855

Swoboda P, Döring TF, Hamer M (2022) Remineralizing soils? The agricultural usage of silicate rock powders: a review. Sci Total Environ 807. https://doi.org/10.1016/j.scitotenv.2021.150976

Tanchum M (2022) Morroco's new challenges as a gatekeeper of the world's food supply: the geopolitics, economics, and sustainability of OCP's global fertilizer exports. Middle East Institute. Available: https://www.mei.edu/sites/default/files/2022-01/Morocco%27s%20New%20Challenges%20as%20a%20Gatekeeper%20of%20the%20World%27s%20Food%20Supply.pdf

Tavares LF, Carvalho AMX, Camargo LGB, Pereira SGF, Cardoso IM (2018) Nutrients release from powder phonolite mediated by bioweathering actions. Int J Recycl Organic Waste Agric 7:1–10. https://doi.org/10.1007/s40093-018-0194-x

Theodoro SH, Leonardos OH, Rocha EL, Macedo I, Rego K (2013) Stonemeal of amazon soils with sediments from reservoirs: a case study of remineralization of the Tucuruí degraded land for agroforest reclamation. Acad Brasil Ciências 85(1):23–34. https://doi.org/10.1590/S0001-37652013000100003

Theodoro SH, Leonardos OH (2006) The use of rocks to improve family agriculture in Brazil. An Acad Bras Ciências 78(4):721–730. https://doi.org/10.1590/S0001-37652006000400008

Theodoro SH, Leonardos OH (2015) Stonemeal: principles, potential and perspective from Brazil. In: Goreau TJ, Larson RW, Campe J (eds) Geotherapy: innovative methods of soil fertility restoration, carbon sequestration and reversing CO_2 increase. CRC Press, USA, pp 403–418

Theodoro SH, Leonardos OH (2021) The unsustainable sustainability. An Acad Bras Ciênc 93(1). https://doi.org/10.1590/0001-3765202120181226

Theodoro SH, Medeiros FP, Ianniruberto M, Jacobson TKB (2021) Soil remineralization and recovery of degraded areas: an experience in the tropical region. J S Am Earth Sci 103014. https://doi.org/10.1016/j.jsames.2020.103014

Theodoro SH, Manning DAC, Carvalho AXM, Ferrão FR, Almeida GR (2022) Soil remineralizer: a new rote to sustainability for Brazil, a giant exporting agro-mineral commodities. In: Yakovleva N, Nickless E (eds) Routledge handbook of the extractive industries and sustainable development, 1st edn, pp 261–281. https://doi.org/10.4324/9781003001317

Theodoro SH (2000) A Fertilização da terra pela Terra: uma alternativa para a sustentabilidade do pequeno produtor rural. Ph.D. Universidade de Brasília, Brasília, DF, p 221. Available: https://repositorio.unb.br/handle/10482/20881. Accessed 24 May 2024

Theodoro SH (2021) Ressignificando a Rochagem. In: Theodoro SH, Martins EdS, Monte MB (eds) Anais do IV Congresso Brasileiro de Rochagem, 1st edn. Autografia, RJ, pp 15–35

Van Straaten P (2002) Rocks for crops: agrominerals of sub-Saharan Africa. International Centre for Research in Agroforestry, Nairobi. ISBN: 0-88955-512-5

Van Straaten P (2006) Farming with rocks and minerals: challenges and opportunities. An Acad Bras Ciências 78(4):731–747. https://doi.org/10.1590/S0001-37652006000400009

Van Straaten P (2022) Distribution of agromineral resources in space and time—a global geological perspective. Pesquisa Agropecuária Brasileira 57. https://doi.org/10.1590/S1678-3921

Vermeulen SJ, Campbell BM, Ingram JSI (2012) Climate change and food systems. Annu Rev Environ Resour 37:195–222. https://doi.org/10.1146/annurevenviron-020411-130608

Vicca S, Goll DS, Hagens M et al (2022) Is the climate change mitigation effect of enhanced silicate weathering governed by biological processes? Glob Change Biol 28(3). https://doi.org/10.1111/gcb.15993

Vidal MC, Dias RP (2023) Bioinsumos a partir das contribuições da agroecologia. Rev Bras Agroecol 18(1):171–192. https://doi.org/10.33240/rba.v18i1.23735

Vidal MC, Saldanha R, Verissimo MAA (2020) Bioinsumos: o programa nacional e a sua relação com a produção sustentável. In: Gindri DM, Moreira PAB, Verissimo MAA (Org.) Sanidade vegetal: uma estratégia global para eliminar a fome, reduzir a pobreza, proteger o meio ambiente e estimular o desenvolvimento econômico sustentável, 1st edn. CIDASC, Florianópolis, pp 382–409

Wenliang W, Yuhui Q, Fanqiao M, Ji L, Yanbin G, Huafen L (2016) Status quo and development policies of organic agriculture in China. In: Shiming L, Gliessman S (eds) Agroecology in China. CRC Press, pp 291–299. ISBN 978-1-4822-4935-4

Wood C, Harrison AL, Power IM (2023) Impacts of dissolved phosphorus and soil-mineral-fluid interactions on CO_2 removal through enhanced weathering of wollastonite in soils. Appl Geochem 148. https://doi.org/10.1016/j.apgeochem.2022.105511

Zalles V, Hansen MC, Potapoy PV, Chavez S (2018) Near doubling of Brazil's intensive row crop area since 2000. PNAS 116(2):428–435. https://doi.org/10.1073/pnas.181030111

Zhang L, Gadd GM, Li Z (2020) Microbial biomodification of clay minerals. In: Advances in applied microbiology. Academic Press. https://doi.org/10.1016/bs.aambs.2020.07.002

An Overview of Research and Innovation in Stonemeal Technology in Brazil

Manoel Batista Viana, Caroline Siqueira Gomide, Fernanda de Paula Medeiros, Claudete Gindri Ramos, and Suzi Huff Theodoro

Abstract Stonemeal, an agricultural practice where rock dust is used as an agricultural input, is a promising technological procedure in Brazil that has ensured its role as a leader in the agricultural sector. This chapter analyzed articles published in four editions of the Brazilian Stonemeal Congresses (BSC), which were held in 2009, 2013, 2016, and 2021. We examined identifying trends, recurring themes, types of rocks, agricultural crops and soils, as well as the main research groups, the most frequently used methodologies in the analyses, and the results listed in the articles. The systematized data were analyzed from a total of 108 articles. Our survey enabled us to establish correlations derived from the data using the IRaMuTeQ software. The results of this analysis show that stonemeal research in Brazil encompasses several objectives, including the extraction/ processing of rocks and their main characteristics, how they are used in different types of soil, and the benefits they offer to plants. Soybeans and maize were the plants that were tested the most. However, several identified challenges need to be addressed, including the expansion of research groups, the concentration of research on a limited number of crops, and the need for public policies that promote the adoption of this technique among a diverse range of farmers. After systematically analyzing the results, we concluded that stonemeal has the potential to contribute to a robust, resilient, and sustainable agricultural system in Brazil.

M. B. Viana · C. S. Gomide · F. de Paula Medeiros · S. H. Theodoro (✉)
Postgraduate Program in Environment and Rural Development (PPG-MADER), FUP-UnB, University of Brasília, Planaltina, DF, Brazil
e-mail: suzitheodoro@unb.br

Stonemeal Technology Research Group, CNPq-National Council for Scientific and Technological Development, Brasilia, Brazil

C. S. Gomide
e-mail: carolinegomide@unb.br

C. G. Ramos
Department of Civil and Environmental, Universidad de La Costa, Barranquilla, Atlántico, Colombia
e-mail: cgindri@cuc.edu.co

É. de Souza Martins and S. Huff Theodoro (eds.), *Soil Remineralizers and Silicate Fertilizers*, Frontier Studies in Soil Science,
https://doi.org/10.1007/978-3-032-14656-4_12

Keywords Soil remineralizer · Tropical agriculture · Rock dust · Textual analysis

1 Introduction

Soil remineralizers, sometimes referred to as rock dust, play a crucial role in Brazilian agriculture (Leonardos et al. 2000). The use of these inputs is based on the suppositions of stonemeal technology, which claims that certain types of crushed rocks become sources of macro and micronutrients for soil and plants (Theodoro 2000). The use of rock dust is regulated by a legal framework consisting of a law, a decree, and normative instructions.

The growing interest in the use of remineralizers can be attributed to several factors, including higher purchase costs for soluble fertilizers, the pursuit of more sustainable agricultural practices, and the realization of direct and indirect benefits from the use of available multinutrient inputs throughout Brazil. Several studies (Ramos et al. 2021a; Theodoro et al. 2022; Medeiros et al. 2025) have listed a series of direct agronomic benefits, including changes in soil pH and CEC, a reduction in aluminum toxicity, increased moisture retention capacity, and products that meet farmers' expectations.

In addition to the direct benefits, there are important indirect benefits from using locally/regionally available mineral sources. Some of these benefits include adopting production practices that align more closely with the Sustainable Development Goals (SDGs) established by the UN (Manning and Theodoro 2020). Another relevant aspect involves achieving higher levels of food and nutritional security (Theodoro 2023), as well as the potential to capture atmospheric carbon (Beerling et al. 2020).

A number of these benefits were published in scientific articles as well in the expanded abstracts in the four editions of the Brazilian Stonemeal Congresses (BSC). Understanding, cataloging, and organizing results from different research groups over time helps systematize the findings on this topic, leading to benefits that can both strengthen and promote a more sustainable agricultural model. Furthermore, understanding the background of the studies presented in the BSCs helps to further our understanding of how this research has evolved and innovated over the years as technology gains more prominence and recognition.

Another important aspect is that the use of mineral sources, with or without organic inputs, achieves better food security and higher rural and agricultural development rates for Brazil, which in turn results in a better quality of life for society as a whole. More than 200 articles and expanded abstracts were published: 40 in the 1st Edition, 43 in the 2nd, 56 in the 3rd, and 69 in the 4th. The 4th Edition also included short summaries on a wide range of topics and results.

Understanding and analyzing the benefits and challenges reported in the articles that comprise the four editions of the BSCs is key to describing the transformative potential of this technological option for agriculture. Our analysis aimed to integrate scientific and technical knowledge about the trajectory of research and the performance of remineralizers to enhance this agricultural practice within the country's agricultural sector. The primary objective of this work is to present the state of the art (historical) regarding stonemeal, as well as to identify the main areas of research, scientific advances, and/or possible knowledge gaps, primarily focusing on the results obtained by the different research groups in the four editions of the BSCs. To achieve this, we systematized the data published in the BSCs, identified the main objectives and methodologies in the articles, and highlighted some of the key results of the research published in the BSC editions. The results obtained from this analysis can benefit and guide future studies on improving management practices and mechanisms to facilitate the adoption of stonemeal technology among farmers (both business/family) and on implementing subsidies to promote the widespread adoption of remineralizers in Brazil.

2 Materials and Methods

The articles were selected based on the following criteria: (I) objective of the work; (II) type of rock used; (III) soil class; (IV) type of crop used; (V); analysis methodology; (VI) research groups; and (VII) main results. This data was compiled and entered into an Excel table. A total of 108 articles and expanded abstracts (65%) were selected from the available 167, with an analysis conducted to identify patterns and particularities related to this area. There were 21 articles from international authors that were excluded from the analysis, as they focused on objectives that differed from those in this analysis.

An additional 40 texts were also not analyzed, as they address more theoretical issues, the construction of the legal framework, or the use of rock dust for other purposes, such as sewage treatment. The data were then systematically organized to group the information into categories, which will be presented throughout this chapter.

The data obtained from each article were used to formulate guiding questions, which were then used to compile a textual corpus. This corpus was analyzed using the IRaMuTeQ software to identify language similarities (Camargo and Justo 2013). The IRaMuTeQ software enables you to perform word frequency analyses, identifying the most frequent words in a text and their relationships. In addition, the software performs word-class analyses, grouping words with similar meanings (and forming word networks), which helps visualize the relationships between them. These analyses identify important themes and concepts in the analyzed texts (Camargo and Justo 2013; Melo and Souza 2023).

The software consists of an R interface code for the multidimensional analysis of texts and questionnaires. Developed at the University of Toulouse in France, this program is versatile as it is used for academic research, media data analysis, and knowledge management (Salviati 2017). This analysis enables us to identify the most explored areas, emerging themes, and ultimately highlight some gaps that need to be addressed in stone research in Brazil.

3 Results

3.1 History of Stonemeal Technology

Stonemeal technology is not new in Brazil. According to Theodoro (2000), the first studies in stonemeal technology were done by Ilchenko and Guimarães in 1955, followed by Leonardos and collaborators in the 1970s and 1980s (Leonardos et al. 1987). It is worth mentioning previous and pioneering studies carried out by James Hutton (1750), who recommended the use of rocks on his properties in Scotland, and by Julius Hensel (1894), in Germany, who wrote the book Bread from Stones, where the first assumptions about the use of this type of rock dust were established.

However, it has only been over the last 25 years that research in this area in Brazil has garnered attention, ultimately leading to the establishment of a regulatory framework related to the topic. Law 12,890/2013 (Brazil 2013) established the concept of soil remineralizers and, subsequently, Decree 8,384/2014 (Brazil 2014) and Normative Instructions (NI) 05/2016 (Brazil/Mapa 2016) defined the minimum conditions and guarantees that these materials must have in order to be converted into agricultural input.

Following Leonardo's pioneering research in the late 1990s, other groups of Brazilian researchers innovated their research on the use of rock dust for crops such as maize, sugarcane and cassava (Theodoro 2000), soybeans and millet (Martins et al. 2008), soybeans (Ferreira et al. 2010), fodder palm (Medeiros 2017), quinoa (Burbano et al. 2022), maize (Krahl et al. 2022) and others.

Studies by Carvalho et al. (2010), Bamberg et al. (2013), and Alovisi et al. (2021), among others, have examined several major crops for agribusiness. These studies are important as they help strengthen large-scale production (commodities) even though they do not address the necessary establishment of their importance for food production, which do meet the ecological and social principles of agroecology. There are, however, counterpoints that deserve mention, such as research on the importance of agroecology, crop diversity, and various management forms by Almeida et al. (2007) and Soares et al. (2022). They report that the use of these materials has grown among family farmers who adhere to agroecological principles.

These studies emphasize the significance of stonemeal technology as a viable option for family farming, with an increasingly prevalent agroecological profile in the lives of farmers. Some of these studies examined the properties of remineralizers and investigated which types of rocks have macro- and micronutrients more readily available for plants (Bergmann 2019; Ramos et al. 2021b). Other studies aim to understand the differences between crops that utilize rock dust and those that do not (Medeiros et al. 2025). Figure 1 presents a historic timeline of initiatives and milestones in the development of stonemeal technology in Brazil.

Considering the historical context and the diverse themes and objectives related to stonemeal technology, the works published in the four editions of the Brazilian Stonemeal Congresses (BSCs) were accessed.

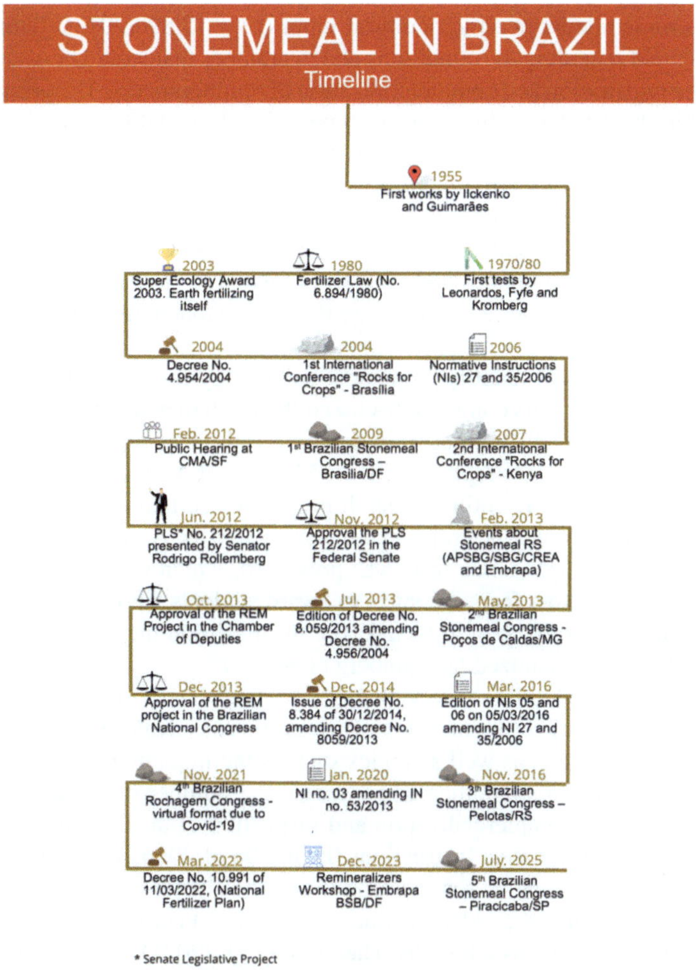

Fig. 1 Development timeline of stonemeal technology in Brazil (modified from Theodoro 2016)

The 1st Edition was held in Brasilia, between September 22 and 25, 2009. The aim was to present the first research on stonemeal technology in Brazil, establish the theme as a technological option for Brazilian agriculture, and present the initial results from the use of rock dust. This Edition sought to establish dissemination strategies through a formal network of researchers to develop the principles and benefits of stonemeal technology. A total of 40 articles were published on the 1st BSC, which were edited by Martins and Theodoro (2010).

The 2nd BSC was held in Poços de Caldas/MG, between May 12 and 17, 2013. The main theme of this congress was the regulation of rock dust in Brazil. Bill No. 212/2012, currently being processed by the National Congress, was discussed as it includes remineralizers as an agricultural input (Law 12,890/2013). Initial discussions were held regarding the conditions and guarantees that would be established to regulate the registration and commercialization of remineralizers. The primary topics addressed in the articles include various types of rocks suitable for agricultural use across different ecosystems, as well as methodologies for chemical analysis and the agronomic efficiency of soil remineralizers. A total of 43 articles were published. This Edition saw the first essay contest for undergraduate students about stonemeal technology. The Edition was edited by Theodoro et al. (2013). Considering the historical context and the diverse themes and objectives related to stonemeal technology, the works published in the four editions of the Brazilian Stonemeal Congresses (BSCs) were accessed.

The 3rd BSC helped identify stonemeal technology as a suitable technique for Brazilian agriculture. This congress discussed the implementation of public policies in the form of decrees and Normative Instructions (NIs) that outlined the criteria that remineralizers must meet to register with the Ministry of Agriculture, Livestock, and Food Supply (MAPA). Scientific research has shown certain aspects related to the interaction between minerals, crops, and microorganisms. A methodological procedure for conducting agronomic tests was presented, which guides the experiments to date. There was also a second contest to award students who wrote essays on the future of soil remineralization in Brazil. The third Edition included 52 articles and expanded abstracts, organized by Bamberg et al. (2017).

The 4th Edition, held remotely due to the COVID-19 pandemic, addressed the characterization of different types of rocks, recent advances, the formation of networks of researchers, as well as topics such as the implementation of a national remineralizer policy to strengthen and increase the number of registered products suitable for the use, commercialization and inspection of this agricultural practice, as provided for in Normative Instructions 05 and 06 of 2016. This 4th Edition also presented the first studies on remineralizers and their potential for capturing CO_2. Thirty-one articles and expanded abstracts were published, in addition to 42 short abstracts. This Edition was edited by Theodoro et al. (2021), and was the first one made available online.

International researchers participated and contributed to the discussions in all the Congresses by presenting experiences from their respective countries. There were 21 articles published in English and/or Spanish: seven in the 1st BSC, five in the 2nd BSC, three in the 3rd BSC, and six in the 4th BSC. Over the course of 11 years, the Congresses have attracted a growing number of participants. There are many interested parties and new research addressing topics such as the consolidation of technology and the creation of an association. They also focus on the regulation of the commercialization of rock dust and the search for sustainable solutions in agriculture. These events have promoted further research, innovation, and the adoption of stonemeal as a suitable agricultural practice for Brazilian tropical agriculture, which can contribute to environmental sustainability, increased productivity, and the promotion of practices that result in more fertile soils. All proceedings from the four editions are available on the Geological Survey of Brazil website (https://www.sgb.gov.br/remineralizadores/anais.html).

4 Data About the Brazilian Stonemeal Congresses

Systematizing the data from the articles and expanded abstracts in the four editions of the BSC revealed a wide diversity of objectives related to the different types of rocks tested as remineralizers, their agronomic effects on various crops, the experimental models and analysis methodologies, the main research groups, and the main results obtained. We were able to determine that the studies are committed to scientific and technological development, as well as exploring different types of rocks in various regions of the country. We observed the evolution of methodological procedures, different types of analysis, and interdisciplinary collaboration, all of which are essential for advancing knowledge and technology.

In terms of the types of rocks (Fig. 2), basalts and schists were the lithotypes that were most used in the tests (24 each), followed by phonolite (11), granites (5), syenites (4), amphibolites, serpentinites and ornamental rock residues (3 each), carbonatites, siltstones, phlogopite and dacites (2), gabbro, diabase, sandstones with zeolites and pegmatite tailings (1). This diversity of lithological types confirms the immense potential that rocks similar to these can be used to increase soil fertility in the different regions of the country where they are found. In any case, the use of basalts and schists can ensure a wider use of these rocks since they are widely available throughout Brazil.

In terms of the crops that were most used in agronomic tests (Fig. 3), maize (16) and soybeans (15) were the most frequent, followed by sugarcane (10), Beans (7), *Brachiaria* (3), coffee, rice and lettuce (2), as well as fruits such as blackberries (1), grapes (1) and oranges (1), cereals such as oats and wheat (1), and sweet potatoes (1) and cassava (1).

As seen from our analyses of the works, there was a predominance of experiments using Ferralsol for Oxisols (32), particularly the Dystrophic Red Yellow Latosol, followed by Argisols (14), Haplic Planosol (5), and Arenosol for Quartzarenic Neosol (3). Despite this frequency, some studies do not specify the types of soil used or do not

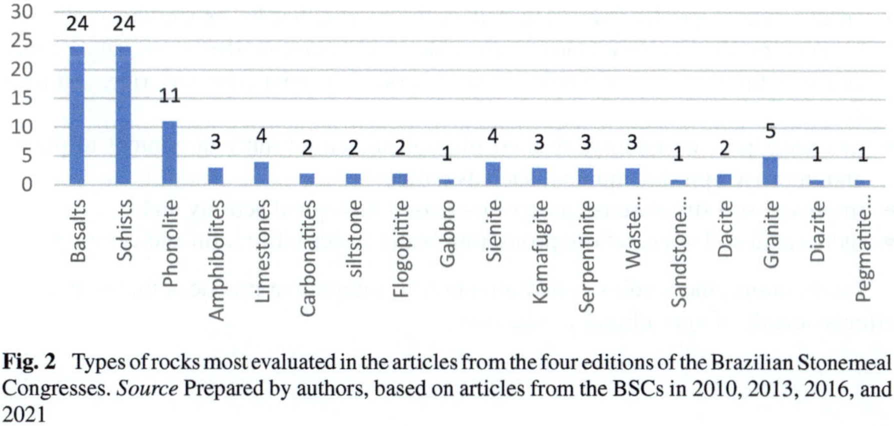

Fig. 2 Types of rocks most evaluated in the articles from the four editions of the Brazilian Stonemeal Congresses. *Source* Prepared by authors, based on articles from the BSCs in 2010, 2013, 2016, and 2021

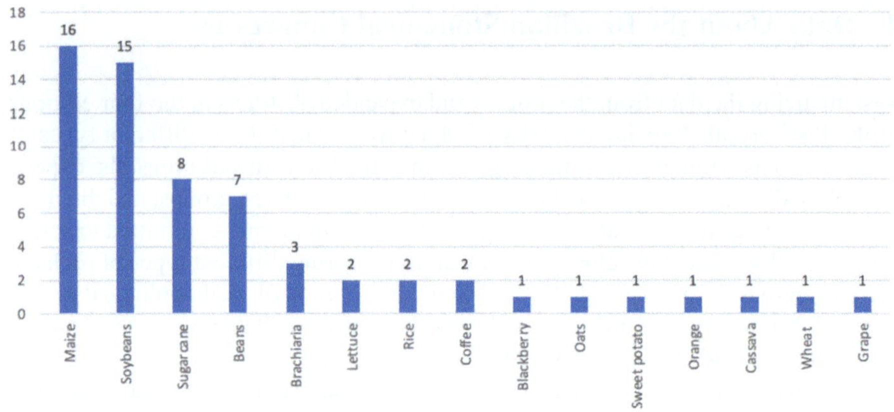

Fig. 3 Main agricultural crops utilized in agronomic tests. *Source* Prepared by authors, based on articles from the BSCs in 2010, 2013, 2016 and 2021

include agronomic experiments. In these instances, articles focused on lithological characterizations of various rock types are particularly noteworthy.

In order to aggregate similar information about the articles and expanded abstracts from the four editions of the BSCs, we divided the articles according to the types of tests that were applied and then gathered the information on the various results listed by the researchers and attributed them to three categories: (i) field experiments, (ii) greenhouse experiments, and (iii) laboratory analyses. Aggregating the results is a way to systematize the broad spectrum of benefits and outcomes of applying ground rocks in agriculture. Overall, the published studies yielded the following results:

- improved soil fertility levels by increasing pH, cation exchange capacity (CEC), and availability of essential nutrients such as calcium, magnesium, potassium, and phosphorus;
- reduction in soil acidity by decreasing exchangeable aluminum levels and potential acidity, making the soil more suitable for plant development;
- an increase in productivity reported in the works that by expanding grain and biomass production in various crops such as soybeans, maize, beans, sugar cane, and sweet potatoes, in addition to other works that reported beans, rice, and fruit crops;
- increased product quality, verified through increased nutrient content in plants and improved product quality, such as coffee;
- improved soil structure because of increased biological activity and
- increase in soil aggregation, promoting better water infiltration and aeration.

Furthermore, the studies called attention to integrative practices that may have effects outside of agricultural production:

- synergistic effect with other agricultural practices, especially with regards to crushed rocks combined with organic matter and other practices that incorporate green manure into the soil;
- the possibility of using waste (discarded products and/or residue) as an agricultural input may, over the medium term, reduce dependence on chemical fertilizers as these waste products can be registered as remineralizers (if they meet the specifications established in NI 05/2016), thus facilitating the partial or total replacement of conventional fertilizers with ground rocks. The articles suggested that there should be a reduction in production costs with respect to using waste materials in this manner and, ultimately, a reduction in the emission of greenhouse gases which can harm the environment. Some works claim that the use of these materials enhances the aggregation of resources and the extraction and processing of mineral assets, which could benefit regions that produce other mineral assets, such as ornamental rocks;
- some works highlighted the multinutrient potential that different types of rocks present in their mineralogical and lithochemical properties, which meet the different nutritional needs of forest crops;
- some results highlighted the handling and dose rates of different rocks, as the efficiency of crushed rocks depends on the dose that is applied, the type of soil that is used, the regional climate, and the crop itself. Adequate management is essential to obtaining the best results;
- microbiological studies showed significant changes in the structure of bacterial communities in soils that were treated with ground rocks compared to soils that were treated with conventional potassium sources (KCl);
- the combination of organic fertilizers and remineralizers proved promising for the initial development of grasses, with emphasis on fresh and dry mass and the number of tillers;
- basic rocks, such as basalts, have faster nutrient release kinetics compared to acidic rocks, such as gneisses and granites. However, the latter also has the capacity for gradual and continuous nutrient release, contributing to the expansion of mineral sources for the agricultural system.

The papers that addressed rock characterization, whether in the laboratory or systematic mapping of the geological resources available in Brazil, revealed that phonolites are important sources of potassium, as are some types of schists. Basalts, particularly their finer fractions, present significant levels of several nutrients, particularly Ca and Mg, and, occasionally, K, in addition to a series of micronutrients that meet the nutritional demand of plants.

Investigating which research groups produced these results is also important as it helps to understand the scope of the studies, as well as the different research institutions that invest in stonemeal innovation. In general, the institutions based in the south-central region of the country had the greatest number of publications included in the Congresses.

Stonemeal technology facilitates synergy and interaction between the agricultural and mineral sectors, and our systematization of the data reveals that the various

Embrapa units, particularly the Temperate Climate and Cerrado units, have published the most research results. Other Units from across Brazil have also participated in the BSC and presented various results. Likewise, the Geological Survey of Brazil (SBG/CPRM) also presented research results, particularly related to surveys of regional rock sources suitable for agricultural use. SBG studies identify and characterize rocks that have the potential to provide nutrients.

Universities from across the country were also important centers for experimenting and developing concepts, methodological procedures and results. There were around 47 articles developed at Brazilian universities, with many coming from ESALQ/USP, UnB, and Viçosa. These institutions are among the top universities in the country and leaders in their commitment to high-level research through innovation, scope, and significant contributions to this topic. Other equally important universities also produced excellent quality work (UNITINS, UFRGS, UFSC, UFPR, UFSB, UFMG, UFMS, and others). In addition to the Mineral Technology Center (CETEM) from the Ministry of Science, Technology, and Innovation (MCTI), private companies have also participated and supported research on the use of ground rocks as fertilizers. State institutions and non-governmental organizations (NGOs) have also been involved with the topic. Figure 4 illustrates the distribution of research by authors and the institutions to which they belong.

The research showed that the most used methodology in the work was chemical analysis of the soil. This type of analysis is essential for obtaining good productivity results and management practices over time since it provides information about the availability of nutrients in the soil, thus allowing farmers to plan their land use more appropriately in terms of the nutritional needs of their plants and the supply of nutrients (a lack or excess of), which has an effect on fertility levels and helps determine which fertilizer management practices and other inputs should be used (Silva 2009).

In addition to the systematized results, the research involved conducting an analysis using the IRaMuTeQ software. The same analytical categories (objectives, types of rocks, soils, agricultural crops, research groups, methodologies and results) were analyzed by the software.

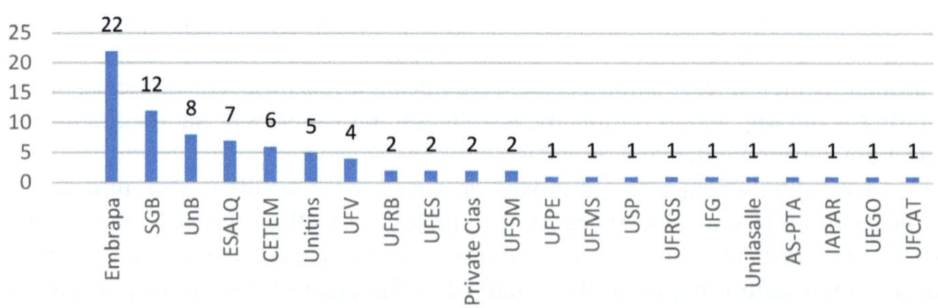

Fig. 4 The main research groups whose articles were published in the Brazilian Stonemeal Congresses. *Source* Prepared by authors, based on articles from the BSCs in 2010, 2013, 2016 and 2021

The data obtained from each article were used to form guiding questions that helped build a text corpus, which was analyzed using the IRaMuTeQ software to identify language similarities. This software analyzes word frequency, identifying the most frequent words and their relationships (word network) to identify the most relevant themes and/or concepts.

The similarity analysis (Fig. 5) performed using the IRaMuTeQ software revealed a rich range of information on stonemeal, including its main attributes and relationships. Examining the word 'network' helped identify the various aspects involved in Stonemeal research. This is why it is a transdisciplinary topic; the analysis revealed the similarities between the words and their connections, providing the context for the research and the main responses.

This analysis allowed us to identify the most explored areas, emerging themes, and highlight some gaps in stonemeal research in Brazil. This tool (IRaMuTeQ) helped qualitatively confirm the previously presented results, thereby aiding in the identification of patterns and trends.

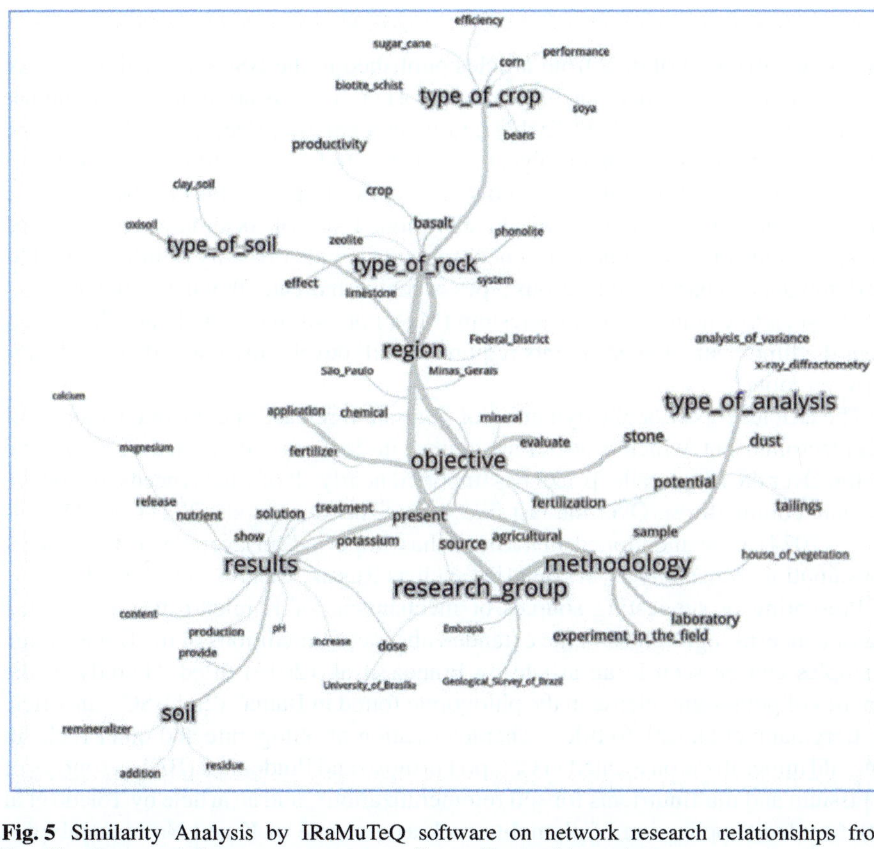

Fig. 5 Similarity Analysis by IRaMuTeQ software on network research relationships from Brazilian Stonemeal Congresses text corpus. *Source* Prepared by authors

To further highlight this information, data from Martins et al. (2024), published in Novo Solo Journal by the Brazilian Association of Producers of Remineralizers and Natural Fertilizers (ABREFEN), were incorporated. This association compiled a summary of Brazilian production of REM and simple mineral fertilizers with silicate composition (FSi) from 2019 to 2023, during which approximately nine million tons of these products were produced. This study revealed that the Brazilian states producing the most REM and Silicate fertilizers (FSi) were Minas Gerais (43%), Goiás (24%), São Paulo (10%), Paraná (9%), and Rio Grande do Sul (6%). Data from the Integrated System of Agricultural Products and Establishments (SIPEAGRO), of the Ministry of Agriculture and Livestock (MAPA), show that 68 products have already been registered in 2024. According to the above-mentioned study, over the last four years, around 10% of the country's agricultural area has been using remineralizers in their production processes.

5 Discussion

The systematization of data from articles published in the BSCs showed a series of achievements reached over an 11-year span. There is legislation governing the use of remineralizers (Law 12,890/2013) and their regulation (Decree 8,384/2016 and NI 05 and 06/2016). The main objectives of the studies were to evaluate different types of rocks for different ecosystems and types of agriculture (agribusiness and family). From an innovative perspective, various types of rocks have been utilized in experiments or agronomic tests, encompassing a wide diversity of lithotypes. The predominance of basalts and various types of schists indicates that the main macronutrients, such as calcium (Ca), magnesium (Mg), potassium (K), and phosphorus (P), meet the fundamental needs plants require for full development and their exchange between roots and soil.

The articles report on the dynamics of potassium release and the main sources of this macronutrient, which is one of the highest in-demand sources in the agricultural sector. Over the last decade, Brazil has imported nearly all of what it needs to produce its main commodities. Over the last five years, Theodoro et al. (2022) and Martins et al. (2024) have mentioned that Brazil has imported an average of 96% of its consumption, especially from countries such as Russia, Belarus, and Canada.

Presenting or suggesting sources or mechanisms of K release was one of the major concerns highlighted in the extended abstract four editions of the BSCs. Some examples can be seen in an article by França et al. (2013) titled "a study of the kinetics of potassium release in the phlogopite found in Bahia" (2nd BSC), an article by Bergmann et al. (2016) titled "characterization of phlogopite and other rocks in emerald mineralization located in Campo Formoso and Pindobaçu (BA) as sources of potassium and multinutrients for soil remineralization", and an article by Toledo et al (2016) titled "application of phonolite rock as a source of K and Na in eucalyptus plantations", both of which are from the 3rd BSC, and an article by Cortes et al.

(2010) titled "Phonolite as a substitute for potassium chloride and/or other sources of potassium in agriculture and livestock in Brazil" (1st BSC).

Although there is a significant quantity and variety of rocks that are possible sources of potassium, the release of mineral compounds that contain K needs to be better understood in order to advance or delay the reactions. Studies (Manning 2010) have shown significant results regarding the release of K from minerals such as leucite, biotite, muscovite, and K-feldspars. This is important because, as mentioned by Churchmann et al. (2020), K solubilization can vary according to the nature and composition of the minerals, their crystalline structure, and the degree of alteration and genesis. Furthermore, as reported by Souza et al. (2019) and Carvalho (2012), the presence of organic acids secreted by plant roots and microorganisms increases potassium solubility, which is a crucial indicator to monitor.

Similarly, phosphorus (P) is another imported macronutrient that Brazil continues to rely on. Recent years have seen a significant portion of the demand being met by importing mixtures that contain P, primarily from Morocco. One of the works that addresses this topic is by Carneiro et al. (2016), titled "Phosphorus availability for the application of a siltstone remineralizer." Hinsinger (2001) notes that phosphorus is probably the least mobile and least available nutrient for plants in most soils, which is a significant factor, and possibly the primary limiting factor, for the growth of agricultural species. The authors studied the dissolution of phosphate rock in the presence and absence of cultivated plants, in a substrate that simulates an acidic phosphorus-fixing soil (without a source of P and Ca). They concluded that, in the absence of plants, the low pH of the substrate resulted in the dissolution of approximately 8% to 30% of the phosphate rock, depending on the dosage applied to the substrate. Similar results were reported in an article by Stamford (2016) titled "Phosphate and potash rocks with microorganisms and organic matter" (3rd BSC).

Calcium (Ca) and magnesium (Mg) were also prominent topics in the articles from the four editions of the BSCs. The following articles stand out: the "effect of limestone and by-product of limestone exploration on soil pH, Ca, and Mg, and on the production of dry mass of corn (Rodrigues et al. 2016) and the "migration of exchangeable Ca^{2+} in the soil after application of limestone and agricultural gypsum in eucalyptus plantations" by Macana et al. (2017). Ca and Mg are usually obtained from liming, which is a particular type of stonemeal technology that uses ground calcitic or magnesian limestone as a source for these nutrients. Over the last 20 years, there has been an increase in the use of these inputs (particularly in soils in the central-west and northern regions) to improve their acidity. This practice has been questioned in recent years because the dissolution of Ca and Mg can increase the effects of climate change, according to Emrich et al. (2013). In any case, silicate rocks rich in Ca and Mg, such as basalts, are highly abundant in Brazil.

Although not addressed in the works, another important point concerns the necessary presence of silicon (Si) in the soil for the full development of most plants (especially wheat, rice, and sugarcane). It is worth noting that sources of this nutrient are present in all minerals that form silicate rocks. Although the conventional model considers Si as a non-essential but beneficial element for plants, Kelland et al. (2020) point out that most crops, especially cereals, accumulate this nutrient in the aerial part, resulting in greater yields and resistance to abiotic (drought, salinity and heat) and biotic (pests) stress.

The types of crops that were subjected to testing and then reported on in the articles and expanded abstracts of the BSCs were predominantly soybeans and maize. This makes sense since these crops are abundant in most agricultural areas in Brazil, with approximately 54.7 million hectares (Mha) for soybeans and 20.6 Mha for maize (IBGE 2017). Similarly, sugarcane was one of the agricultural species that was researched the most. Sugarcane cultivation is a historical practice in Brazil and is one of the three most important crops in agribusiness, occupying an area of approximately 9 million Hectares (IBGE 2017). According to data from the National Fertilizer Plan (PNF-2050, Brazil 2022b, a), soybeans, corn, and sugarcane accounted for 72% of fertilizer consumption in Brazil. Another crop with large-scale production is coffee, which was also the subject of research in the studies. Wheat and oats (which are some of the more common crops in southern Brazil) were also tested using rock dust. This information suggests that research and studies are focusing on strengthening and supporting agribusiness. Even still, we were able to verify the presence of several crops produced by family farming, including cassava, beans, sweet potatoes, and various vegetables. These choices may indicate a concern with essential crops because they are used in food hampers throughout Brazil.

There is a concentration of experiments and analyses being conducted in institutions located in the central and southern regions of Brazil. Not only do these regions have greater access to public resources for research, but the South, Southeast, and Midwest are the regions that consume the most fertilizers in the country. Approximately 73.5% of all fertilizers consumed (ANDA 2022) are used in these regions. As a result, access to new sources of fertilizers is, theoretically, an alternative to maintaining the country's current production model. The use of remineralizers, therefore, becomes a necessity to reduce dependence on soluble inputs.

The data revealed that some of Embrapa's Units (Temperate Climate, in Rio Grande do Sul, and Cerrados, in the Federal District) are the groups with the largest number of published works and, therefore, research being conducted. The characterization of rocks suitable for use as remineralizers was mainly produced by scientists at SGB/CPRM. Just as important as these institutions, Brazil's public universities have made important contributions to boosting research, particularly ESALQ/USP and UnB. However, it is worth noting that other universities and research centers have also contributed important articles to these events. Some examples are the Federal Universities of Southern Bahia and UFPE. UNITINS (Tocantins State University) was also present at three of the events, contributing reports on the use of rocks in the Northeastern and Northern states.

With regards to the reported results, the systematization of the data showed that, in general, the application of different rocks promoted the development of several crops, improved the chemical attributes of the soil such as pH, CEC, and availability of essential nutrients (Ca, Mg, K, P), reducing acidity and exchangeable aluminum. Furthermore, the use of ground rocks altered the physical properties of the soil by reducing erosion processes and forming aggregates within the soil. Some articles reported important changes in the soil bacterial community, which enhanced plant development. From a lithogeochemical point of view, studies have shown that basalts and schists are the primary types of rocks used in agriculture, either because they contain the essential nutrients required by plants or because they are widely distributed throughout the country. The more gradual release of nutrients from rocks was mentioned in several studies, which drew attention to the fact that, unlike soluble fertilizers, the residual and long-term effects are a comparative advantage of REM over soluble fertilizers.

Several international researchers, who are known for their research on rock dust, also shared their knowledge and participated in discussions on the topic. Some of these researchers are international references in research on the use of rock dust. University representatives from other countries also participated in the BSCs, including those from Angola, Argentina, South Africa, Colombia, Uganda, Portugal, the Netherlands, the United States, and Germany.

All these advances in research, as described in the articles, demonstrate the importance of the Brazilian Stonemeal Congresses in disseminating research and fostering innovation in Brazilian sciences. These efforts have led to the establishment of a legal framework that provides security for the use of this input throughout the country. The research presented at Stonemeal Events also helped establish procedures and best practices for the use of remineralizers, as well as protocols that facilitate the observation of agronomic results in various types of crops across distinct agroecosystems.

6 Conclusions

The analysis of the four editions of the BSCs, published between 2010 and 2021, provided a detailed mapping of the state of the art of stonemeal technology in Brazil, and helped identify advances, trends, and knowledge gaps on this topic. The results showed an evolution in research and stonemeal development, which includes a substantial growth in the number of works on the subject, demonstrating the growing interest and the diversification of the theme, ranging from basic studies to practical applications in different regions and agroecosystems.

The systematization of the data revealed that the studies primarily focus on agribusiness crops. Even still, we identified some works that focused on agricultural crops produced by family farming. What is also evident is that the use of remineralizers helps recover degraded areas and increases interest in other areas. The challenges and prospects are similar; therefore, it is essential to strengthen research in order to improve understanding of the mechanisms involved in soil modifications and the assimilation of nutrients by plants. Developing more efficient management methods and combining remineralizers with organic sources such as manure, straw, and bioinputs tends to improve results.

Although little is mentioned about it in the analyzed texts, one of the biggest challenges to expanding the use of rock dust in Brazil may stem from the commercialization of products that have already been registered with the Ministry of Agriculture and Livestock. Knowledge of the productive potential and benefits of remineralizers among farmers and professionals remains limited. Furthermore, it became clear from reading the articles (particularly those that present a more general overview of the production system in Brazil) that there is a need to invest in public policies that encourage family farms and other levels of farms, as well. To use stonemeal technology, particularly since the cost of soil remineralizers can be significantly lower compared to the costs when using the chemical model. Appropriating stonemeal technology has the potential to help build a fairer, more resilient, and sustainable agricultural system for Brazil.

The challenges and achievements identified in the analyzed literature can serve as a reference for researchers in other countries who are investigating the use of crushed rocks in agriculture. Although several barriers have been overcome, the most critical one was the establishment of a legal framework that regulates the application of these materials in a safe and economically viable manner. Thus, adapting regulatory standards to the reality of each country appears to be the most effective way to promote the adoption of remineralizers. This measure is particularly relevant for nations that aim to prioritize climate change mitigation through soil carbon sequestration, as proposed by the enhanced rock weathering mechanism.

Acknowledgements We are grateful for the financial support of the Grantham Foundation for the Protection of the Environmental and we extend our thanks to Foundation for Scientific and Technological Enterprises (Finatec) for management of resources and the support of the National Council for Scientific and Technological Development (CNPq) providing scholarships for part of research group

Declaration of Competing Interest The authors declare that they have no known competing financial interests or personal relationships that could have appeared to influence the work reported in this publication.

References

Almeida E, Silva F, Jr P, Ralisch R (2007) Revitalização dos solos em processos de transição agroecológica no sul do Brasil. Agriculturas 4(1):7–10

Alovisi AMT, Gomes WL, Alovisi AA, Silva JAM, Silva RS et al (2021) Produtividade e componentes de produção da soja adubada com pó de basalto. In: Theodoro MM (org.) Anais do IV Cong. Bras. de Rochagem. Rio de Janeiro, RJ, pp 217–278

ANDA (2022) Principais indicadores do setor de fertilizantes. [ANDA–National Association for the Diffusion of Fertilizers. Main Indicators of the Fertilizer Sector], São Paulo. 2022. Available: https://anda.org.br/wp-content/uploads/2023/03/Principais_Indicadores_2022.pdf. Accessed 06 Feb 2024

Bamberg AL, Silveira CAP, Martinazzo R, Bergmann M, Toniolo JÁ, Grecco MF (2013) Desempenho agronômico de fontes minerais e orgânicas de nutrientes para as culturas de milho e trigo. In: Theodoro et al. (org.) Anais do II Cong. Bras. de Rochagem, pp 24–31

Bamberg A, Silveira CAP, Martins ÉS, Bergmann M, Martinazzo R, Theodoro SH (2017) Anais do III Cong. Bras. de Rochagem. Triunfal Gráfica e Editora, p 455

Beerling DJ, Kantzas E, Lomas M, Wade P, Eufrasio RM, Renforth P et al (2020) Potential for large-scale CO_2 removal via enhanced rock weathering with croplands. Nature 583(7815):242–248

Bergmann M, Blaskowski A, Silveira CAP, Camargo MA, Simas MW, Cavalcante O (2016) Caracterização de flogopititos e outras rochas encaixantes das mineralizações de esmeralda de Campo Formoso e Pindobaçu (BA) como fontes de potássio e multinutrientes para remineralização de solos. In: Bamberg et al. (org.) Anais do III Cong. Bras. de Rochagem, pp 103–108

Bergmann M (2019) Grupo Serra Seral no Estado do Rio Grande do Sul: agrominerais para uso em remineralização e correção de solos. In: Simpósio Nacional de Estudos Tectônicos, 17/ International Symposium on Tectonics, 11. Anais

Brazil (2013) Lei 12.890/2013 de 10 de dezembro de 2013 - Altera a Lei no 6.894, de 16 de dezembro de 1980. 2013. Available: https://www.planalto.gov.br/ccivil_03/_ato2011-2014/2013/lei/l12890.htm. Accessed 10 Sept 2023

Brazil (2014) Decreto n.° 8.384 de 29 de dezembro de 2014. Altera o anexo ao decreto n.° 4.954 de 14 de janeiro de 2004, que aprova o regulamento da lei n.° 6.894 de 16 de dezembro de 1980

Brazil (2016) MAPA. Instrução Normativa N.° 05/ 2016. Available: https://www.in.gov.br/materia/-/asset_publisher/Kujrw0TZC2Mb/content/id/21393137/do1-2016-03-14-instrucao-normativa-n-5-de-10-de-marco-de-2016-21393106. Accessed: 22 Mar 2023

Brazil (2022) Decreto n.° 10.991 de 11 de março de 2022. Institui o Plano Nacional de Fertilizantes 2022–2050. 2022. Available: https://in.gov.br/web/dou/-/decreto-n-10.991-de-11-de-marco-de2022385453056#:~:text=D%20E%20C%20R%20E%20T%20A%20%3A,data%20de%20publica%C3%A7%C3%A3o%20deste%20. Accessed 26 Nov 2023

Brazil (2022) MAPA. Plano Nacional de Fertilizantes Estatísticas do Setor. 2022. [Internet] Available: https://www.gov.br/agricultura/pt-br/assuntos/insumos-agropecuarios/insumosagricolas/fertilizantes/plano-nacional-de-fertilizantes/estatisticas-do-setor. Accessed 20 Jul 2023

Burbano DFM, Theodoro SH, Carvalho AMX, Ramos CG (2022) Crushed volcanic rock as soil remineralizer: a strategy to overcome the global fertilizer crisis. Nat Resour Res 31:2197–2210. https://doi.org/10.1007/s11053-022-10107-x

Camargo BV, Justo AM (2013) Iramuteq: um software gratuito para análise de dados textuais. Temas Em Psicologia 21(2):513–518

Carneiro YI, Santos RA, Ieda JJC, Sermarini RA, Azevedo AC (2016) Disponibilidade de fósforo em função da aplicação de um remineralizador de siltito. In: Bamberg et al. (org.) Anais do III Cong. Bras. de Rochagem, pp 311–315

Carvalho AMX, Deliberali DC, Cardoso IM (2010) Efeito da rochagem no crescimento e nutrição de plantas de soja sob manejo agroecológico. In: Theodoro M (org.) Anais I Cong. Bras. de Rochagem. Planaltina/DF, pp 183–189

Carvalho AMX (2012) Rochagem e suas interações no ambiente solo: contribuições para aplicação em agroecossistemas sob manejo agroecológico. Tese de Doutorado, Universidade Federal de Viçosa, p 116. Available: https://locus.ufv.br//handle/123456789/1631. Accessed 17 Nov 2022

Churchman GJ, Singh M, Schapel A, Sarkar B, Bolan N (2020) Clay minerals as the key to the sequestration of carbon in soils. Clays Clay Min 68(2):135–143

Cortes GP, Ferreira RC, Cortes G, Rampazzo L, Ferreira LC (2010) Fonolito como substituto do cloreto de potássio e/ou outras fontes de potássio na agricultura e pecuária no Brasil. In: Theodoro M (org.) Anais I Cong. Bras. de Rochagem. Planaltina–DF, pp 75–83

Emrich EB, Silva DRG, Chagas WFT, Silva CSC, Caputo ALC, Silveira MTP (2013) Estimativas de emissões de CO_2 pela aplicação de calcário para o Brasil e seus estados. Anais do XXXIV CBCS

Ferreira LHG, Silveira CAP, Pillon CN, Santos LC (2010) Efeito da combinação de calcário de xisto e calcário dolomítico com diferentes fontes de fósforo sobre a produtividade da cultura da soja. In: Theodoro M (org.) Anais I Cong. Bras. de Rochagem. Planaltina–DF, pp 219–224

França SCA, Silva AAS, Ronconi CM (2013) Estudo da cinética de liberação de potássio contido no flogopitito da Bahia. In: Theodoro et al. (org.) Anais do II Cong. Bras. de Rochagem, pp 200–210

Hinsinger P (2001) Bioavailability of inorganic P in the rhizosphere as affected by root-induced chemical changes: a review. Plant Soil 237:173–195. https://doi.org/10.1023/A:1013351617532
IBGE-Instituto Brasileiro de Geografia e Estatística (2017) Censo Agropecuário 2017 -resultados definitivos. Rio de Janeiro. Available: https://sidra.ibge.gov.br/pesquisa/censo-agropecuario/censo-agropecuario-2017. Accessed 3 Feb 2024
Kelland M, Wade P, Lewis A, Taylor L, Sarkar B, Andrews M (2020) Grace et al. Increased yield and CO_2 sequestration potential with the C_4 cereal Sorghum bicolor cultivated in basaltic rock dust-amended agricultural soil. Glob Change Biol. https://doi.org/10.1111/gcb.15089
Krahl LL, Marchi G, Paz SPA, Angelica RS, Sousa-Silva JC et al (2022) Increase in cation exchange capacity by the action of maize rhizosphere on Mg or Fe biotite-rich rocks. Pesquisa Agropecuária Tropical 52:e72376. https://doi.org/10.1590/1983-40632022v5272376
Leonardos OH, Fyfe WS, Kronberg BI (1987) The use of ground rocks in laterite systems: an improvement to the use of conventional soluble fertilizers. Chem Geol 60(1–4):361–370
Leonardos OH, Theodoro SH, Assad MLL (2000) Remineralization for sustainable agriculture: a tropical perspective from a Brazilian viewpoint. Nutr Cycl Agroecosyst 56(1):3–9. https://doi.org/10.1023/A:1009855409700
Macana YAM, Toledo FHSF, Ferraz AV, Ferreira D et al (2017) Migração de Ca^{2+} trocável no perfil do solo após aplicação de calcário e gesso agrícola em plantio de eucalipto. In: Bamberg et al. (org.) Anais do III Cong. Bras. de Rochagem, pp 417—420
Manning DAC (2010) Mineral sources of potassium for plant nutrition. A review. Agron Sustain Dev 30:281–294. https://doi.org/10.1051/agro/2009023
Manning DAC, Theodoro SH (2020) Enabling food security through use of local rocks and minerals. Extract Ind Soc 7(2):480–487. https://doi.org/10.1016/j.exis.2018.11.002
Martins ÉS, Theodoro SH (2010) Anais I Cong. Bras. de Rochagem. Planaltina–DF, p 322
Martins ÉS, Oliveira CG, Resende ÁV, Matos MSF (2008) Rochas silicáticas como fontes minerais alternativas de potássio para a agricultura. Rochas e Minerais Industriais - CETEM. 2.ª Edição
Martins ÉS, Theodoro SH, Bernardez FFG, Luchese AV, Bergmann M, Siqueira DS, Teixeira A, Azevedo A, Curtis JCD (2024) Produção Brasileira de Remineralizadores e Fertilizantes Naturais: 2019 a 2023. Revista Novo Solo. Available: https://abrefen.org.br/2024/11/08/artigos-producao-brasileira-de-remineralizadores-e-fertilizantes-naturais-2019-a-2023/. Accessed 24 Dez 2025
Medeiros FP, Carvalho AMX, Ramos CG, Dotto GL, Cardoso IM, Theodoro SH (2025) Rock powder enhances soil nutrition and coffee quality agroforestry systems. Sustainability 16:354. https://doi.org/10.3390/su16010354
Medeiros FP (2017) Uso dos remineralizadores associados a policultivos para a produção de palma forrageira no semiárido baiano. Ph.G/PPG-MADER. Universidade de Brasília, p 130. 2017. Available: https://repositorio.unb.br/jspui/handle/10482/24660. Accessed 10 Mar 2023
Melo UMBF, Souza LO (2023) Os potenciais do IRaMuTeQ para análise de conteúdo de decisões judiciais. Braz J Dev 9(1):4886–4911
Ramos CG, Dalmora AC, Kautzmann RM, Hower J, Dotto GL, Oliveira LFS (2021a) Sustainable release of macronutrients to black oat and maize crops from organically-altered dacite rock powder. Nat Resour Res 30:1941–1953. https://doi.org/10.1007/s11053-021-09862-0
Ramos CG, Hower JC, Blanco E, Oliveira MLS, Theodoro SH (2021b) Possibilities of using silicate rock powder: an overview. Geosci Front. https://doi.org/10.1016/j.gsf.2021.101185
Rodrigues M, Silveira CAP, Vahl LC (2016) Efeito da aplicação de calcário e subproduto da exploração de calcário sobre o pH, Ca e Mg do solo e na produção de massa seca do milho. In: Bamberg et al. (org.) Anais do III Cong. Bras. de Rochagem, pp 293–298
Salviati ME (2017) Manual do Aplicativo IRaMuTeQ: compilação, organização e notas. Iramuteq.org. Planaltina, DF, v 31
Silva FC (2009) Manual de análises químicas de solos, plantas e fertilizantes. Brasília–DF: Embrapa Informação Tecnológica; Rio de Janeiro: Embrapa Solos.

Soares G, Theodoro SH, Carvalho AMX, Burbano D, Ramos CG (2022) Remineralizadores de suelos y sistemas agroforestales: una opción para la captura de CO_2. Revista Agroecológia 15:35–45

Souza MEP, Cardoso IM, Carvalho AM, Lopes AP, Jucksch I (2019) Gneiss and steatite vermicomposted with organic residues: release of nutrients and heavy metals. Int J Recycl Organic Waste Agric 8:1. https://doi.org/10.1007/s40093-019-0244-z

Stamford NP (2016) Rochas fosfatadas e potássicas com micro-organismos e matéria orgânica. In: Bamberg et al. (org.) Anais do III Cong. Bras. de Rochagem, pp 189–204

Theodoro SH, Martins ÉS, Fernandes MM, Carvalho AMX (2013) Anais II Cong. Bras. de Rochagem. Poços de Caldas–MG, p 399

Theodoro SH, Monte MBM, Martins ÉS (2021) Anais IV Cong. Bras. de Rochagem. Rio de Janeiro–RJ, p 320

Theodoro SH, Manning DAC, Carvalho AMX, Ferrão FR, Almeida GR (2022) Soil remineralizer: a new rote to sustentability for Brazil, a giant exporting agro-mineral commoditites. In: Yakovleva, N. Nickless, E. (eds.) Routledge Handbook of the extractive industries and sustainable development, 1st edn, pp 261–281. https://doi.org/10.4324/9781003001317

Theodoro SH (2000) A Fertilização da Terra pela Terra: Uma Alternativa de Sustentabilidade para o Pequeno Produtor Rural. Ph.D tesis, Universidade de Brasília. p. 231. 2000. Available: https://repositorio.unb.br/handle/10482/20881. Accessed 5 May 2024

Theodoro SH (2023) Remineralizadores de solos para fortalecer a segurança alimentar e nutricional. Revista Novo Solo. Available: https://abrefen.org.br/2023/06/28/remineralizadores-de-solos-para-fortalecer-a-seguranca-alimentar-e-nutricional/. Accessed 14 Aug 2024

Toledo FHSF, Macana YAM, Nascimento GA, Teixeira ACCG, Ferraz AV, Ferreira EVO et al. (2016) Aplicação de rocha fonolito como fonte de K e Na em plantio de eucalipto. In: Bamberg et al. (org.) Anais do III Cong. Bras. de Rochagem, pp 323–328

Optimizing Research Protocols for Evaluating the Agronomic Efficiency of Soil Remineralizers

Carlos Augusto Posser Silveira, Adilson Luis Bamberg, Rosane Martinazzo, Éder de Souza Martins, Magda Bergmann, Andréa Sander, Luís Henrique Gularte Ferreira, Ivan dos Santos Pereira, Giuliano Marchi, and Gisele Freitas Vilela

Abstract This chapter explores research strategies for evaluating soil remineralizers (REM), emphasizing their nature as insoluble, multi-nutrient sources that release nutrients gradually and in low concentrations. It highlights the importance of biological processes and microorganism interactions, specifically bioweathering processes,

---

C. A. P. Silveira · A. L. Bamberg · R. Martinazzo · I. dos Santos Pereira
Empresa Brasileira de Pesquisa Agropecuária—Embrapa Clima Temperado, Pelotas, RS, Brasil
e-mail: augusto.posser@embrapa.br

A. L. Bamberg
e-mail: adilson.bamberg@embrapa.br

R. Martinazzo
e-mail: rosane.martinazzo@embrapa.br

I. dos Santos Pereira
e-mail: ipereira@emater.tche.br

É. de Souza Martins (✉) · G. Marchi
Agrogeological Laboratory, Empresa Brasileira de Pesquisa Agropecuária—Embrapa Cerrados, BR 020, km 18, Planaltina, DF 73310970, Federal District, Brasil
e-mail: eder.martins@embrapa.br

G. Marchi
e-mail: giuliano.marchi@embrapa.br

M. Bergmann
Serviço Geológico Do Brasil, SUREGE-PA, Porto Alegre, RS, Brazil
e-mail: magda.bergmann@sgb.gov.br

A. Sander · L. H. G. Ferreira
Instituto Federal de Educação, Ciência e Tecnologia do Rio Grande do Sul—Campus, Rio Grande, RS, Brazil
e-mail: andrea.sander@sgb.gov.br

L. H. G. Ferreira
e-mail: luis.ferreira@riogrande.ifrs.edu.br

I. dos Santos Pereira
EMATER-RS, Porto Alegre, Brazil

É. de Souza Martins and S. Huff Theodoro (eds.), *Soil Remineralizers and Silicate Fertilizers*, Frontier Studies in Soil Science,
https://doi.org/10.1007/978-3-032-14656-4_13

in enhancing mineral weathering and increasing REM effectiveness. The chapter discusses the importance of experimental design, ranging from simple chemical extraction systems to complex field trials, in order to accurately assess the REM potential for improving soil fertility and crop nutrition in tropical conditions. It emphasizes the necessity of characterizing REM based on geochemical, mineralogical, and granulometric properties, and tailoring experiments to specific soil types and crop needs. Here, we advocate multi-faceted approaches, considering soil type, crop species, application rates, physical properties, and experiment duration to optimize REM use. It also critically evaluates the limitations of traditional controls in experimental setups, recommending strategies to understand REM's agronomic contributions better. Ultimately, this chapter provides a comprehensive framework to guide future research, aiming to promote sustainable practices through the effective use of silicate mineral-based soil amendments.

Keywords Research approaches · Multifunctional agrominerals · Complexity

1 Introduction

Research strategies involving soil remineralizers (REM) are grounded in the understanding that these products are insoluble in water and serve as multi-nutrient sources (Van Straaten 2007). They release nutrients slowly and in relatively low concentrations. Consequently, specific experimental setups—such as those with short durations, fast-growing plants responsive only to soluble fertilizers, or direct comparisons with highly soluble alternatives—are inadequate for accurately evaluating REM efficacy.

Biological processes are essential for studying the functioning of REM, just as they are vital for transforming rock into soil. In addition to the differential efficiency of crops in nutrient extraction, the direct interaction of microorganisms with minerals in REM enhances the weathering rate. This process, also referred to as bio-weathering, plays a fundamental role in increasing efficiency.

The primary aims of these research strategies considering: 1: the intrinsic characteristics of REM; 2: agronomic conditions and 3: types of experiments for the agronomic evaluation of REM) should be to evaluate the potential of REM to improve specific response variables in target crops or soil, and to illustrate the impact of these products on plant nutrition and growth along with their role in improving soil quality.

Furthermore, it is crucial to adhere to the core principles of agronomic experimentation, ensuring these principles are consistently applied across all agronomic research endeavors. REM, sometimes referred to as rock dust, play a crucial role in Brazilian agriculture (Leonardos et al. 2000). The use of these inputs is based on

G. F. Vilela
Empresa Brasileira de Pesquisa Agropecuária—Embrapa Territorial, Campinas, SP, Brazil
e-mail: gisele.vilela@embrapa.br

the suppositions of stonemeal technology, which claims that certain types of crushed rocks become sources of macro and micronutrients for soil and plants (Theodoro 2000). The use of rock dust is regulated by a legal framework consisting of a law, a decree, and normative instructions.

The growing interest in the use of remineralizers can be attributed to several factors, including higher purchase costs for soluble fertilizers, the pursuit of more sustainable agricultural practices, and the realization of direct and indirect benefits from the use of available multinutrient inputs throughout Brazil (Brazil 2013, 2016, 2018). Several studies (Ramos et al. 2021; Theodoro et al. 2022; Medeiros et al. 2024) have listed a series of direct agronomic benefits, including changes in soil pH and CEC, a reduction in aluminum toxicity, increased moisture retention capacity, and products that meet farmers' expectations.

In addition to the direct benefits, there are important indirect benefits from using locally/regionally available mineral sources. Some of these benefits include adopting production practices that align more closely with the Sustainable Development Goals (SDGs) established by the UN (Manning and Theodoro 2018). Another relevant aspect involves achieving higher levels of food and nutritional security (Theodoro et al. 2022), as well as the potential to capture atmospheric carbon (Beerling et al. 2020).

A number of these benefits were published in scientific articles and expanded abstracts in the five editions of the Brazilian Stonemeal Congresses (BSC). Understanding, cataloging, and organizing results from different research groups over time helps systematize the findings on this topic, leading to benefits that can both strengthen and promote a more sustainable agricultural model. Furthermore, understanding the background of the studies presented in the BSCs helps to further our understanding of how this research has evolved over the years as technology gains more prominence and recognition.

Another important aspect is that the use of mineral sources, with or without organic inputs, achieves better food security and higher rural and agricultural development rates for Brazil, which in turn results in a better quality of life for society as a whole. 347 congress papers and expanded abstracts were published: 40 in the 1st edition, 43 in the 2nd, 56 in the 3rd, 69 in the 4th, and 139 in the 5th. The 4th and 5th editions also included short summaries on a wide range of topics and results.

Understanding and analyzing the benefits and challenges reported in the articles that comprise the four editions of the BSCs is key to describing the transformative potential of this technological option for agriculture. Our analysis aimed to integrate scientific and technical knowledge about the trajectory of research and the performance of remineralizers to enhance this agricultural practice within the country's agricultural sector. The primary objective of this work is to present the state of the art (historical) regarding stonemeal, as well as to identify the main areas of research, scientific advances, and/or possible knowledge gaps, primarily focusing on the results obtained by the different research groups in the four editions of the BSCs. To achieve this, we aimed to systematize the data published in the BSCs between 2009 and 2025, identify the primary objectives and methodologies in the articles, and highlight some of the key research findings from the BSC editions.

We anticipate that the results obtained from this analysis can benefit and guide future studies on improving management practices and mechanisms to facilitate the adoption of stonemeal technology among farmers (both business/family) and thus inform public policies to promote the more extensive adoption of REM in Brazil.

2 Research Approaches for Soil Remineralizers

REM should be studied in various interaction systems to describe their agronomic effects (Fig. 1). The ultimate focus should be on the real situation in the agricultural landscape, which is more complex, where climate, soil, cultivated plants, and management interact with REM. On the other hand, this situation presents the least control over variables and is the most complex possible. This complex scenario entails processes that are challenging to assess directly due to the various potential pathways for transforming low-solubility silicate minerals within agricultural contexts.

A more accurate understanding of the processes and mechanisms involved in the weathering of minerals in agricultural soils can be obtained through simpler experimental systems that allow greater control over variables and conditions. Since bioweathering processes are essential for the efficiency of REMs, studies on the biological interaction of the rhizosphere and solubilizing microorganisms should form the focus of evaluating the processes and mechanisms involving weathering

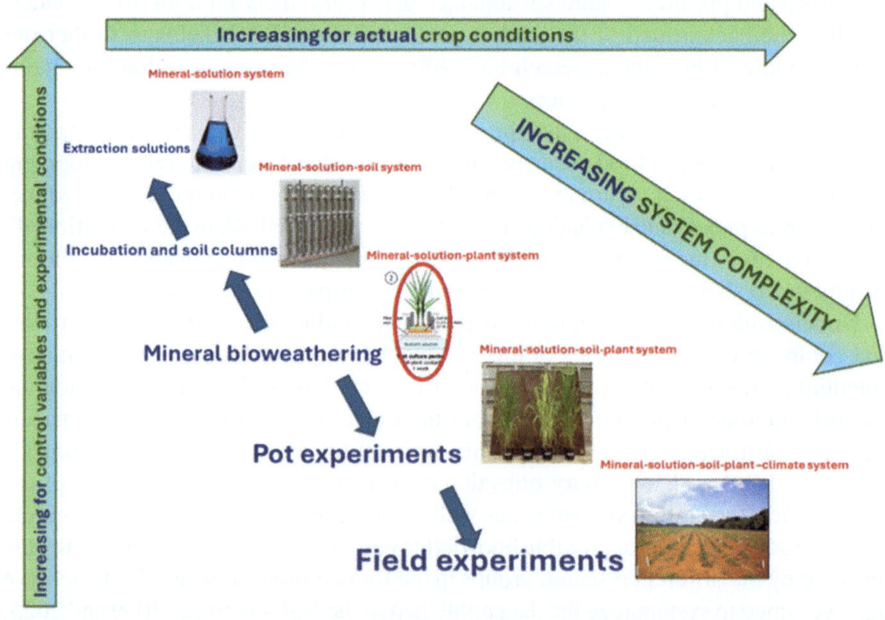

Fig. 1 Agronomic experiments with REM, from simple to complex

in mineral-solution-(soil)-plant-(microorganisms) systems. In this case, studies can be conducted on the interaction between the plant rhizosphere and REMs, either with or without soil, and with or without the application of microorganisms. At the same time, it is important to consider that different soil types can promote different transformation and weathering pathways of minerals present in REMs.

The simplest experimental system with the greatest control over variables is the one that uses chemical extractants (mineral-solution system) to evaluate nutrient availability and weathering directly in REMs. In this case, the main objective of this study is to describe the potential of REM to release plant nutrients.

The second type of evaluation involves interactions from incubating REM in soil (mineral-solution-soil system) with or without leaching of released elements. It increases the complexity of the processes, but it is possible to evaluate physico-chemical modifications in soils, such as changes in pH and nutrient availability.

The other two types of experiments, conducted in a greenhouse or in the field, involve the interaction of REM with soil and plants under controlled conditions.

3 Key Questions on the Agronomic Potential of Soil Remineralizers

Before outlining the strategies to be used in a research protocol for evaluating the REM, it is important to address some fundamental questions. These questions concern the product and its evaluation conditions. Before evaluating an REM, consider these questions:

a. Regarding the product: What are the nutrient levels of the product? What are the main nutrient-bearing minerals and their composition? What is the granulometry of the final product?
b. Regarding the experimental or agronomic conditions: How long will the experiments last? In areas with low natural fertility, degraded areas, or in areas with high natural or constructed fertility? As a total replacement of soluble fertilizers or as a complement to them? Alone or in combination with other REM and/or bioinputs? To increase productivity or improve food quality? In monoculture systems, short-cycle systems, or crop rotation systems with long-cycle crops? On the surface, with or without incorporation or in the cultivation furrow? In improved species that respond to soluble fertilizers or in rustic, unimproved species?
c. From these conditions, practical questions arise, such as: What dose to use? Is this dose based on the total nutrients present in the REM or a particular one? When to reapply?

3.1 *Product Characteristics*

For an accurate agronomic evaluation of a REM, intended for future agronomic classification and use, it is essential to understand its chemical, mineralogical, and granulometric characteristics (Table 1).

3.1.1 Geochemical Composition

Examining the geochemical composition of a REM (Table 1) as an example, there are notable differences in the total contents of oxides and sums of bases. This is one of the first pieces of information necessary for properly planning strategies for agronomic evaluation and the use of REM as potential nutrient sources. For example, mica schist and olivine melilitite show the lowest total K_2O contents when compared to concentrated biotite, phonolite, and glauconitic siltstone. Conversely, olivine melilitite presents the highest CaO and MgO contents, while phonolite has the lowest. Thus, based solely on these data, olivine melilitite (as well as basalts) can be classified as sources of CaO and MgO, and biotite, phonolite, and glauconitic siltstone as sources of K_2O.

3.1.2 Mineralogy

In greenhouse or field experiments, rocks composed of feldspathoids, certain types of micas (biotite and phlogopite), as well as carbonates and sulfates, require a short incubation period in humid soil before the experiment begins, typically lasting up to 60 days. For rocks containing these minerals, the final product's granulometry has little impact on agronomic efficiency; as a rule, powder-class granulometry (100% < 2.0 to > 0.3 mm) is sufficient for positive responses.

For rocks composed of minerals with crystalline structures (feldspars and plagioclases), longer incubation periods are needed—120 days or more; for rocks with these minerals, the granulometry of the final product has a significant impact on agronomic efficiency. Ultra-grinding is generally required (100% < 0.3 mm, or even smaller: < 0.105 to < 0.075 mm); even with ultra-grinding, there may be a need to incorporate other physicochemical or biological agents to increase their weathering (e.g.: thermal process, acidification, the use of elemental sulfur and organic matter) and bio-weathering (use of solubilizing microorganisms) processes.

For rocks composed of crystalline minerals, even after prolonged incubation periods, the magnitude of agricultural responses may vary; thus, it is suggested to conduct experiments in greenhouses using plants that require the nutrients present in the REM.

Table 1 Rock type, geochemical composition (total major oxide contents), predominant potassic minerals, minerals dissolution rate, biological reactivity, and granulometry

Rock type	Total contents (%)						Predominant potassium mineral	Dissolution rate (susceptibility to weathering) of potassium mineral groups present in the rock[2]	Biological reactivity of minerals (predisposition to bioweathering)[3]	Final product granulometry
	SiO_2	K_2O	CaO	MgO	P_2O_5	SB[1]				
Kamafugite	35.0	4.00	4.00	4.00	3.00	12.0	Sanidine (feldspar)/ kalsilite an leucite (feldspathoide)	Low/medium	High	100% < .00 mm; 80% < 0.84 mm; 70% < 0.30 mm
Olivine melilitite	38.2	2.80	13.55	15.05	1.11	31.4	Flogopite (mica) and amorphous minerals	Medium/high	High	100% < 0.3 mm
Concentrated biotite	49.6	6.85	1.70	3.66	0.44	12.2	Biotite (mica)	Medium	High	100% < 0.3 mm
Phonolite	53.7	8.40	1.30	0.20	0.07	9.9	Sanidine (potassic feldspars)/Nefeline (feldspathoide)	Low/high	High	100% < 0.105 mm
Mica-schist	57.9	2.10	1.80	4.80	0.20	8.7	Biotite/muscovite (Mica)	Medium/low	High	100% < 0.3 mm

(continued)

Table 1 (continued)

Rock type	Total contents (%)						Predominant potassium mineral	Dissolution rate (susceptibility to weathering) of potassium mineral groups present in the rock[2]	Biological reactivity of minerals (predisposition to bioweathering)[3]	Final product granulometry
	SiO_2	K_2O	CaO	MgO	P_2O_5	SB[1]				
Glauconitic siltstone	60.2	10.70	0.01	2.48	0.07	13.2	Glauconite (argilomineral)/biotite/ muscovite (mica)/ microcline (potassic feldspars)	High/medium/ low	High	100% < 2.00 mm; 90% < 0.3 mm
Monzogranite	70.3	4.33	2.42	0.99	0.15	7.7	Microcline and orthoclase (potassic feldspars)/biotite (mica)	Low/medium	Low	100% < 0.3 mm

[1]SB—Sum of Bases (CaO + MgO + K2O)
[2]Oliveira (2024)
[3]Martins et al. (2023)

3.1.3 Granulometry

Another specific characteristic of REM to consider is the granulometry of the final product. Since REM undergo solely grinding processes, with their reactivity significantly influenced by this factor, it is suggested to adopt the same granulometric ranges as those specified for soil acidity correctives (IN 35/2006). Particles passing through an ABNT no. A 50-mesh sieve (0.30 mm aperture) has a theoretical reactivity of 100% over a period of 12–36 months. Particles passing through an ABNT no. 20 sieve but retained in an ABNT no. 50 sieve (0.30–0.84 mm) show a reactivity of 60%, while coarser particles (retained in ABNT no. 20 sieve and passing through ABNT no. 10 sieve (0.84–2.00 mm) exhibit a reactivity of 20%, and finally, particles larger than 2.00 mm show no effect until 36 months.

Considering that most minerals constituting REM exhibit low solubility in water and reactivity, grinding is the simplest and most effective technique for enhancing the product. When finely ground, minerals are broken, increasing the contact surface area between particles and the soil. A good example of this process is soil acidity correctives, where particles smaller than 0.053 mm (passing through ABNT no. 270 sieve) react completely in less than a month, whereas particles between 2.00 and 0.84 mm (retained in ABNT no. 20 sieve and passing through ABNT no. 10 sieve) require more than 60 months for complete reaction.

For REM with K present in minerals that are less susceptible to weathering (generally potassium feldspars and muscovite), it is suggested to adopt an ultrafine granulometry (<0.105 mm or < 0.075 mm). However, to determine the best granulometry, mineralogical analyses should be conducted to identify the minerals, their proportion in the rock matrix, and their integrity. If there is uncertainty regarding the optimal granulometry to be employed, it can also be considered a treatment variable for testing. For example, the following granulometric fractions of a particular REM should be tested: (1) between 2.0 and 0.84 mm; (2) between 0.84 and 0.30 mm; (3) between 0.30 and 0.105 mm; (4) between 0.105 and 0.053 mm; (5) < 0.053 mm.

3.2 Agronomic Conditions

3.2.1 Soil Type and Fertility Level

Agronomic responses are highly influenced by soil type. Therefore, it is suggested to include at least two characteristic soil types from each region, which contrast in terms of texture and organic matter content. Soils with low nutrient levels, low CEC, low pH, low organic matter content, and high exchangeable Al content allow faster and more robust responses to REM application.

3.2.2 Plant Type

Plant species should be chosen based on their ability to extract nutrients in REM or by the economic importance of the crops in the region where REM are marketed. It is recommended to carry out successive crops, including several plant species, especially those from the Poaceae family, with a robust root system. Grain-producing species (maize, rice, wheat, and soybean) have been used as benchmarks for nutrient requirements, especially P and K; among forages, brachiaria, (*Brachiaria ruziziensis*, *B. brizantha*, *B. decumbens*), elephant grass (*Cenchrus purpureus*) and ryegrass (Lolium multiflorum) are the most indicated; for cover crops, millet (*Pennisetum glaucum*), Sudan grass (*Sorghum sudanense*), and black oat (*Avena strigosa*) are recommended.

The grasses have a high demand for K, as do some vegetables (such as garlic, beet, carrot, tomato, potato, and strawberry) and fruit tree species (including banana, citrus, and other Rosaceae). Considering the cycle of crops, it is essential to foresee the inclusion of long-cycle species, such as perennial or semi-perennial crops (like elephant grass and sugarcane), that can withstand regular cutting, as well as perennial species (like fruit trees). Thus, considering this information, REM containing higher levels of K could yield better responses when species that require large quantities of K are grown, especially if the soils used are classified as low or very low in this element, particularly if they are sandy soils with a low CEC. In contrast, for those soils with low levels of Ca and Mg, the more suitable REM would be those with higher CaO and MgO contents. In this case, the most indicated species would belong to the leguminous and Rosaceae families, as they are recognized for their high demand for Ca and Mg.

When selecting species, it is crucial to consider the differences between improved and unimproved plant species. Conventional plant breeding has led to significant increases in productivity for most grain-producing agricultural species. However, these cultivars require optimal conditions throughout their growth cycle to express their full productive potential. Among these conditions is the high number of soluble fertilizers. Since REM generally have low solubility, presenting a gradual release of nutrients, cultivars that demand immediate-effect inputs are not the best suited for assessing the agronomic efficiency of REM. On the other hand, species that have not undergone conventional breeding are often more resilient, adapted, and tolerant to variations in chemical, physical, and biological soil fertility, making them more suitable, especially in tropical environments. Examples of these species include cassava, sweet potato, fruit trees from the *Myrtaceae* family, and native pastures.

3.2.3 Doses of REM to Use

Another important aspect involves defining doses. Since this type of input contains several minerals with different nutrient release rates (ranging from immediate to slow and gradual) and, currently, calculations for doses to be used are based on total contents, it is suggested that the first dose in the experimental treatment design provide 100% of the soil requirements (determined through soil analysis) and the crop of interest (considering information on crop demand and, when available, historical productivity data). However, when the REM presents easily weathered minerals, it is possible to suggest a dose based on the theoretical availability of the nutrients of interest. However, since chemical characterization usually informs the total nutrient content, which is released slowly/gradually, subsequent doses can be defined at 200%, 300%, 400%, and even 500% of the recommendation. This approach allows the determination of an appropriate dose for the specific soil condition and plant species being evaluated. Until adequate information on available nutrient levels, via organic acids, is validated and provided in the geochemical characterization reports, the REM should base its dose recommendations on total nutrient contents. At the same time, it is essential to identify other important parameters for recommendations, such as cation exchange capacity (CEC), soil pH interference, water holding capacity (WHC), and physical, chemical, and/or biological soil conditioning.

3.2.4 Experiment Duration

Regarding the experiment's duration, a maximum of 36 months is recommended for greenhouse experiments, and a minimum of 24 months is recommended for field experiments. If the REM of interest contains minerals with medium to high dissolution rates and a high predisposition to bio-weathering, effects can often be observed in less than 6 months.

3.2.5 Mineral Weathering and Bioweathering

The widespread presence and constant action of carbonic acid are the main factors affecting mineral weathering. However, for this process to occur meaningfully, a significant timescale (geological, on the order of thousands to millions of years) is usually required. At the same time, conditions can be created to accelerate this phenomenon through the grinding of minerals, which increases their surface area exposed to water, organic acids, plants, and microorganisms. In this context, soils with high organic matter content and the application of organic materials (such as poultry litter, manure, vegetable cakes, and organic compost) help speed up the weathering of minerals. Additionally, some silicate minerals are inherently less stable than others and can undergo structural changes that facilitate the quicker release of nutrients. Examples include olivines, pyroxenes, calcium plagioclases, biotite, phlogopite, and glauconite.

Regarding mineral bio-weathering, potassium-solubilizing microorganisms act within the rhizosphere of plants, transforming insoluble potassium (found in minerals such as muscovite, orthoclase, biotite, feldspar, illite, and mica) into soluble forms suitable for plant absorption. This process involves various bacteria (e.g., *Bacillus, Acidothiobacillus, Paenibacillus spp.*) and fungi (e.g., *Aspergillus spp.*). For example,

Paris et al. (2015) highlight the role of ectomycorrhizal species in the mineralogical alteration of phlogopite, suggesting that these microorganisms play a crucial role in mineral weathering through increased oxalic acid production. Similarly, Raghavendra et al. (2016) emphasize the contributions of bacteria and fungi in dissolving and solubilizing potassium from various potassium minerals. These reactions occur through the production of acetic and gluconic acids, as well as polysaccharides, in addition to oxalic, tartaric, and citric acids, and $H+$ ions (Kour et al. 2020).

3.2.6 Reapplication of REM

In experiments, it is important to consider reapplying REM in plots. Therefore, it is recommended that the experimental plot dimensions be adjusted to account for this factor.

Generally, field experimental plots have variable dimensions, but an adequate size may be 5.0 × 5.0 m (25 m^2), allowing the useful area—after excluding borders—to provide sufficient biomass for conducting nutritional evaluations (in both leaves and grains).

To account for the reapplication of REM, for instance, if planning for two reapplications, it is advisable to double the plot size. This adjustment would result in dimensions of 5.0 × 10.0 m (50 m^2). With this adjustment, it will be possible to evaluate both the immediate and residual effects of both the reapplication and dose factors simultaneously. The time between applications, as mentioned in controlled condition trials, will be determined based on foliar nutrient levels or the nutrient availability in the soil provided by the REM.

3.2.7 Evaluations (Response Variables)

Evaluations should consist of chemical, physical, and biological analyses of the soil, foliar tissue analysis, dry matter of aerial parts, and assessment of production/productivity. Soil analyses may be performed only at the end of the last successive cultivation; however, it is preferable to conduct them at the conclusion of each crop cycle.

3.2.8 Combining REM with Highly Soluble and Concentrated Fertilizers

REM can help increase the efficiency of conventional fertilizers that are highly soluble and concentrated in nutrients, either by reducing the applied doses or by applying full doses under adverse conditions (under low soil moisture, in sandy soils, for example). A good example is the effect of REM on soil P. The level of available P in the soil is typically increased through the combined effects of increased pH, the formation of monosilicic acid, and the stimulation of solubilizing microorganisms. This makes it possible to reduce the total dose of P to be applied. Generally, when REM release nutrients into the soil, they form secondary minerals with high CEC and WHC. These properties interact synergistically with soluble fertilizers, resulting in reduced losses through leaching, adsorption, and volatilization. Therefore, if one is interested in testing the combination of REM with soluble sources, it is important to know the mineralogy of the REM, the secondary products formed by the weathering of the predominant primary minerals, as well as the mineralogy of the soils where the REM will be applied. Minerals such as biotite produce 2:1 minerals with a permanent negative charge, allowing the temporary retention of hydrated cations, while sandy soils generally exhibit minerals with low CEC and low organic matter content. Therefore, such conditions would theoretically provide the optimal environment for evaluating the combination of REM with soluble sources.

3.3 Types of Experiments for the Agronomic Evaluation of REM

3.3.1 Incubation Experiments

Incubation involves mixing soil with the REM to be tested, employing desired doses and granulometries. This method allows for numerous combinations of treatment factors (types of soil, type, granulometry, and dose of the REM, as well as incubation time). For an initial approach prior to conducting experiments with plants in greenhouses or fields, it is recommended to include the most representative soil types for each region (at least two types, one sandy and another clayey), selected based on their agricultural representativeness.

The duration can be up to 12 months, with soil samples taken monthly, quarterly, or semiannually during this period. The first sampling can occur one week after incubation. Thus, at the end of the incubation period, it will be possible to determine pH, nutrient release into the soil solution, the most appropriate dose, and the best granulometry for each REM in the different tested soils.

The quantity of soil (as well as the size of the containers) depends on the frequency of sampling. It is suggested to start with a minimum of 300 g of dry soil. The minimum sample quantity for comprehensive soil analysis (macro and micronutrients, pH, clay, and organic matter) is 100 g; therefore, with 300 g, it would be possible to carry out up to three samplings staggered in time. For monthly sampling experiments lasting up to 12 months of incubation, approximately 1.2 kg of dry soil would be necessary.

The moisture of the incubated soil should be maintained at 80% of the water retention capacity for each soil type. It is recommended to use distilled water.

3.3.2 Pot Experiments

After determining the type, reactivity, dose(s), and granulometry of the REM in the incubation study, pot experiments using plants should be conducted. At least two of the most representative soils from each region should be used, preferably with different textures (one sandy and one clayey), as these characteristics are crucial in the dynamics of chemical, physical, and biological reactions in the soil, as well as in agronomic responses.

The sizes of the pots may range from 2 dm^3 to 30 dm^3, depending on the plant species and the duration of the experiment (Novais et al. 1991). Evaluations should involve chemical and biological analyses of the soil, foliar tissue analysis, dry matter of aerial parts, and productivity (when possible). Soil and foliar tissue analyses may be conducted only at the conclusion of the cultivation cycle or sequence; however, it is preferable to perform them at the end of each crop cycle.

As previously mentioned, it is essential to define the type of plant for the experiments, as some species can also absorb nutrients in their less soluble or non-exchangeable forms due to mechanisms such as acidification in the root environment via the release of organic acids, which can alter the structure of minerals. Therefore, to evaluate the fertilizing (REM as a K source, for example) or soil conditioning effect of REM, it is suggested that, in addition to evaluating the immediate effects, the experiment should continue with successive crops of plant species that are known to be sensitive to the nutrients provided by the tested REM, that grow rapidly under controlled conditions and can withstand frequent successive cuttings (species of the Poaceae family, for example, brachiaria (*Brachiaria ruziziensis*, *B. brizantha*, *B. decumbens*), elephant grass (*Cenchrus purpureus*) and ryegrass (*Lolium multiflorum*). Generally, fertilization studies in greenhouses last a maximum of 60 days for grain-producing species. However, for the evaluation of REM, it is suggested to conduct trials with successive cultivations lasting from 12 to 36 months.

The successive cultivation trial will also help determine the need for reapplication of the REM by monitoring foliar nutrient levels (plant absorption) and the available levels of the nutrients supplied by the REM in the soil. The values found can be compared with existing reference tables (Pauletti 2004; CQFS-RS/SC 2016; SBCS/NEPAR 2017; Boletim 100 2022). At the same time, it is important to consider the accumulation of nutrients in the aerial parts of plants. Leaf content is a widely used indicator for monitoring the nutritional status of plants, but when evaluated in isolation, it may not be a reliable indicator. Therefore, for REM, it is suggested to evaluate the biomass of the aerial parts (dry weight) in combination with the nutrient levels of the aerial parts to quantify the accumulation of nutrients. This variable may be the appropriate complement to evaluate the efficiency of a REM.

Additionally, in cases of REM that are K sources, it is also possible to perform standard soil analyses (such as Melich 1) in conjunction with soil K fractionation to determine the total, structural, non-exchangeable, exchangeable, and dissolved fractions. (Melo et al. 2009) of this element in response to treatment factors (type and doses of REM; soil type, etc.). This will enable the determination of the contribution of the REM to the mineral reserve of the soil.

In addition to the fertilizing effect, REM consists of primary minerals, either intact or altered. Some of them, during the processes of weathering and bio-weathering, may form 2:1 clay minerals with high cation exchange capacity, thereby increasing the soil's nutrient and water retention capacity (Martins et al. 2010).

Given the controlled condition experiments (incubation and pot experiments), which allow for considering representative and contrasting soil types, plant species, granulometries, and doses of REM simultaneously, as well as assessing both immediate and residual effects (through successive cultivations), thus enabling the conduct of experiments with multiple treatments.

3.3.3 Field Experiments

Based on the results observed in incubation and pot experiments, field experiments should be conducted to validate the findings obtained under controlled conditions. However, special attention must be given to the selection of sites where experiments will be established to ensure adequate control over heterogeneous variations in soil types, fertility levels, compacted soils, and conditions of excess or lack of moisture. When large-scale experiments (covering an area of 1.0 hectare or larger) are implemented, attention should be given to the correct allocation to avoid having different soil types or patches of varying soil quality within the same experiment (Fig. 2). It is recommended to conduct a soil analysis in advance to correctly allocate experiments or their blocks to a single type of soil later.

Another possibility for controlling variations in soil fertility is to delineate the experimental area and conduct soil analysis in advance. Based on the chemical analysis reports, blocking can be performed based on fertility levels, rather than solely on slope or solar exposure. In this case, blocking considers the fertility levels of the factor of interest (K content, for example), and categories can be adopted for the blocks based on the K content of the soil (Very Low, Low, Medium, High, and Very High) (Fig. 3a). Thus, considering an experiment to evaluate an REM as a source of K, in a certain type of soil, the traditional blocking for the design in randomized blocks with four replications would appear as shown in Fig. 3b, while blocking according to the soil K contents would be as described in Fig. 3c.

For field experiments, it is recommended that they be conducted in at least two locations with different soil types over a period of at least two years. If possible, the chosen areas should have chemically and physically distinct but uniform soils, as emphasized earlier. The greater the number of locations in which the REM is evaluated, the more robust and representative the results will be.

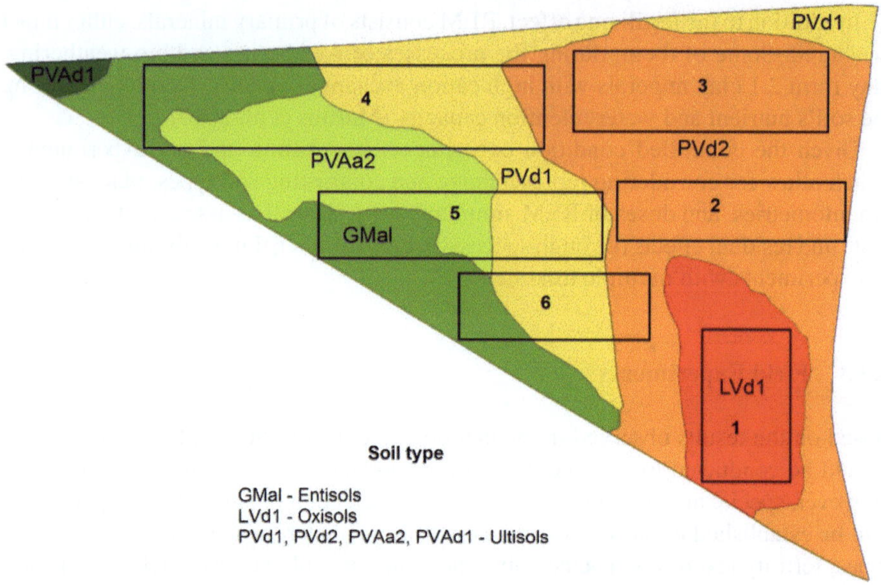

Fig. 2 Soil map of a 5-hectare experimental site. Note the significant variability in soil types and implications for experimental design (blocking in experiments 4, 5, and 6 was unable to control for location adequately)

4 Examples of Treatment Designs for Experiments with rem

Considering the specificities of REM previously mentioned, below are some suggested examples of treatment designs that can be used for both incubation studies and plant studies in controlled and/or field conditions.

4.1 Example 1. Experiment with a Quantitative Treatment Factor Tested in a Single Soil Type

Quantitative Factor—REM dose as a source of total K_2O content, with five levels: 0%, 100%, 200%, 300%, and 400% of the K_2O recommendation. The total number of treatments will be equal to the factor levels, and after the variance analysis, a polynomial regression model should be adopted. In this case, the hypotheses of interest refer to the linear and quadratic effects. If the latter is significant, it is possible to obtain the maximum technical efficiency dose of REM. In this design, an equivalent K source can be included as a positive control for comparison at the desired dose (it is suggested that this control be a registered product equivalent to REM) and a negative control, representing the native soil content without REM application (Table 2).

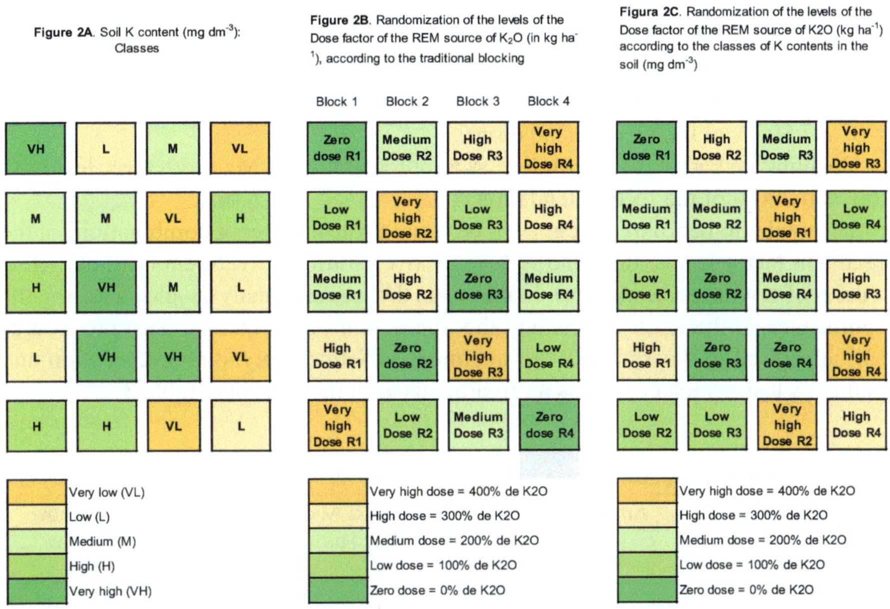

Fig. 3 Classes of soil K contents and treatment distribution (randomization) for an experiment with five treatments and four replications based on soil K content (**a**) and randomization of K_2O REM doses per randomized block design (traditional blocking—**b**) and randomization of treatments based on soil K contents (**c**)

Table 2 Example of treatment design for an experiment with one quantitative treatment factor (REM dose as a source of K_2O) with an additional treatment (registered K_2O source) at the recommended dose (140 kg ha^{-1} of K_2O)

Dose factor levels	Treatment description	Amount of K_2O (kg ha^{-1})	REM doses how to provide K_2O	
			%	kg ha^{-1}
1	Negative control	0	0	0
2	Recommendation	140[1]	100	978
3		280	200	1.955
4		420	300	2.933
5		560	400	3.911
6		700	500	4.888
REM K[2]	Positive control	140	100	1.750

Recommended dose (140 kg ha^{-1} of K_2O)

[1]Recommended dose of K_2O (100%), considering the following principles: (a) Soil with K_2O content in the low class (31–60 mg dm^{-3}) and CEC at pH 7.0, 7.6–15.0 cmol$_c$ dm^{-3}; (b) Correction and maintenance fertilization (CQFS SC/RS 2016) and (c) expected productivity of > 10 Mg ha^{-1} of corn grains

[2]REM source of K_2O

4.2 Example 2. Experiment with Two Quantitative Treatment Factors in a Single Soil Type

Quantitative Factor 1—REM dose as a source of K (with six levels: 0, 100, 200, 300, 400, and 500% of the recommendation).

Quantitative Factor 2—REM dose as a source of Ca and Mg (with three levels: 0, 100, and 200% of the recommendation).

The total number of treatments will be equal to the number of combinations of the treatment factor levels, that is, $3 \times 6 = 18 + 1$ additional treatment (without REM sources of K_2O and Ca and Mg). In this case, the variance analysis allows testing the significance of the main effects and the interaction of the factors. Two procedures can be applied to discriminate the treatment variation: polynomial regression and Dunnett's test, which compares all treatments with the controls (Table 3).

Table 3 Example of treatment design for an experiment with two quantitative treatment factors

Treatments	Amount of K_2O (kg ha^{-1})	REM doses how to provide K_2O		REM doses how to provide Mg	
		%	kg ha^{-1}	%	kg ha^{-1}
T1—negative control	0	0	0	0	0
T2—recommendation	140[1]	100	978	100[2]	688
T3	280	200	1.955	100	688
T4	420	300	2.933		
T5	560	400	3.911		
T6	700	500	4.888		
T7	140[1]	100	978	200	1.376
T8	280	200	1.955		
T9	420	300	2.933		
T10	560	400	3.911		
T11	700	500	4.888		
T12	140[1]	100	978	300	2.064
T13	280	200	1.955		
T14	420	300	2.933		
T15	560	400	3.911		
T16	700	500	4.888		

[1]Recommended dose of K_2O (100%), considering the following principles: (a) Soil with K_2O content in the low class (31–60 mg dm^{-3}) and CEC at pH 7.0, 7.6–15.0 $cmol_c$ dm^{-3}; (b) soil with low Ca and Mg contents (< 2.0 and < 0.5 $cmol_c$ dm^{-3}, respectively); (c) correction and maintenance fertilization (CQFS SC/RS 2016); (d) expected productivity of > 10 Mg ha^{-1} of corn grains; (e) the dose of remineralizer B was based on Mg export, considering the export of 4.4 kg of Mg per ton of grains produced

4.3 *Example 3. Experiment with One Qualitative Treatment Factor and One Quantitative Treatment Factor in a Single Soil Type*

Qualitative Factor—Granulometric ranges (with five levels: 2.0–0.84 mm; 0.84–0.30 mm; 0.30–0.105 mm; 0.105–0.053 mm; < 0.053 mm).

Quantitative Factor—REM dose as a source of K_2O (978, 4888 g ha-1, equivalent to 100 and 500% of the recommendation + one additional treatment: without REM).

If there are doubts regarding the most suitable granulometries for the REM, for example, one possibility is the design proposed in Table 5, which would consist of two levels of the REM, one providing 100% and another 500% of the K2O requirement, both combined with five granulometric ranges of the REM; plus one additional treatment (zero dose, soil without remineralizer), totaling eleven treatments (Table 4).

Table 4 Example of treatment design for an experiment with two treatment factors, one qualitative (Granulometric ranges of the REM) and the other quantitative (REM dose)

Treatments	Amount of K_2O (kg ha^{-1})	REM doses how to provide K_2O		REM particle size range (mm)
		%	kg ha^{-1}	
T1—negative control	0	0	0	Without remineralizer
T2	140[1]	100	978	100% between 2.0 and 0.84
T3				100% between 0.84 and 0.30
T4				100% between 0.3 and 0.105
T5				100% between 0.105 and 0.053
T6				100% < 0.053
T7	700[2]	500	4888	100% between 2.0 and 0.84
T8				100% between 0.84 and 0.30
T9				100% between 0.3 and 0.105
T10				100% between 0.105 and 0.053
T11				100% < 0.053

[1]Dose of 100% of K_2O considering the following principles: (a) Soil with K_2O content in the low class (31–60 mg dm^{-3}) and CEC at pH 7.0, 7.6–15.0 $cmol_c$ dm^{-3}; (b) correction and maintenance fertilization (CQFS SC/RS 2016); (c) expected productivity of > 10 Mg ha^{-1} of corn grains. [2]Dose of 500% of the recommended K_2O

Table 5 Example of treatment design for an experiment with two treatment factors, both qualitative (Granulometric ranges of the REM and Soil Type)

Treatments	Amount of K_2O (kg ha^{-1})	REM doses how to provide K_2O		Soil type	REM particle size range (mm)
		%	kg ha^{-1}		
T1—control 1					Without remineralizer
T2	140[1]	100	978	Sandy soil	100% between 2.0 and 0.84
T3					100% between 0.84 and 0.30
T4					100% between 0.3 and 0.105
T5					100% between 0.105 and 0.053
T6					100% < 0.053
T7—control 2					Without remineralizer
T8	140[1]	100	978	Clayey Oxisol	100% between 2.0 and 0.84
T9					100% between 0.84 and 0.30
T10					100% between 0.3 and 0.105
T11					100% between 0.105 and 0.053
T12					100% < 0.053

[1]100% K_2O dose considering the following principles: (a) Soil with K_2O content in the Low class (31–60 mg dm^{-3}) and CEC at pH 7.0, 7.6–15.0 $cmol_c$ dm^{-3}; (b) correction and maintenance fertilization (CQFS SC/RS 2016); (c) expected productivity of > 10 Mg ha^{-1} of corn grains

The responses from the analysis of variance would provide practical information, such as the optimal REM particle sizes that promote increases in soil K levels and whether the REM dose factor can overshadow the particle size factor. It is important to emphasize that negative control should only lack the nutrient being evaluated and should be used whenever possible.

Variations of this design may involve replacing the REM dose factor with a qualitative factor (such as soil type, for example) or transforming both factors into qualitative ones: (1) Biological weathering sources (e.g., plant species and/or microorganisms) and (2) Presence and absence of the REM. Thus, in the first scenario, the REM dose could be kept fixed (for example, to provide 100% of the K_2O recommendation for a determined soil fertility level). The combinations of levels for these factors in the first example are presented in Table 5, while those for the latter are presented in Table 6.

For experiments with such designs (many treatments), they could first be conducted under controlled conditions, beginning with incubation experiments and

Table 6 Example of treatment design for an experiment with two treatment factors, both qualitative (Sources of biological weathering and Presence/Absence of the REM)

Treatments	Amount of K_2O (kg ha^{-1})	REM doses how to provide K_2O			Biological agents of mineral weathering	
		%	kg ha^{-1}		Microrganisms	Plant
T1—control 1	0^1	0	0	Absence	Absence	Absence
T2					Absence	Presence
T3					Presence	Absence
T4					Presence	Presence
T5—control 2	140^2	100	978	Presence	Absence	Absence
T6					Absence	Presence
T7					Presence	Absence
T8					Presence	Presence

[1,2]Doses of 0 and 100% of K_2O considering the following principles: (a) Soil with K_2O content in the Low class (31–60 mg dm^{-3}) and CEC at pH 7.0, 7.6–15.0 $cmol_c$ dm^{-3}; (b) correction and maintenance fertilization (CQFS SC/RS 2016); (c) expected productivity of > 10 Mg ha^{-1} of corn grains

subsequently moving to greenhouse studies, and finally to field trials after identifying the best treatments.

4.4 *Example 4. Experiment with Three Treatment Factors, Two Qualitative and One Quantitative*

Qualitative Factor 1—Type of REM (with two levels, for example, REM as a source of K_2O and REM as a source of Ca and Mg).

Qualitative Factor 2—Type of soil (with two levels: sandy soil and clayey soil; or soil with low levels of K, Ca, and Mg and soil with high levels of K, Ca, and Mg; or soil of low and high fertility).

Quantitative Factor—REM dose (with five levels for the potassium REM: 0, 100, 200, 300, 400, and 500% of the recommendation; and two levels for the REM as a source of Ca and Mg: 0 and 688 kg ha^{-1}—the dose to provide 100% of the Mg required by a specific crop).

In this case, the total number of treatments will equal the combinations of the three treatment factors, that is, $2 \times 2 \times 5 = 20$. The variance analysis could indicate significant effects of the individual factors or even the interaction among the factors (Table 7).

The responses from the variance analysis would yield practical information such as the efficiency of the REM in promoting increases in Ca, Mg, and K levels for each of the soils. For experiments with this design (containing many treatments), they should first be conducted under controlled conditions, initially through incubation

Table 7 Example of treatment design for an experiment with three treatment factors: two qualitative (Type of REM as a source of Ca and Mg and Type of Soil) and one quantitative (REM dose as a source of K_2O)

Treatments	Soil type	REM doses how to provide Mg		Amount of K_2O (kg ha^{-1})	REM doses how to provide K_2O	
		%	kg ha^{-1}		%	kg ha^{-1}
T1	Sandy soil	0	0	0	0	0
T2				140[1]	100	978
T3				280	200	1.955
T4				420	300	2.933
T5				560	400	3.911
T6				700	500	4.888
T7		100	688	0	0	0
T8				140[1]	100	978
T9				280	200	1.955
T10				420	300	2.933
T11				560	400	3.911
T12				700	500	4.888
T13	Clayey oxisol	0	0	0	0	0
T14				140[1]	100	978
T15				280	200	1.955
T16				420	300	2.933
T17				560	400	3.911
T18				700	500	4.888
T19		100	688	0	0	0
T20				140[1]	100	978
T21				280	200	1.955
T22				420	300	2.933
T23				560	400	3.911
T24				700	500	4.888

[1]Recommended dose of K_2O (100%), considering the following principles: (a) Soil with K_2O content in the low class (31–60 mg dm^{-3}) and CTC pH 7.0, 7.6–15.0 $cmol_c$ dm^{-3}; (b) soil with low Ca and Mg contents (< 2.0 and < 0.5 $cmol_c$ dm^{-3}, respectively); (c) correction and maintenance fertilization (CQFS SC/RS 2016); (d) expected productivity of > 10 Mg ha^{-1} of corn grains; (e) the dose of remineralizer B was based on Mg export, considering the export of 4.4 kg of Mg per ton of grains produced, that is, 44 kg of Mg for an expected 10 Mg ha^{-1} of corn grains

methods and then in a greenhouse. After identifying the best treatments, they would be incorporated into an experiment to be conducted under field conditions.

5 Concluding Remarks

Considering the nature of REM—multi-element nutrient sources with low and/or gradual solubility and relatively low nutrient concentration—this chapter proposes a specific agronomic protocol. Existing agronomic protocols were developed for evaluating sources with high solubility and concentration. The proposed protocol is based on different research strategies for evaluating REM. This approach takes into account the basic principles of agronomic experimentation, knowledge of soil fertility, and plant nutrition, in order to simultaneously meet the following objectives: (a) testing the ability of REM to positively alter one or more response variables of the target crop or soil, and (b) demonstrating that the product contributes to plant nutrition and/or development, either directly or indirectly, or to soil conditioning.

The suggested strategies help define treatment factors based on the intrinsic characteristics of REM. They aim to conduct experiments to evaluate agronomic efficiency. This enables the identification that this category of agricultural input alters soil fertility indices by adding macro-and micronutrients for plants. Additionally, it promotes improvements in the physical, physicochemical, or biological properties of the soil. Ultimately, it will be possible to obtain practical information for application (effective use), such as the most appropriate granulometries, the most efficient doses, responsive crops, and soil types with the most significant probability of response. At the same time, the experiments should aim to adequately evaluate REM, whether alone or in combination with highly soluble mineral sources, organic sources, and microorganisms, observing how their efficiency is improved in different soil and climate conditions. Furthermore, the use of positive controls of high solubility in experimental designs with REM has not proven to be an adequate strategy for evaluating the efficiency of REM.

Acknowledgements The authors gratefully acknowledge the financial support provided by Embrapa, by Embrapa and Ministry of Science, Technology and Innovation Contractn. 1AAFMI/2021 'Adequações em protocolo de eficiência agronômica em função das particularidades do remineralizadores de solo', and by FINEP—Fundo Setorial Mineral through the project 'Rede de PD&I sobre remineralizadores e agrominerais (REMAg)' (Project No. 1541/2022), which made possible the development of this book chapter.

Author's Contribution Carlos Augusto Posser Silveira, Adilson Luis Bamberg and Rosane Martinazzo conceived the research protocols and wrote the chapter. Éder de Souza Martins, Magda Bergmann, Andrea Sander, Luis Henrique Gularte Ferreira, Ivan dos Santos Pereira, Giuliano Marchi and Gisele Freitas Vilela contributed to the interpretation of protocol design and critically revised the final version of the chapter.

Declaration of Competing Interest The authors declare that they have no known competing financial interests or personal relationships that could have appeared to influence the work reported in this paper.

References

Beerling DJ et al (2020) Potential for large-scale CO2 removal via enhanced rock weathering with croplands. Nature 583:242–248. https://doi.org/10.1038/s41586-020-2448-9

Boletim 100 (2022) Cantarella H, Antonio Quaggio J, Jr. Mattos D et al (eds) Recomendações de adubação e calagem para o estado de São Paulo. instituto Agronômico, Campinas, p 489

Brazil (2013) Ministério da Agricultura, Pecuária e Abastecimento. Instrução Normativa no 53, de 23 de outubro de 2013. Das definições, da classificação, do registro e renovação de registro de estabelecimentos, do cadastro e renovação de cadastro de prestadores de serviços, fornecedores de minérios e geradores de materiais secundários, das alterações e cancelamentos

Brazil (2016) MAPA. Ministério da Agricultura, Pecuária e Abastecimento. Instrução Normativa no 5, de 10 de março de 2016. Regras sobre definições, classificação, especificações e garantias, tolerâncias, registro, embalagem, rotulagem e propaganda dos remineralizadores e substratos para plantas, destinados à agricultura

Brazil (2018) MAPA. Ministério da Agricultura, Pecuária e Abastecimento. Instrução Normativa no 39, de 08 de agosto de 2018. Relação de fertilizantes minerais simples e fertilizantes contendo novos aditivos, quelantes ou complexantes, novos materiais para serem utilizados como carga e novos minérios para a fabricação de fertilizantes

Kour D, Rana KL, Kaur T, Yadav N, Halder SK, Yadav AN, Sachan SG, Saxena AK (2020) Potassium solubilizing and mobilizing microbes: biodiversity, mechanisms of solubilization, and biotechnological implication for alleviations of abiotic stress. In: Rastegari AA, Yadav AN, Yadav N (ed). New and future developments in microbial biotechnology and bioengineering: trends of microbial biotechnology for sustainable agriculture and biomedicine systems: diversity and functional perspectives. Elsevier, Amsterdam, Cap. 11, pp 177–202

Leonardos OH, Theodoro SH, Assad ML (2000) Remineralization for sustainable agriculture: a tropical perspective from a Brazilian viewpoint. Nutr Cycl Agroecosyst 56: 3–9. https://doi.org/10.1023/A:1009855409700

Manning DAC, Theodoro SH (2018) Enabling food security through use of local rocks and minerals. Extr Ind Soc 7(2):480–487. https://doi.org/10.1016/j.exis.2018.11.002

Martins ÉS, Martins ES, Hardoim PR (2023) Princípios geoquímicos, mineralógicos e biológicos do manejo de remineralizadores de solos. Informe Agropecuário, EPAMIG, 44(321):26–39. Available: https://www.alice.cnptia.embrapa.br/alice/bitstream/doc/1156783/1/Eder-principios-geoquimicos-mineralogicos.pdf

Martins ÉS, Resende AV, Oliveira CG, Furtini Neto AE (2010) Materiais silicáticos como fontes regionais de nutrientes e condicionadores de solos. In: Fernandes FRC, da Luz AB, Castilhos. (Org.) Remineralizadores para o Brasil, vol 1, 1st edn. CETEM, Rio de Janeiro, RJ, p. 89–104

Medeiros FP, Carvalho AMX, Gindri Ramos C, Dotto GL, Cardoso IM, Theodoro SH (2024) Rock powder enhances soil nutrition and coffee quality in agroforestry systems. Sustainability 16:354. https://doi.org/10.3390/su16010354

Melo VF, Castilhos RMV, Pinto LFS (2009) Reserva mineral do solo. In: Química e mineralogia do solo. Edts.: Vander de Freitas Melo & Luís Fernando Ferracciú Alleoni. SBCS, Viçosa, MG, pp 251–332

Novais RF, Neves JCL, Barros NF (1991) Ensaio em ambiente controlado. In: de Oliveira AJ, Garrido WE, de Araújo JD, S Lourenço (eds) Métodos de pesquisa em fertilidade do soloBrasília, Embrapa-SEA, Embrapa-SEA, Documentos 3, p 392

Pauletti V (2004) Nutrientes: teores e interpretações, 2nd edn. Fundação ABC para Assistência e Divulgação Técnica e Agropecuária, Castro, p 86

Comissão de Química e Fertilidade do Solo—CQFS (2016) Recomendações de adubação e calagem para os estados do Rio Grande do Sul e Santa Catarina, 11th edn. Sociedade Brasileira de Ciência do Solo, Porto Alegre, Brasil

Raghavendra MP, Chandra Nayara S, Nuthan BR (2016) Role of Rhizosphere Microflora in Potassium Solubilization, Cap.4. In: Meena VS, Maurya BR, Verma JP, Meena RS (eds) Potassium solubilizing microrganisms for sustainable agriculture, pp. 43–59. Springer India, New Delhi

Ramos CG, Dalmora AC, Kautzmann RM, Hower J, Dotto GL, Oliveira LFS (2021) Sustainable release of macronutrients to black oat and maize crops from organically-altered dacite rock powder. Nat Resour Res 30:1941–1953. https://doi.org/10.1007/s11053-021-09862-0
Ramos CG, Hower JC, Blanc E, Oliveira MLS, Theodoro SH (2022) Possibilities of using silicate rock powder: an overview. Geosci Front 13(1). https://doi.org/10.1016/j.gsf.2021.101185
Sociedade Brasileira de Ciência do Solo (2017) Núcleo Estadual Paraná. Manual de adubação e calagem para o estado do Paraná. SBCS/NEPAR, Curitiba, p 482
Theodoro SH (2000) A Fertilização da terra pela Terra: uma alternativa para a sustentabilidade do pequeno produtor rural. 221 p. Ph.D. Universidade de Brasília, Brasília, DF. https://repositorio.unb.br/handle/10482/20881
Theodoro SH, Manning DAC, Carvalho AXM, Ferrão FR, Almeida GR (2022) Soil remineralizer: a new rote to sustentability for Brazil, a giant exporting agro-mineral commoditites. In: Yakovleva N, Nickless E (eds) Routledge Handbook of the Extractive Industries and Sustainable Development, 1st Edn, 261–281. https://doi.org/10.4324/9781003001317
Van Straaten P (2007) Agrogeology: the use of rocks for crops. Enviroquest Ltda., Ontario, CA, p 440

MIX
Papier aus verantwortungsvollen Quellen
Paper from responsible sources
FSC® C105338

If you have any concerns about our products,
you can contact us on
ProductSafety@springernature.com

In case Publisher is established outside the EU,
the EU authorized representative is:
Springer Nature Customer Service Center GmbH
Europaplatz 3, 69115 Heidelberg, Germany

Printed by Libri Plureos GmbH
in Hamburg, Germany